CURRENT ORNITHOLOGY

VOLUME 2

Current Ornithology

Editorial Board

A Continuation Order Plan is available for this series. A continuation order will bring delivery of each new volume immediately upon publication. Volumes are billed only upon actual shipment. For further information please contact the publisher.

CURRENT ORNITHOLOGY

VOLUME 2

Edited by
RICHARD F. JOHNSTON
University of Kansas
Lawrence, Kansas

PLENUM PRESS • NEW YORK AND LONDON

ISBN 0-306-41780-4

© 1985 Plenum Press, New York
A Division of Plenum Publishing Corporation
233 Spring Street, New York, N.Y. 10013

Printed in the United States of America

CONTRIBUTORS

MYRON CHARLES BAKER, Department of Zoology/Entomology, Colorado State University, Fort Collins, Colorado 80523

LUIS F. BAPTISTA, Ornithology and Mammalogy, California Academy of Sciences, Golden Gate Park, San Francisco, California 94118

GEORGE F. BARROWCLOUGH, Department of Ornithology, American Museum of Natural History, New York, New York 10024

ABBOT S. GAUNT, Department of Zoology, The Ohio State University, Columbus, Ohio 43210

SANDRA L. L. GAUNT, Department of Zoology, The Ohio State University, Columbus, Ohio 43210

FRANCES C. JAMES, Department of Biological Science, Florida State University, Tallahassee, Florida 32306

NED K. JOHNSON, Museum of Vertebrate Zoology and Department of Zoology, University of California, Berkeley, California 94720

DONALD E. KROODSMA, Department of Zoology, University of Massachusetts, Amherst, Massachusetts 01003

BERND LEISLER, Max-Planck-Institute for Behavioral Pediatrics, Radolfzell Ornithogical Station, Am Obstberg, D-7760 Radolfzell-Möggingen, Federal Republic of Germany

CHARLES E. McCULLOCH, Biometrics Unit, Department of Plant Breeding and Biometry, Cornell University, Ithaca, New York 14853

ALBERT H. MEIER, Department of Zoology and Physiology, Louisiana State University, Baton Rouge, Louisiana 70803

KENNETH MEYER, Department of Biology, University of North Carolina at Chapel Hill, Chapel Hill, North Carolina 27514

HELMUT C. MUELLER, Department of Biology, University of North Carolina at Chapel Hill, Chapel Hill, North Carolina 27514

LEWIS PETRINOVICH, Department of Psychology, University of California, Riverside, California 92521

ROBERT J. RAIKOW, Department of Biological Sciences, University of Pittsburgh, Pittsburgh, Pennsylvania 15260, and Carnegie Museum of Natural History, Pittsburgh, Pennsylvania 15213

ALBERT C. RUSSO, Department of Zoology and Physiology, Louisiana State University, Baton Rouge, Louisiana 70803

JARED VERNER, Forestry Sciences Laboratory, Fresno, California 93710

HANS WINKLER, Institute for Limnology of the Austrian Academy of Sciences, A-5310 Mondsee, Austria

ROBERT M. ZINK, Department of Ornithology, American Museum of Natural History, New York, New York 10024 Present Address: Museum of Zoology, Louisiana State University, Baton Rouge, Louisiana 70803

PREFACE

It is not often that a century of scholarly activity breaks conveniently into halves, but ornithology of the first half of the 20th century is clearly different from that of the second half. The break actually can be marked in 1949, with the appearance of Meyer and Schuz's *Ornithologie als Biologische Wissenschaft*. Prior to this, ornithologists had tended to speak mostly to other ornithologists, experiments (the testing of hypotheses) were uncommon, and a concern for birds as birds was the dominant thread in our thinking. Subsequent to 1949, ornithologists have tended to become ever more professional in their pursuits and to incorporate protocols of experimental biology into their work; more importantly perhaps, they have begun to show a concern for birds as agencies for the study of biology. Many of the most satisfying of recent ornithological studies have come from reductionist research approaches, and have been accomplished by specialists in such areas as biochemistry, ethology, genetics, and ecology. A great many studies routinely rely on statistical hypothesis testing, allowing us to come to conclusions unmarred by wishful thinking. Some of us are ready to tell the world that we are a "hard" science, and perhaps that time is not so very far off for most of us.

Volume 2 examines several solid examples of late 20th-century ornithology. James and McCulloch look into many aspects of experimental design and data analysis currently available to ornithologists, and make certain value judgements elaborating on the subject of paragraph 1 of this Preface. The Gaunts provide a close-up view of the structure and functioning of the syrinx—for decades the black box of avian anatomy. Raikow presents a review of the disturbing and exciting problems in avian classification. Kroodsma, Baker, Baptista, and Petrinovich attempt to resolve some of the more operationally intractable problems characteristic of studies in vocal dialects of songbirds. Mueller and Meyer present results of statistical tests of a series of hypotheses designed to illuminate so-called reversed sexual size dimorphism in falconiform birds.

Barrowclough, Johnson, and Zink examine the apparently non-Darwinian variation of allele electromorph in birds. Leisler and Winkler provide an extended examination of techniques in and results of work on covariation of morphological and ecological variables for a number of avian species. Meier and Russo summarize recent advances in the study of behavioral and physiologic regulation of the annual cycle of birds. Verner provides a critique of the current status of the ever-evolving set of censusing techniques for free-living birds. A diverse set of papers such as this is not readily summarized, yet there is an evident unitary message: whole-organism biology is alive and well and living in a reductionist world.

We are pleased to have received a good number of solid suggestions over the past year for topics to be addressed in this series. We will present as many of these as possible in forthcoming volumes. As always, we are open to suggestions for both topics and authors that can pursue the further development of ornithology.

Richard F. Johnston

Lawrence, Kansas

CONTENTS

CHAPTER 2

THE EVOLUTION OF REVERSED SEXUAL DIMORPHISM IN SIZE:
A COMPARATIVE ANALYSIS OF THE FALCONIFORMES OF THE
WESTERN PALEARCTIC

HELMUT C. MUELLER AND KENNETH MEYER

CHAPTER 3

VOCAL "DIALECTS" IN NUTTALL'S WHITE-CROWNED SPARROW

DONALD E. KROODSMA, MYRON CHARLES BAKER, LUIS F. BAPTISTA,
AND LEWIS PETRINOVICH

CHAPTER 4

ON THE NATURE OF GENIC VARIATION IN BIRDS

GEORGE F. BARROWCLOUGH, NED K. JOHNSON, AND ROBERT M. ZINK

CHAPTER 8

ASSESSMENT OF COUNTING TECHNIQUES

   JARED VERNER

CHAPTER 9

CIRCADIAN ORGANIZATION OF THE AVIAN ANNUAL CYCLE

ALBERT H. MEIER AND ALBERT C. RUSSO

CHAPTER 1

DATA ANALYSIS AND THE DESIGN OF EXPERIMENTS IN ORNITHOLOGY

FRANCES C. JAMES and
CHARLES E. McCULLOCH

1. INTRODUCTION

One familiar statement one hears about data analysis in ornithology is that traditional ornithologists accumulated facts, but did not make generalizations or formulate causal hypotheses (Emlen, 1981; Wiens, 1980). The approach of modern ornithologists, on the other hand, is said to be more theoretical. Modern ornithologists formulate hypotheses, make predictions, check the predictions with new data sets, perform experiments, and do statistical tests. Proponents of the modern approach frequently state their position in a condescending tone, and this is guaranteed to arouse the ire of those more committed to empirical research (Olson, 1981). In this essay, we hope to lessen the differences of opinion on this subject as they were recently expressed in the commentary section of *The Auk* (Austin et al., 1981). We agree with Popper (1959, 1972, 1983) that the goal of science is to develop better explanatory theories about processes in nature, but it does not follow that advances are made only by hypothesis testing. Many philosophers think

FRANCES C. JAMES • Department of Biological Science, Florida State University, Tallahassee, Florida 32306. CHARLES E. McCULLOCH • Biometrics Unit, Department of Plant Breeding and Biometry, Cornell University, Ithaca, New York 14853.

that Popper has exaggerated the role of falsification in scientific inquiry (Lieberson, 1982), and several reviewers of theoretical ecology (Southwood, 1973; Levandowsky, 1977; MacFadyen, 1975; McIntosh, 1980; Simberloff, 1980; Strong, 1980; Wiens, 1983) have questioned the validity of the way in which hypothesis testing has been conducted in the last 30 years. Peters (1976, 1978) and Gould and Lewontin (1979) claim that natural selection theory is untestable and therefore unscientific and that sociobiology is just story telling. If these charges are true, and descriptive work is trivial, is the entire field of ornithology nonscientific except for laboratory experiments in physiology and behavior? Rather than take such an extreme view, we want to explore how data analysis might be improved in both empirical and theoretical research.

Our original objective was to review methods of data analysis and the design of experiments in ornithology from a statistical point of view. In the course of organizing our ideas on this topic, we found that it was also necessary to try to clarify the relationship between statistical hypothesis testing and the type of hypothesis testing advocated by Karl Popper. As a result, our paper begins with sections on models and scientific methods and then proceeds to a discussion of formal inductive methods of data analysis. First, we define modeling very broadly and give examples of the development of specific empiric and theoretic models. Empiric models simply describe observed relationships among variables. They describe relationships that result from processes but they are not about processes themselves. Theoretic models are about processes. The research hypotheses of Karl Popper are verbal forms of major theoretic models (theories), but it is important to distinguish between Popperian "testing" of research hypotheses and statistical hypothesis testing. Statistical hypothesis testing involves only evaluating predictions made from models (empiric or theoretic) when data are for samples from clearly defined populations. It uses inductive inference, the process by which one proceeds from particular to general propositions. It is not directly involved in decisions about whether to retain theories. Popper is still adamantly opposed to induction as a method of gaining reliable knowledge (Popper and Miller, 1983). He discounts its practical importance and logical interest. Nevertheless, both theory choice and statistical hypothesis testing use principles of refutation, so criteria can be defined for rejecting models, theories, and their predictions.

The general field of statistics has been moving away from what is now viewed as an overemphasis on statistical hypothesis testing. For a clear argument documenting the misplaced prestige currently enjoyed by statistical hypothesis testing, see Guttman (1977) on "What is not what in statistics." The current view of many statisticians is that in many cases researchers should spend more time in exploratory data analysis (Tukey, 1977), the stage that precedes formulation of formal models and testing of predictions based on them. We give examples of exploratory methods and references to literature on this subject. In Section 4, we discuss common problems with the design of field experiments and we show how principles of experimental design can be incorporated into the analysis stage of purely observational studies. "Natural experiments" can play an important role in deriving weak inferences from data, even though they have many drawbacks compared with well-controlled experiments in which the effects of treatments can be more directly tested. In Section 5, on the strength of inference, we emphasize that violations of the assumptions of models can lead to serious misinterpretations. Our view of how the steps of data analysis and the design of experiments lead from observations to conclusions is summarized in Fig. 1.

In Section 6, four particular topics are discussed in relation to their effect on how ornithologists interpret data. First, we think that continuous models, such as direct gradient analysis, are excellent ways to describe ecological patterns of distribution. They need not be tied to community ecology or the Hutchinsonian niche. Second, using an example from morphometrics, we discuss the relationship among correlation, regression, and principal components analysis. Third, we present Mosimann's (Mosimann, 1975a; Mosimann and James, 1979; Mosimann *et al.*, 1978) direct methods of size and shape analysis. Finally, we discuss multivariate distance measures in resource space. In an appendix, we give a glossary of terms with references to full explanations. Our major conclusion is that if ornithology is unscientific, it is not so much because we have too much description and too little hypothesis testing, but rather because we have too many untestable speculative hypotheses and too many instances of misapplied inference. Unfortunately, pressure from editors and granting agencies to cast research in terms of hypothesis tests has been encouraging biologists to conduct many premature and even inappropriate statistical tests.

2. MODELS: VERBAL, EMPIRIC, THEORETIC

We use the term modeling very broadly. Models are simply statements about relationships among variables. Everyone who interprets data has a model (e.g., regression, gradient analysis, cluster analysis) by which the data are organized and, even before that stage, a verbal

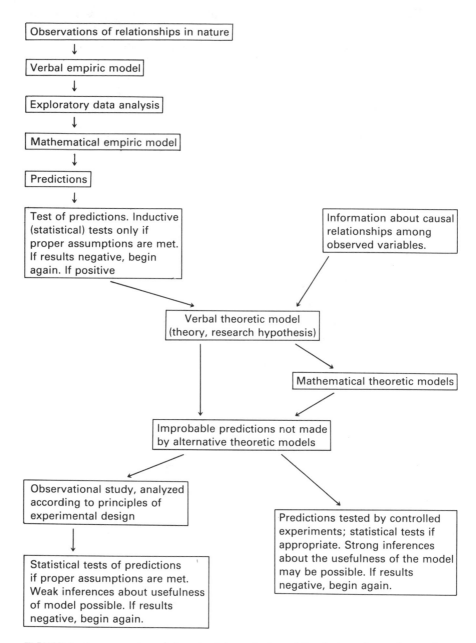

FIGURE 1. A progression of steps in data analysis and the design of experiments from observations to conclusions.

model is used. Attention to models can help the investigator define variables precisely and express their relationships quantitatively. It also can clarify what assumptions are being made and what predictions can be tested. Possibly even more important, modeling allows formal comparisons with alternative constructs that may make the same predictions but use other variables, relationships, and assumptions. Models are never proved true with statistics. They are simply tentatively considered statements that, if framed properly, permit testable predictions.

2.1. Verbal Models

The process of formulating models is probably the most creative part of science (Hanson, 1958); so, clearly, conscious attention to modeling is warranted. Sometimes only verbal (nonmathematical) models are possible, and these are very useful (May, 1981a,b). At the most general level population biology, community ecology, physiology, biogeography, morphology, and evolution are each verbal models for how information can be organized and interpreted. When verbal models are expressed in mathematical terms, they can be classed as either empiric or theoretic. The level of mathematical formality of a model should suit the level of understanding of the process of interest, and the first step toward quantification is expression of the verbal relationship among variables in mathematical terms. When the relationship is derived from observed data rather than from theoretical constructs about processes, the result is an empiric model.

2.2. Empiric Models

Empiric models are symbolic expressions of relationships among variables derived from data. Examples are ratios, correlations, regression equations, coefficients of variation, diversity and similarity indices, and multivariate equations. The assumptions of an empiric model should be tested to try to determine whether the model is useful in practice.

Below are examples of the natural progression from verbal to empiric models. Each example starts with a verbal model, and we try to show how it has evolved. In example 3 we argue that a model should be discarded.

1. Larger birds have lower gram-specific metabolic rates. Aschoff and Pohl (1970) expressed this relationship for passerines as a regression of standard metabolism (kcal/day) on body weight (kg):

$$M = 140.9W^{0.704}$$

This estimate was based on metabolic rates of resting birds under constant laboratory conditions during daylight hours. Calder and King (1974) review the literature on the development of allometric (power function) models in avian energetics and Calder (1983) explores their broader implications.

King and his associates (Mahoney and King, 1977; Walsberg and King, 1978a,b; Mugaas and King, 1981) have developed a new type of model that is useful for studying the energetics of active birds in field situations. This approach considers the various avenues of heat exchange between the bird and its environment under a variety of conditions by using the principle of "operative environmental temperature" (equivalent black-body temperature). Bakken (1980) uses data from heated taxidermic mounts to integrate theoretical developments with laboratory analyses. King's empirical methods provide links among biophysical ecology (Porter and Gates, 1969), physiological ecology, and behavioral studies of time and energy budgets of birds. Walsberg (1983) summarizes the most recent work.

2. Altricial birds grow faster than precocial birds. Ricklefs (1967, 1968, 1969, 1973, 1979) has used a graphical method to fit growth data to three nonlinear growth curves, the Gompertz, and logistic, and the von Bertalanffy. Johnson et al. (1975) review available techniques. An alternative to Ricklefs' method would be fitting the more general Richards equation (Richards, 1959; Johnson et al., 1975; White and Brisbin, 1980), which can assume the form of many standard growth curves. The Richards equation is

$$W(t) = [A^{1-m} + \beta e^{-kt}]^{1/(1-m)}$$

where W is the quantity subject to growth, t is time, A is the asymptotic value of W, m is a shape parameter, β is a parameter related to the time of origin, and k is growth rate.

3. Bird species diversity increases with foliage height diversity. In this case the verbal model is generally acceptable, but both terms in the model are so complex that the expression of their relationship as a linear regression of one information theoretic statistic on another (MacArthur and MacArthur, 1961) has not been very useful. In addition to problems of sampling procedures discussed by Wiens (1983), there are other difficulties. First, because very different combinations of species richness and evenness can give the same value of species diversity, any particular value of the index is difficult to interpret. Second, in the same community, bird species estimates increase with the size of the area sampled, even though area sampled is not a term in the model

(James and Rathbun, 1981). Third, foliage height diversity is a misnomer because the index increases with increasing evenness of foliage distribution rather than increasing variation in the distribution of the foliage in height categories (James and Wamer, 1982, p. 169). James and Wamer (1982) show how principal components analysis and contour plots can be used to express nonlinear relationships between species richness or density and vegetation structure.

4. Mayfield (1961) proposed a model for estimating survivorship that helps to overcome biased estimates caused by the tendency for those nests found to misrepresent survivorship in all nests built. This was a vast improvement over techniques used at the time, but the method has also been found to have drawbacks. Refinements of the original empiric model have led not only to more reliable measurements, but also to the ability to make statistical inferences such as confidence intervals and hypothesis tests. Green (1977) showed that Mayfield's method would give biased estimates when nests had different survival rates. Dow (1978) pointed out problems with tests of statistical significance proposed by Mayfield (1975). Miller and Johnson (1978) and Klett and Johnson (1982) also noted situations in which the Mayfield estimator can give biased estimates. Bart and Robson (1982) state that the Mayfield estimator often gives good estimates, but that it is difficult to set confidence intervals or perform significance tests with the estimator because of its skewed sampling distribution. To alleviate this problem, Bart and Robson (1982) proposed a likelihood analysis of survival rates. This approach solves many problems inherent in Mayfield's method; it allows confidence intervals and hypothesis tests. It is not influenced to a great extent by different survival rates, and it reduces biases of the estimates.

5. Mayr's (1963) statement of Bergmann's Rule is an informal verbal model stating that "races from cooler climates tend to be larger in species of warm-blooded vertebrates than races of the same species living in warmer climates." Exceptions to the model have been described as altitude effects and aridity effects by Snow (1954), Moreau (1957), and Hamilton (1961). By comparing patterns of various climatic variables with patterns of geographic size variation, James (1970) found that several variables that are functions of both temperature and humidity, along with their altitudinal correlates, are more highly related to size variation than are variables expressing dry-bulb temperature alone. These variables (wet-bulb temperature, absolute humidity, vapor pressure deficit, actual evapotranspiration) are all functions of the evaporative power of the air, which varies with altitude and aridity as well as with temperature. This result suggests that a physiological interpre-

tation relating the ratio of body surface to respiratory surface is the simplest adaptive explanation for the pattern of size variation and climatic variation. This is a verbal theoretic model akin to Bergmann's original formulation of the problem. When Rosenzweig (1968) studied data for geographic size variation in mammals, he also noted that actual evapotranspiration is positively correlated with body size. But, because actual evapotranspiration is negatively correlated with plant productivity, he constructed a different verbal theoretic model stating that large size variation is an adaptation to low food availability.

This is a case in which we do not know how to differentiate among theoretic models about processes causing patterns. Alternative models fit the observed pattern equally well. With the present lack of knowledge about the causal relationships, it is best to work with Mayr's empiric model, modified at least for birds, to say that those species that have geographic variation in size tend to be larger where the environment is either cooler or drier.

Even though empiric models are not about processes, the models themselves (e.g., regression; see Section 6.2.) may have underlying assumptions that affect the interpretation of data. Alternate empiric models with other assumptions may fit the data equally well. Empiric modeling may use statistical hypothesis testing in those situations in which inferences are made about relationships in populations from relationships in samples, but inferences about processes causing the relationships are not reliable (Romesburg, 1981). Even if an empiric model makes good predictions, one is not justified in concluding anything about the process that caused the relationships. Two types of multivariate empiric models will be discussed in later sections. Multivariate gradient analysis is discussed in Section 6.1 and morphometrics is discussed in Section 6.3.

2.3. Theoretic Models

Theoretic models should be derived from prior understanding of processes. Their advantage over empiric models is that they can be used to study processes, whereas empiric models can only be used to study the results of processes. A good theoretic model simplifies the process being studied while retaining its essence. It allows the researcher to conceptualize the process more easily. Theoretic modeling also allows the estimation and testing of meaningful model parameters (May, 1981a,b). For example, theoretic mathematical models can be manipulated to make predictions outside the realm in which they were developed. This form of mathematical theoretic modeling can help guide the choice of experiments or data gathering, and predictions can then be tested (Pielou, 1977). The way to investigate a theoretic model is not to fit one's observations to it. A good fit may mean only that the model is not very discriminating. It is better to perform experiments designed to discriminate among competing models.

Here are five examples of verbal theoretic models (models of processes) with comments on their mathematical development and usefulness.

1. The fauna of an island is a function of the invasion rates of new species, extinction rates of those already present, the size of the island, and the distance to the source of invading species (MacArthur and Wilson, 1967). This is the equilibrium model of island biogeography. The model is being applied to the design of refuges and is widely accepted in spite of serious challenges that it is far too simplistic (Simberloff, 1983a,b). Simberloff (1983c) and Keough and Butler (1983) contend that there is no objective definition of what constitutes an equilibrium, and Lynch and Johnson (1974) question the extent to which there is any connection between extinction and immigration that could conceivably exert control over species turnover. Lack (1969) insisted that the failure of birds to establish populations on islands comes mainly from the birds' failure to find the right conditions rather than a failure to disperse there. Williamson (1981) provides the best overall summary of the literature about island biology. He points out that too often processes are inferred when the data are simply lists of species. Even more recently Haila (1983) has summarized his extensive quantitative analyses of bird populations on Finnish islands. He contends that the theory of island biogeography is a useful construct even if its usefulness is merely as a departure point for studying the dynamics of island faunas. But only experiments will permit strong inferences about the processes involved, and these are very difficult with birds. Lovejoy (1980) is directing a large-scale, long-term experiment in Brazil that should eventually provide guidance for the design of reserves in neotropical rain forests. Abbott (1980) suggests specific ways that field experiments could clarify our understanding of the regulation of bird populations on islands. (See Section 3.1 for comments on the argument between Diamond and Simberloff on island biogeography.)

2. So many offspring are produced each year in excess of the carrying capacity of the environment that winter mortality or removal by hunters of a number equivalent to the excess does not disrupt the population dynamics of the species. This is a statement of Errington's

(1945) threshold-of-security hypothesis. Romesburg (1981) charges that this theoretic model, which is so central to the philosophy of wildlife management (Wagner, 1969), has never really been tested by experimentation. It is an example of a theoretic model that is frequently used to interpret observations, but Romesburg insists that this practice produces unreliable knowledge. He suggests that critical experiments could be conducted with introductions of wild turkeys *(Meleagris gallopavo)* to see whether predictions of the model are supported.

3. Interspecific competition is an important ecological force in regulating the structure of communities (MacArthur and Levins, 1967; Cody, 1974; Schoener, 1982; Diamond, 1975, 1978; Hutchinson, 1958, 1965, 1968; Roughgarden, 1983). The widespread use of this theoretic model, and its underlying Lotka-Volterra competition equations, in analyses of ecological niches is another case of the dangers of assuming that the congruence of an observed pattern with predictions of a theoretic model is evidence that the model expresses a real process in nature (Brown, 1981; Birch, 1979; Pielou, 1977; Connell, 1980; Simberloff, 1982; Wiens, 1973, 1983). Cody (1974, 1978), Diamond (1978), Noon (1981), and Alatalo (1982) say that observations of behavioral or habitat differences when species co-occur with different combinations of other species are evidence of interspecific competition. But this argument fails to sort out all of the other causes that influence distribution (Connell, 1975; James et al., 1984b). Schoener (1982) and Strong et al. (1984) present the full spectrum of opposing views on this topic.

James et al. (1984b) try to show that modern methods of data analysis can be applied to the less competition-oriented and more autecological model of the niche originally proposed by Joseph Grinnell (1917). With the neo-Grinnellian approach, one studies resource use throughout the geographic range of one species at a time. The idea is that by looking at many situations, one may be able to sort out potential factors regulating distribution. Even so, either experiments or blocking in the analysis stage will be required if inferences about causes are to be made. (See Section 4.1 on blocking in observational studies.)

4. Character displacement is an important evolutionary force in adaptation (Brown and Wilson, 1956). Experiments are not possible here. Grant (1972) reviewed the literature about morphological character displacement and found few if any unequivocal examples of even the pattern that such a process would produce. This is a good example of a generally accepted and intuitively pleasing theoretic model for which there is virtually no ornithological support. See Abbott et al. (1977) and Grant and Schluter (1984) for a possible exception in the Galápagos finches.

5. The neo-Darwinian theory of natural selection is a theoretic model stating that genetic changes in populations occur by the gradual accumulation of alleles that are associated with higher fitness. By this process, populations develop behavioral, physiological, and morphological traits that result in successively better adaptations to environments. Examples of observational studies of birds that are consistent with this theoretic model could be taken from the fields of population genetics, sociobiology, kin selection, parent–offspring conflict, mating systems theory, life history strategies, and optimal foraging. Much of the new work in these areas is experimental, but usually the experiments are not designed to test the theoretic model of natural selection per se.

A. Optimal Foraging Theory. Foraging animals make decisions based on the ratio of benefit accrued from food items to the cost in energy of obtaining the food (Pyke et al., 1977; Schoener, 1979). The original presentation of these ideas in the form of a graphic optimization model by MacArthur and Pianka (1966), as a hypothesis for testing, has developed into the most elaborate example of modeling in the ornithological literature. In a very thorough recent review, Krebs et al. (1983) emphasize that the selectionist model itself is not being tested. They insist that the objective of optimal foraging research is to understand the decision rules by which animals forage, not whether they are optimizing anything or even whether within-population differences have a genetic basis. Nevertheless, the models are based on optimality theory.

Optimal foraging models are being used to investigate how foragers modify their behavior in relation to factors such as distribution of prey (Cowie, 1977; Oaten, 1977) and the extent to which prey size is a function of distance from a central place (Orians and Pearson, 1978). Sometimes the theoretical considerations seem to get bogged down in "thickets" of algebra (May, 1977), but when predictions can be tested by quantification of costs and benefits, the connection with reality is more clear. A good example is the empirical study by Carpenter and MacMillen (1976) on feeding territoriality in a Hawaiian honeycreeper. Another is the experiments with territoriality in migrant Rufous Hummingbirds by Kodric-Brown and Brown (1978). Krebs et al. (1978) used a stochastic model in an aviary experiment with the Great Tit (Parus major) in which the birds were allowed to sample two patches of prey. Oaten (1977) advocates the stochastic models of Stephens and Charnov (1982) because in nature the animal must respond to ever-changing resources. Myers (1983) prefers the approach of Janetos and Cole

(1981), which uses minimum threshold criteria for behavioral tactics rather than maximization of net benefit. Schluter (1981), who found little support for the optimality prediction that animals take more specialized prey when food is abundant and diverse, would probably agree.

B. Life History Strategies. The literature comparing life history "strategies" is usually cast in terms of natural selection theory (May, 1977; Stearns, 1976, 1977). This theme started with Deevey's (1947) empirical life tables and a classic paper by Crook (1964) on ecological correlates of the life history differences among weaver finches (ploceids). Maynard Smith (1974, 1978) has developed more and more highly mathematical models for what he terms evolutionarily stable strategies. Pielou (1977) shows that the mathematical modeling of birth and death processes has intrinsic value because seemingly unlikely predictions of models can then be checked with real data. But continued analysis of life tables is not recommended (Eberhardt, 1972). Brownie *et al.* (1978), in a thorough analysis of band recovery data, show that because estimates of population parameters have such extremely wide confidence intervals, newer stochastic models have important advantages over life table approaches. Ricklefs (1983) summarizes the literature interpreting comparisons of life table statistics for birds.

There are two common misunderstandings about models. The first is that only theoretic models permit predictions (Fretwell, 1975; Emlen, 1981; MacArthur, 1962). As we have seen, both empiric and theoretic models make predictions. The index of leading economic indicators used to forecast the performance of the United States economy is an empiric model derived from much past experience about which economic variables precede changes in the general economy. It makes fairly reliable short-term forecasts. Of course, the success of predictions based on empiric models depends on constancy of the system being modeled. If the system is the same as when the empiric model was formulated and the associations used to derive the model are strong ones, then the predictions will be accurate. However, if the system has changed, or the empiric model is used to make predictions for a completely different system, then predictions may be poor. Poole (1978) emphasizes that empirical statistical models are useful for forecasting population fluctuations, regardless of whether the causes of the fluctuations are understood.

Poole (1978) and Pielou (1977) define both models and predictions more narrowly than we do. They treat only regression and time series modeling as predictive and all other modeling as theoretic or explanatory. Pielou (1981) restricts the term modeling to strictly mathematical and theoretical modeling of the functioning of an ecosystem. She thinks this area of research in ecology is unlikely to produce real scientific advances.

The second common misunderstanding about models is that if their predictions are fulfilled, the model has been "tested." The model that the sun revolves around the earth makes precise and repeatable predictions, but it is incorrect. The number of storks and the human birth rate are both declining in Europe, but the model that storks bring babies is incorrect. Similarly, statistically significant confirmations of predictions about the number of birds on islands, the size ratios of coexisting congeners, the extent of kin recognition in ground squirrels, or niche shifts in European warblers do not give information about the validity of the theoretic models that led to the predictions. If predictions of a model are supported, all we have learned is the extent to which the model makes correct predictions. It seems obvious that to investigate a process, you have to study the process, not its consequences. Thus, when Wilson (1975) says that progress in sociobiology will depend on how well data can be fitted to models, he should also confess that this approach will never reveal whether natural selection is producing adaptive behavior. Unfortunately, evolutionary biology contains a lot of theoretic models based on vague ideas about natural selection (Dayton, 1979; Cain, 1977). Stearns (1977, 1980) gives a pessimistic view of whether their continued application is the best use of research time.

The complaint that theories like natural selection do not make *unique* predictions (Peters, 1978) is an important point in light of the fact that the entire field of sociobiology is based on predictions. Peters means that congruence of observations with predictions is not justification for accepting a theory. Brady (1979) makes this point beautifully. "It would be naive to suppose that testing by generating and checking predictions one could either prove or disprove a theory. When a hypothesis predicts that, given specified conditions, a specified consequent will result, the fact that the prediction holds does indeed extend the congruence between the anticipated pattern and the observed pattern, but it does not tell us whether this congruence is coincidental or necessary." We will try to show in Section 4 that theoretic models can only be studied by some form of an experiment.

3. CONFIRMATORY VERSUS EXPLORATORY STUDIES

In this section we discuss the differences between confirmatory studies (those that are made to test statistical hypotheses or to make significance tests) and exploratory studies (those that are made to explore data and thereby facilitate the formulation of models and their predictions). Any good research program uses both approaches, but for the different purposes stated above. Exploratory analyses have received far less attention than they deserve. They allow the detection of patterns in data before specific models have been formulated. They permit an interplay between ideas and data before one uses formal inference. Furthermore, they can prevent premature hypothesis formulation, preventing the researcher from testing unrealistic hypotheses and then concluding that a worthwhile contribution has been made.

We will discuss the current influence of Popper's philosophy on biology. Then we will explore the relationship between Popper's philosophy of conjecture and refutation and the field of statistics. Finally, we will describe the important new subfield of statistics called "exploratory data analsis."

3.1. Popper's Hypothetico-deductive Method

Popper's (1959, 1968, 1972) central idea is that scientific research should be viewed not as the accretion of information, but rather as the activity of solving problems. He recommends proposing bold research hypotheses (theoretic models), deducing their consequences, and specifying seemingly improbable predictions that can be checked by observations or experiments. If predictions are not supported, a research hypothesis can either be modified to accommodate the new information or discarded, but the refutationist approach must continue. Because experience may not yet have included observations that will eventually falsify the hypothesis, it can only be tentatively entertained. The hypothesis must always be phrased in a way that its parameters are clearly defined, and its causal relationships must be clearly specified. Otherwise rejection would never be possible, and the whole endeavor would become pseudoscientific (see also Platt, 1964). For Popper the objective of science is to develop universal theories (research hypotheses) that have not yet been falsified. These hypotheses are the equivalent of the verbal theoretic models in Section 2.3. They may be but are not necessarily probabilistic. We will describe in Section 3.2 how adamantly Popper has been opposed to inductive reasoning as a way of gaining reliable knowledge. Biologists do not seem to be aware of the fact that Popperian science does not require statistical hypothesis testing.

The challenge from Popper that our activities may be "pseudoscientific" has elicited a scramble to justify what we do in Popperian terms. This justification is particularly important because natural selection theory, which is the major theory of whole-organism biology, is being exposed to severe scrutiny on these grounds. Ghiselin (1969) says that Darwin's own bold and critical methods fully qualify as Popperian. He used logic to formulate bold conjectures independently of the general biological thinking of his time, and he used deductive reasoning to make unlikely predictions that might falsify his ideas. How is neo-Darwinian research being conducted in the 1980s? Are our explanatory theories refutable? Even Dobzhansky (1975) confessed that no biologist "can judge reliably which 'characters' are neutral, useful, or harmful in a given species." If it is impossible to test the fitness of a trait, how can one study adaptation that optimizes its fitness? The theory as stated is irrefutable, and "irrefutability is not a virtue of a theory (as people often think) but a vice" (Popper, 1965).

The ornithological literature is particularly weak on this score. In fact, most of the literature on life history strategies (Stearns, 1980), ecological adaptations (Lewontin, 1972), and community ecology (Wiens, 1983) stands guilty of leaving the implication that the selectionist model has been confirmed because our observations can be interpreted in its terms. Many optimization models such as those for life histories (Horn, 1978), mating strategies (Parker, 1978), and territorial defense (Myers et al, 1981) take natural selection theory as a given. Someone should be figuring out how to make the model more operational (Brady, 1979; Lewontin, 1972; Gould and Lewontin, 1979). Because birds are probably the best-known class of animals, ornithology may provide a major new contribution.

In the effort to defend disciplines as Popperian, there has been an interesting debate among systematists. Of the three camps—phenetics, evolutionary systematics, and cladistics—Bock (1973) claims that the second is hypothetico-deductive, and Wiley (1975), Platnick and Gaffney (1977), Nelson (1978), and others claim that the third is hypothetico-deductive. In this case, the argument is over differences in criteria for making classifications and whether systematic statements are refutable (Settle, 1979; Kitts, 1977, 1980; Cracraft, 1974) by statistical tests. The data in cladistics are observations of character variation among taxa. Hennig (1966) constructed criteria by which he thought such data could be used to reconstruct phylogenetic relationships, but Patterson (1982) shows that this goal is probably too optimistic.

In ecology, MacArthur was certainly Popperian in the sense that he made bold conjectures. His major contribution was the expression of his ideas in qualitative analytical and graphic models. But MacArthur did not apply the critical refutational part of Popper's method. In MacArthur's (1962) own words, a good ecologist "arranges ecological data as examples testing the proposed theories and spends most of the time patching up theories to account for as many data as possible." There were no ingenious attempts to refute.

The current controversy over how to interpret the distribution of birds on islands is an example of how a refutationist approach can contribute to the methods of science. It is in this spirit that Strong *et al.* (1979), Connor and Simberloff (1979), and Simberloff (1983a,b,c) have taken species lists for sets of islands and asked whether the patterns of distribution can be distinguished from what might occur if the null model of independent allocation were operating. The objective is to propose an alternative to the reasoning of Diamond (1975) that by observing the pattern of species coexistence on islands, one can infer interspecific competitive effects. Connor and Simberloff (1983, 1984) acknowledge that testing for a completely random distribution would be trivial, so they ask whether the distribution is random given the constraints that each island supports the number of species actually found there and each species occurs on the same number of islands actually observed. Notice that in this case the hypothesis is a statistical one about the existence of a pattern, not about its cause. Connor and Simberloff (1983, 1984) summarize the status of this continuing debate. The Popperian aspect of the controversy between Diamond and Simberloff is not that both sides are now doing statistical tests, although this is a positive spin-off. The Popperian part is that a bold theoretical model—assembly rules for the occurrence of birds on islands—is being subjected to refutation. Connor and Simberloff show that their hypothesis about distribution cannot be rejected for some archipelagoes. Following Popperian principles, Diamond's model would be rejected because some of its predictions are unfalsifiable. For others, the Diamond predictions and Connor-Simberloff predictions are so similar that real data will probably not allow model discrimination. However, the statistical issue of whether Simberloff's randomized model is able to make predictions that have realistic alternatives is important. See Harvey *et al.* (1983), Grant and Abbott (1980), and Alatalo (1982) on the need for model refinement here and also Abbott's (1980) suggestions for field experiments.

Rationalist philosophers of science such as Popper feel, as do most biologists and statisticians, that science moves progressively toward developing more inclusive and precise explanatory theories. The rationalist philosophers insist that advances in this direction are made solely by conjecture and refutation. Their position is a reaction against the idea that just by gathering evidence one can prove a theory. The charge is that because observations are inextricably theory-laden, this reasoning amounts to little more than story-telling. Nevertheless, most verbal models have resulted from generalization from repeated case analyses (Allee *et al.*, 1949; Harper, 1977, 1982) rather than from a method that involves hypothesis testing.

3.2. Rationalist Philosophy, Statistics, Hypothesis Testing, and Induction

We think there is confusion among some of the biologists who advocate Popper's hypothetico-deductive method. Part of the problem is that many authors do not distinguish between empiric and theoretic models and between research hypotheses and statistical hypotheses. Popper's method is about the refutation of theoretic models (research hypotheses). This is only a limited part of what scientists are trying to do (Ghiselin, 1969; May, 1981a). For example, Popper (1959, 1965, 1968, 1972; Popper and Miller, 1983) has insisted that gathering evidence by induction (generalizing from a sample to a population) is not a legitimate part of science. Nevertheless, induction is what statistical hypothesis testing is all about.

Newton-Smith (1981) takes a less extreme position than Popper on how scientists should proceed. He advocates what he calls "temperate rationalism," the idea that choice among theories can be facilitated by repeated statistical tests of the predictions of alternative theories. It is no wonder that biologists have trouble distinguishing between the advice of philosophers of science and the advice of statisticians, because both are advocating formal methods for rejecting hypotheses. But in the first case the objective is to evaluate whether a theoretic model represents reliable knowledge. The objective of statistics is to decide whether reliable predictions can be made based on models. The information of whether or not predictions are reliable can aid in choosing among alternative theoretic models, but it does not tell us to what extent the preferred model is correct. Statistics are not appropriate for measuring the reliability of a theoretic model. "All theories can be is approximately true, and we can't judge how approximately true by probability theory" (Newton-Smith, 1981, p. 216).

Should Popper's ideas be totally adopted in biological research? As Dolby (1982) states, Popper seems to have had physics or some other "hard" science in mind when he formulated his ideas. In the stochastic, biological world, probabilistic research hypotheses are the most interesting kind (Quinn and Dunham, 1983). For example, the probabilistic statement that "most swans transmit cholera to ducks" is more interesting than "only swans transmit cholera to ducks." Yet, probabilistic hypotheses are inductive and basically unfalisfiable (Kitts, 1977; Dolby, 1982; Quinn and Dunham, 1983). Popper himself (1965, p. 191) has realized that some modification of his original thinking was needed here (see Simberloff, 1983d). We agree with Dolby (1982) that confirmatory statistics has an important role in the advancement of theories, but also that this role is limited. Its role is to test statistical hypotheses formulated from the predictions of theories. Thus, statistical hypothesis tests do not test theories (research hypotheses) per se, but only their predictions and only those predictions that can be tested by sampling from clearly defined populations.

Operationally, predictions are derived from tentatively entertained empiric or theoretic models. Next, these predictions are stated in the form of a statistical hypothesis, traditionally called the alternative hypothesis. The logical complement of the alternative hypothesis is called the null hypothesis. After data are gathered, there is a search for evidence that the null hypothesis is false. If there is sufficient evidence, then the null hypothesis is rejected and the predictions from the model are deemed correct. Note that following Popperian logic, all that has been shown is that a prediction has not been falsified. If the null hypothesis cannot be rejected, then the prediction is deemed incorrect and the model not useful.

As an example, suppose the model is that large harem size in Redwinged Blackbirds causes reproductive success to be higher. A prediction from this model is that the reproductive success of males with larger harems will be higher than the reproductive success of males with smaller harems. This statement can be translated into a statistical hypothesis if the hypothesis is stated as follows:

> The mean reproductive success (fledglings/season) of male Red-winged Blackbirds is a strictly increasing function of harem size.

The null hypothesis would be that mean reproductive success is not a strictly increasing function of harem size. Note that if data were gathered and the null hypothesis were rejected in favor of the alternative hypothesis above, we would not be able to conclude that the model was correct. In particular, we would not be able to say that harem size causes an increase in reproductive success. That would be equivalent to concluding that the model is true. Instead all we can say is that the model has not been falsified. Some indication of factors affecting the number of young fledged might be sorted out (Lenington, 1983). Effects of weather would be obvious. The failure of entire nests might be attributed to predation. The number of young not fledged from otherwise successful nests might indicate starvation. But clearly if males with large harems have higher reproductive success, it would take more work to find out the cause of this relationship. If, on the other hand, the null hypothesis were accepted, then the prediction would be deemed incorrect. The model probably would require modification or replacement. Of course when one uses statistical tests to make decisions there is always the possibility of making an error because the power of the test is too low and because of the random nature of the data.

Statistical hypothesis tests are designed to control the probability of falsely concluding that a prediction has been upheld. This probability, called the type I error rate or the significance level, is controlled at some predetermined, acceptably small level, traditionally 0.1, 0.05, or 0.01. Requiring this probability to be very small can sometimes have the tradeoff of making it difficult to detect when the model predictions actually are upheld.

Popper's idea of making improbable predictions also fits this framework. Although we can never prove a model correct, it nevertheless gains credence as more and more of its predictions are upheld in statistical hypothesis tests. Thus, it is important to choose improbable predictions—predictions that competing models would not make. This way, if the prediction is upheld, it simultaneously indicates that competing models are less valid. Such evidence would lend credence to the proposed model.

An interaction between ideas and data all through the scientific procedure should be pursued. Because observations are necessarily theory-laden, the objective should be to scrutinize this interaction so that ideas and data move along together from informal ideas and exploratory analysis toward more formal models paired with the appropriate level of experimental design. Popper suggests starting anew by frequent challenges to the underlying concepts. But this alone would certainly not have yielded what we understand today about birds. In fact, most of what we know about birds seems to have come from observation and inductive generalization. To say that descriptive work has no underlying concepts is to fail to appreciate that induction is an essential part of science. Popper (1972) is adamant that induction cannot yield reliable knowledge about the validity of research hypotheses, but that does not mean that inductive methods are not useful for evaluating predictions.

The rationalist philosophy of science discourages another important part of the scientific method in biology—that of exploratory methods of hypothesis formulation. Much of what has worked in science has not followed the rationalist formula (May, 1981a,b, Ghiselin, 1969; Newton-Smith, 1981). We certainly do not agree that non-Popperian science has produced only sterile description (Fretwell, 1975). Rather, we agree with Hull (1979) that more scientific effort should be devoted to developing plausible theories. The most creative part of science, the step between observation and the formulation of the research hypothesis, is missing from Popperian methods. According to Hanson (1958, p. 70), "By the time the hypothetico-deductive system has been defined the original thinking is over." Active research disciplines do not just invent logical models and refine them. They discover new patterns.

The formulation of the theory of natural selection, which developed in Darwin's diaries over several years, occurred to Wallace in a delirium. But both men had spent many years observing nature. To deny that gathering evidence and generalizing from particular cases (inductive reasoning) was also part of their scientific procedure in either case would be ridiculous. Toward the end, Darwin said, "Looking back, I think it was more difficult to see what the problems were than to solve them" (quoted in Andrewartha and Birch, 1954).

3.3. Exploratory Data Analysis

"Exploratory data analysis" is the term used by Tukey and others for a set of techniques designed to find patterns in data. Often the original ideas are vague, and we want to know what the data indicate. Several new books give a variety of suggestions for ways to simplify descriptions and make them more effective (Tukey, 1977; Mosteller and Tukey, 1977; Velleman and Hoaglin, 1981; Chambers et al., 1983). Just making simple graphs can force the researcher to notice unexpected relationships. The idea is to avoid premature refinement of the problem because at each step the flexibility of alternate solutions becomes narrower. The magnitude of differences is emphasized rather than their statistical significance.

As one example, Tukey (1977) recommends that samples be displayed by box plots that show the median, range, and distribution of the data. The distribution of a sample is expressed by the upper and lower quartiles, so the box encloses 50% of the observations. Figure 2 gives data for wing length of samples of adult male Red-winged Blackbirds taken in winter in the southern United States ordered generally by increasing latitude. Larger birds seem to occur farther north. Because skewness in the data is displayed, a box plot gives more information than would a more conventional Dice-Leraas diagram (which gives the mean, range, standard deviation, and standard error of the mean). To judge the statistical significance of differences between any two samples by confidence limits (two standard errors on each side of the mean;

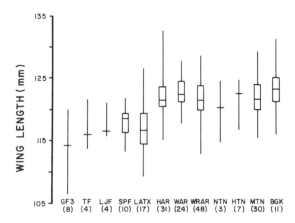

FIGURE 2. Box plots for samples of more than ten wing lengths of adult male Red-winged Blackbirds taken in winter at 12 localities in the southern United States, arranged in order of generally increasing latitude. The southernmost locality, GF3, is Gainesville, Florida; the northernmost, BGK, is Bowling Green, Kentucky. Sample sizes are given in parentheses. From James *et al.* (1984a). Box plots give the median, the range, and the upper and lower quartiles of the data.

one standard error of the mean is the standard deviation divided by the square root of n) would not be appropriate for these data because there was no clearly defined population from which the data would represent independent, identically distributed observations. Dice-Leraas diagrams are particularly inappropriate in cases like the one in the paper by Gill and Wolf (1975) in which the mean minus one standard deviation is below zero, but all data are positive. The basic aim of exploratory data analysis is to look at patterns to see what the data indicate. McGill *et al.* (1978) give variations on box plots that provide additional information such as "notches" to judge statistical significance between groups, when it is appropriate.

Another aspect of exploratory data analysis is explained in new books by Ehrenberg (1978, 1982). He suggests looking for empirical generalizations by examining several sets of data simultaneously, just for the sake of comparison. Examine scatter, medians, and modes, not always the mean. Ehrenberg shows that by simply rounding numbers and ordering the rows or columns of tables remarkable clarity can be obtained. For example, in tables of correlation coefficients the second digit does not really have much meaning. If a table contains a pattern, its impact is enhanced by reordering numbers and reporting only one digit (Table I). Such simple forms of data reduction show that statistical inference is far less important than is usually made out. Even when the researcher starts with a clear conceptual purpose in mind, there

TABLE I

Correlation Coefficients for Tested Variables Possibly Affecting Calling of
Whip-poor-wills and Chuck-will's-widows.[a]

Version A Variables	Singing birds		
	Whip-poor-wills	Chuck-will's-widows	Both species
Calendar date	− 0.3899	− 0.1747	− 0.2877
Sine curve date	0.1681	0.3616	0.2591
Temperature	0.0557	0.3246	0.1825
Relative humidity	− 0.0126	0.2325	0.1031
Wind velocity	0.2229	0.0264	0.1049
Visibility	− 0.0241	− 0.2469	− 0.1292
Moon phase	0.6372*	0.3841	0.5167**

*Significant at p = 0.01 level
**significant at p = 0.001 level

Version B Variables	Singing birds		
	Whip-poor-wills	Chuck-will's-widows	Both species
Moon phase	.6	.4	.5
Sine curve date	.2	.4	.3
Temperature	.1	.3	.2
Wind velocity	.2	.0	.1
Relative humidity	.0	.2	.1
Visibility	-.0	− .2	− .1
Calendar date	− .4	− .2	− .3

[a]The version as published (A) could have been clarified by rounding, omission of three decimal places, and reordering of the variables in the order of decreasing values of the correlation coefficients for both species (B). The significance levels should not be reported. The (A) version is the form of the table that was published (Cooper, 1981).

should be an exploratory stage to the research during which assumptions are as few as possible and observations are merely organized to look for underlying patterns (Mosteller and Tukey, 1977). See Tufte (1983) on the visual display of quantitative information.

Multivariate statistics is an area where exploratory methods are more promising than confirmatory methods (Cooley and Lohnes, 1971; Gnanadesikan, 1977; Johnson, 1981; see also Everett, 1978; Capen, 1981). The assumptions inherent in many multivariate confirmatory procedures are rarely achieved, and violation of the assumptions may invalidate the procedure. Especially useful exploratory techniques are principal components analysis and discriminant analysis. These procedures reduce the dimensionality of data to a few uncorrelated linear combinations of variables. Principal components analysis seeks linear combinations of variables that explain maximal variance, and discriminant analysis searches for new variables that maximally separate pre-existing populations. Both of these techniques are exploratory, because they suggest new variables and/or interpretations without needing a priori hypotheses. The use of these techniques can often suggest hypotheses to be tested in future experiments.

Williams (1976) summarizes a basic difference between statistical hypothesis testing and exploratory analysis. Statistical hypothesis testing is most powerful when there are only a few variates of approximately known distribution. In these situations the methods of exploratory pattern analysis are weak. Conversely, pattern analysis methods are strongest in multivariate cases when the number of attributes is large, and in such cases hypothesis testing is weak.

4. OBSERVATIONAL VERSUS EXPERIMENTAL STUDIES

The difference between an observational and an experimental study is simply that in the experimental one the researcher controls the levels of certain variables. Most of what we know about birds has been learned by observational studies. Experienced observers with informal conceptual models have used case analysis or have conducted surveys, compiled results and interpreted them. If further evidence has lent support to the interpretations, empirical generalizations have developed. In this way observational research has resulted in knowledge of the distribution and abundance of birds, their behavior, habitat affinities, migration habits, and life histories. Subsequent laboratory and museum research has resulted in knowledge of morphological and paleontological relationships.

If the overall objective of science is to develop explanatory theories about causal relationships among variables, and the study of processes requires experimentation (involving processes as well as patterns), then, in addition to purely observational work, an understanding of the methods of experimental design is crucial. Statistical inferences about causes are strongest when they are based on evidence from manipulative experiments, but they can also be made from observational studies that use principles of experimental design in the analysis stage (Eberhardt, 1976, 1978). If it is totally impossible to sort out the possible factors causing a pattern, the researcher should only describe patterns with empirical models.

4.1. Randomization, Replication, and Independence

The idea of experimental design, originally developed in agricultural experiments, is based on the random assignment of treatments to plots (experimental units) and the use of variation among plots given the same treatment as a standard on which to judge whether real differences result from two or more treatments (Box *et al.*, 1978; Eberhardt, 1976). The most essential principles of good experimental design are that:

1. Variables under study should be fixed at more than one level during the experiment. A combination of variable levels allows estimation of their effects separately and simultaneously.
2. The researcher must introduce randomization to break links with possible "lurking" variables that affect the process but are not observed, or even known to exist, but that could cause biases under systematic allocation schemes (Box *et al.*, 1978). Steps should be taken to assure that treatments are interspersed in space (Hurlbert, 1984). If randomization of treatments is not possible or advantageous, a systematic layout may be acceptable.
3. The conditions under which the experiment is conducted must be representative of or similar to the conditions under which the conclusions will be drawn. This condition would be met if the observations in the experiment were randomly selected from the population of interest.
4. Treatments must be replicated. Replicates (units to which the treatments are applied) need not be similar to one another, but they must be statistically independent draws from the population.

Consider the following experiment in which the objective is to determine the effect of fire on the breeding population of Bobwhite (*Colinus virginianus*) in the pine forests of the Apalachicola National Forest of northern Florida. Principle number three says to select study sites at random in the pine forests. Principle number two says to select at random half of the sites to be burned. If the sites vary appreciably according to soil type, tree size, understory, and other factors, consider pairing similar sites and then randomly choosing one from each pair to be burned. This procedure (blocking) is in fact a fifth principle of experimental design. The variable of interest is fire. According to principle number one, it will be manipulated to two levels, fire and no fire. The experimental design outlined here might be impractical, but if it could be carried out, it would have the following advantages:

1. The inference of causation of any observed results due to fire would be strong, because the levels of that variable would have

been manipulated and variation in other variables would have been controlled.

2. By random allocation of experimental units (sites to be burned), systematic error would have been eliminated. If nature had selected which sites were burned, those more susceptible to burning might have differed in other important ways from less susceptible sites.

3. In this experiment, results could be generalized appropriately to all sites in the forest, because the sites were chosen at random from all possible sites in the forest.

Sometimes, true randomization or even a stratified design in which treatments are regularly interspersed with controls is impractical. The U. S. Forest Service has its own criteria for deciding which sites should be burned, and the researcher may be restricted to studying previously burned sites. Or, even if the experiment could be conducted, it might cost much more than a well-conducted observational study. If more data could be gathered in an observational study for the same cost, it might yield more information. In the observational study, variables could be controlled by blocking in the analysis stage. One could block by searching out sites with values of key variables that are thought to influence the results and are similar to those on burned sites, match each burned site with a nearly identical nonburned site, and make comparisions within matched pairs. Or key variables such as acreage burned could be included as a covariate in an analysis of covariance. In a sense, differences in acreage burned would be accounted for. Clearly, the choice of variables to include in the analysis stage is crucial. The analysis of strictly observational studies can be improved through application of the principles of experimental design. By adherence to these principles, one can increase the possibilities for inferences of causality.

4.2. Field Experiments

Recent reviews of field experiments in biology (Hayne, 1978; Underwood, 1981; Hurlbert, 1984; Romesburg, 1981) have documented widespread misapplications of statistical inference. Underwood (1981) found more than 140 papers in marine biology in which analysis of variance was incorrectly reported. Hurlbert (1984), after a major review of the literature on ecological field experiments in which inferential statistics were used, found that in 48% of the papers researchers used what he calls "pseudoreplication." This is the invalid use of inferential statistics to test for treatment effects with data from experiments where either treatments are not replicated (although samples may be) or replicates are not independent (see Section 5).

In field experiments, when treatments are not replicated or replicates are not independent, differences cannot be assumed to be due to treatments. This pitfall also applies to the use of Dice-Leraas diagrams giving paired nonoverlapping confidence intervals. These leave the impression that statistically significant differences are present, but in many cases the treatments were not replicated.

For an example of a well-designed field experiment, see Andersson's (1982) study of Darwin's idea that male secondary sexual ornaments evolve through the preference of females. Andersson experimentally lengthened the tails of male Long-tailed Widowbirds, *Euplectes progne*. These males had higher mating success than males having normal or reduced tails. Because males with shortened tails were unable to maintain their territories, the results suggested that the extreme tail length of normal males is maintained in this species by female mating preferences (Fig. 3). Andersson used blocking on initial tail size and controls to see whether the treatment (cutting and pasting feathers) had an effect.

In another good experiment by Cronmiller and Thompson (1980), the brood size of Red-winged Blackbirds was manipulated. By comparisions with birds in unmanipulated and control nests, they showed that enlarged broods showed less weight gain. Pseudoreplication was avoided because the analysis was done on nests, not individual birds.

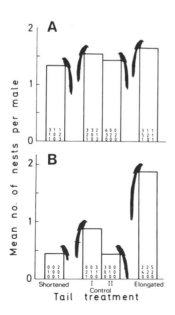

FIGURE 3. Mating success in male Long-tailed Widowbirds subjected to different tail treatments. (A) Mean number of active nests per territory for the nine males of each of four treatment categories before the experiment. (B) Number of new active nests in each territory after treatment of males. Matched comparisons between the number of nests on each male's territory before and after the experiment showed that males with elongated tails became significantly more successful than control males or males with shortened tails. Reprinted by permission from *Nature*, Macmillan Journals Limited. See Andersson (1982).

The weight gain of individual birds within nests would tend to be more similar than that of birds in different nests. Hence measurements on sibling birds were treated as correlated, not independent, replicates.

The main difference between an observational and an experimental study is that the latter uses manipulation of variables. The difference between an exploratory and a confirmatory study is that the latter employs inferential techniques, usually hypothesis or significance tests. Whether such tests are appropriate depends on the important assumptions discussed in this and the next section. Table II gives an example of each of these four combinations as they refer to a particular experiment and the factors underlying patterns of geographic variation in the size and shape of Red-winged Blackbirds (Agelaius phoeniceus) (James, 1983). The data are measurements of nestlings in three groups at each of two localities. One group is nestlings from normal (unmanipulated) eggs. The second group is nestlings from control eggs that had been held in an incubator and then returned to local nests. The third group is nestlings from experimental eggs that had been transported to nests at the other locality. Inferences about causation should only be made from the tests in Section D of Table II, and in this particular case the factors that caused the observed differences are un-

TABLE II

Examples of the Various Aspects of a Research Project Designed to Study the Environmental Component of Geographic Variation in the Morphology of Red-winged Blackbirds Breeding in Florida[a]

	Exploratory (no tests)	Confirmatory (tests)
Observational (no manipulation)	(a) Discriminant function analysis of shape differences among normal (unmanipulated) nestlings in northern and southern Florida.	(b) Univariate or multivariate tests of shape differences among normal (unmanipulated) nestlings in northern and southern Florida.
Experimental (manipulation)	(c) Discriminant function analysis of shape differences among transplanted (manipulated) nestlings in northern and southern Florida (not reported).	(d) Univariate or multivariate tests of differences between control nestlings (eggs held in incubator, then placed in other local nests) and experimental group (nestlings from transplanted eggs).

[a]From James (1983).

known. Hand rearing under controlled conditions (another experiment) might give some clues.

4.3. "Natural Experiments"

The term "natural experiment" has been used by Cody (1974), Cody and Mooney (1978), and others to characterize the apparent adaptive convergence of biotas that occupy similar but widely separated environments. But as Connell (1975) and Abbott (1980) have pointed out, such comparisons in purely observational studies have no real controls, even if they can be considered to be replicates. Mares (1976) attempted to allow for this problem in his comparison of a desert assemblage of rodents in Argentina, another desert assemblage in the Sonoran Desert of the Southwestern United States, and a nearby assemblage in a coniferous forest. He was able to show that the degree of desert adaptation is apparently less in Argentina than in the Sonoran Desert, as would be expected from the known histories of the faunas. In this "natural experiment" there were only two replicates of the treatment (desert faunas) and one replicate of the control (coniferous fauna), so no statistical inferences were possible about desert faunas in general; but this study began to apply the methods of experimental design to the analysis of ecomorphological adaptations in vertebrates. James (1982) reviews the literature on these and other ecomorphological studies. It is possible to expand the concept of controls to "natural experiments" as we showed in the previous section in the discussion of observations of areas that had been burned and areas that had not been burned.

"Natural experiments" play an important role in deriving inferences from data even though they have many drawbacks compared with well-controlled, randomized experiments in which treatments have been manipulated. Experiments in a laboratory may not reflect the conditions under which we desire to make inferences. Although causation may be established under controlled conditions in the laboratory, the results may not generalize to natural settings. A thorough program to demonstrate cause and effect would be to perform a largely experimental study in the laboratory and confirm the results with either field experiments or an observational study in natural settings with controls incorporated at the analysis stage.

In his study of morphological convergence in the avifaunas of New World and Old World peatlands, Niemi (1983) identified genera of birds (*Zonotrichia*, *Spizella*, *Geothlypis*, and *Dendroica* in Minnesota and *Emberiza* in Finland) that had pairs of species with one species in shrubby and the other in coniferous habitats. Each genus was considered to be a replicate, and the shrubs versus conifers were considered to be the "treatments." In each genus the species that occupied the shrubby habitat was smaller and had relatively longer legs and a wider sternum than its congener in coniferous vegetation. Except for the pair of species in *Emberiza*, the bill lengths were relatively longer in the species in the shrubby habitat. These parallel patterns among different genera show that ecomorphological convergence is operating at a very subtle level, one that would not have been detected had the comparisions not been made without matching the genera. This analysis not only demonstrates intrageneric morphological adaptation to habitat differences, it suggests that behavioral experiments conducted in aviaries to test foraging efficiency of the species in these two types of vegetation would be worthwhile. Leisler (1980) has reported similar work on sylviid warblers. See Leisler and Thaler (1982) for an insightful combination of ecomorphological comparisons and behavioral observations with captive birds.

5. STRENGTH OF INFERENCE

The techniques of statistical inference, including hypothesis and significance tests, are tools whose purpose is to draw conclusions about a large set of individuals (the population) based on a small subset of those individuals (the sample). Thus it is not surprising that one of the assumptions necessary for using statistical inference is that the sample represent the population. This assumption usually takes the form of an assumption about random sampling of the population or that the measurements are independent and drawn from a probability distribution that is the same as that of the population. The inferences that can be made desperately depend on these assumptions about how the data were gathered. Unfortunately, the assumptions are often violated in the application of statistical inferential techniques (see Hurlbert, 1984).

Some models are robust, which means that inferences made from them are not sensitive to minor deviations from the assumptions (or at least some of the assumptions). However, when models are not robust, violation of the assumptions can lead to serious misinterpretations of the data. We will discuss statistical models and assumptions and how they affect inferences made from them. Statistical assumptions are usually of two types: assumptions about how the data were collected (random sampling, independence, etc.) and assumptions about the distributions from which the data were collected. Each of these types of assumptions is important, but assumptions about the distribution of the data (for example, that they are normally distributed) are perhaps less important than assumptions about sampling. Nonparametric statistical techniques make less restrictive assumptions about distributions of the data but require nearly the same assumptions about data collection as do parametric techniques. Thus nonparametric statistics are often not a cure for the problems of making statistical inferences.

If the population is defined as those animals or items about which we wish to use data to draw conclusions, then roughly speaking, if statistical inferences are to be drawn about the population, sampling must be random. A typical assumption for statistical inference is that of random sampling or a closely related assumption, that observations are independent, identically distributed, and selected from the population of interest. Here, "independent" means that knowing the value of any observation does not give any information about other observations, and "identically distributed" means that each observation is drawn from the same probability distribution. Violation of the assumption of random sampling or independent, identically distributed observations can lead to two possibly serious consequences: biased estimators and misleading standard errors. Some systematic sampling schemes may give independent measurements.

When samples are selected judgmentally rather than randomly, statistical inferences cannot be used. For example, suppose study sites are judgmentally selected throughout the southeastern United States to represent "typical" sites. Suppose further that random sampling is done *within* each site to measure some quantity, such as density of a certain bird species. Because sites have not been randomly chosen from any population, we cannot use statistical inference to draw conclusions about the United States or even the southeastern United States. Instead, the quality of any inferences made would depend on the data analyst's judgment as to how well the selected study sites represent the population of interest. These inferences may be perfectly valid, but they would be subjective inferences, not statistical inferences. Unless sampling had been random, confidence intervals would not necessarily have the stated confidence probabilities, hypothesis tests would not have the stated significance levels, and so on. On the other hand, if observations within a site had been randomly made, statistical inferences about those particular sites could be made. For example, statistical procedures could be used to test for differences between, say, species densities at site 1 and site 2.

For instance, in a study of whether nest predation is higher in edge habitats than in forest habitats, differences between observations from several parts of a single forest and several parts of an edge habitat should not be assigned a significance level. That would imply that all forest habitats and edge habitats had been randomly sampled or at least that independent observations in each of the two habitats had been made. Only subjective inferences, depending on how representative the two study sites are judged to be, are justified. In fact, it is very important to describe the selected sites carefully in the materials and methods section so that readers can make their own subjective inferences based on how representative of all forest and edge habitats they think the sites are. Note that if the observations had been randomly made within each of the selected sites, then statistical inferences could be made comparing the two sites, but not comparing forest and edge habitats.

The same principle is true with the simplest kind of generalization, such as an association between the size of Downy Woodpeckers (Picoides pubescens) and mean annual wet-bulb temperature in different parts of their geographic range. James (1970) used the product-moment correlation coefficient to express the association between these two variables. When the asterisks symbolizing statistical significance were added to the table of correlations, that implied that the averages for size and temperature by locality were random samples from larger populations of both birds and temperatures. In this case a population of temperatures is an absurdity. Because populations about which the inference was being made could not be defined, James (1970) should not have attached significance levels to a table of such correlations. There are many cases like this in the ornithological literature in which all available data are considered and interdependence within subsamples was probable.

The use of regressions or correlations is often helpful to describe relationships succinctly, but significance should not be assigned to the relationship unless the data meet the criteria of random sampling or independent, identically distributed observations. Even when experiments are being reported, the same rule holds. The simple presentation of a graph can allow the reader the chance to judge the linear dependence of the variables and does not leave a misleading impression of statistical significance.

Another common consequence of sampling that is not random is inaccuracy of variance estimates. The most typical situation is that in which similar units are selected for the sample and regarded, falsely, as independent draws from the population of interest. Hurlbert (1984) has described many situations in which it may occur. Roughly, variance estimates tend to be too small when similar units are selected, but they are regarded as independent. This error leads to the overestimation of the statistical significance of effects and to a loss of control of type I error. Some examples will serve to illustrate the consequences of violations of the assumptions.

Robertson (1972) selected study sites in marshy areas and upland areas and observed the fledging success of Red-winged Blackbirds at each site. Each nest was classified as successful (one or more young fledged) or unsuccessful (none fledged). The nests were then lumped into two general categories (marshy and upland) and a chi-square test was used to test for differences in the proportion fledging. The chi-square test requires that each observation (nest) within a general category be an independent measurement with the same distribution, which is a very dubious assumption in this case. In fact, a chi-square test applied to the marsh sites or the upland sites separately shows that the sites themselves have statistically significantly different fledging rates. The consequences are that inferences must be restricted to the sites actually sampled, and the chi-square test is too likely to indicate differences falsely. If inferences about marshy versus upland sites are desired then marshy and upland sites must be sampled independently. Robertson should have used sites as replicates, not nests. Of course, Robertson carefully described the study sites and their selection, which makes it easy for the reader to make subjective inferences based on the data collected. Thus the reader can make judgments as to the group of nests to which the data collected may be safely extrapolated.

Dow (1978) picked up a potential violation of the assumptions in a proposed method of Mayfield (1975). Mayfield suggested using a chi-square test to detect differences in mortality rates between nests in the incubation period and those in the nestling period. His unit of observation was the nest-day. A chi-square test for this situation would require that nest-days be independent observations. Although this may be a reasonable assumption, he also suggests that the sample size could be multiplied if the nest-hour were considered as the unit of observation. This seeming paradox of generating more observations can be resolved if we consider the independence of the observations. Indeed, as Dow points out, why not consider nest-minutes or nest-seconds? The answer is that as finer and finer divisions of time are considered, the observations become more and more dependent. Given that a nest was without losses in one second, it is very easy to predict (with high certainty) what the state will be in the next second. Thus as more pseudo-observations are generated, they become more and more dependent, and the assumptions for the chi-square test are farther and farther from being satisfied.

The violations of assumptions described above having to do with random sampling and independence can occur in all types of statistical tests including chi-square tests, t-tests, ANOVAs, and regressions. Besides these types of assumptions, there are also assumptions specific to particular statistical methods. Usually there are assumptions about the form of the distributions involved. For example, the F-test in a linear regression of Y on X requires that the observations, Y_i, be normally distributed with mean related to X by the formula $a + b * X_i$ and constant variance. We discuss this topic further in Section 6.

6. SPECIAL TOPICS

In this section we discuss several special topics in data analysis and, at the same time, use them to emphasize some of the issues raised in earlier sections. In particular, we discuss ordination techniques, regression and correlation and their relation to principal components analysis, size and shape analysis, and the use of multivariate distance measures.

6.1. Ordination and Other Continuous Models

In the most general sense, ordination means simply an arrangement of units into a linear order. It is a form of empiric modeling. The first application of ordination methods to ecology is attributed to Ramensky (1930), who constructed a system for ordering plant associations on the basis of a geometrically increasing scale of the percent of cover for the member species. The result was not a classification of associations into communities, but rather their display on a continuous axis. The subsequent development of phytogeography in Russia has extended this approach (Sobolev and Utekhin, 1973). By contrast, the leading early ecologists in the United States (Clements, 1904) interpreted compositional variation in plant associations as discrete communities and constructed classification schemes. Although methods of classification are often convenient and conceptually simple ways to describe variation, biologists who classify continuously varying phenomena risk forgetting the inherent limitations of forcing complex systems into discontinuous categories (Noy-Meir and Whittaker, 1977). Witness the biological reality still credited to many subspecies of birds that are in fact arbitrary divisions of clinal patterns of geographic character variation (Brown and Wilson, 1954; Gould and Johnston, 1972; James, 1970). A typological approach to ecology that began with schemes for classifying plant communities persists today in animal ecology in the form of theoretic models for community-level processes. These processes are thought to maintain structure in communities, but even in relatively stable environments, bird communities may exist as functional units only in the eyes of the community ecologist.

Harper (1977) charges that the dominance of classificatory thinking in the American approach to plant ecology has prevented the field from developing biologically realistic models for evolutionary processes such as adaptation. "If evolution is about individuals and their descendents . . . we should not expect to reach any depth of understanding from studies that are based at the level of the superindividual" (Harper, 1977, p. 148). He advocates a population-level approach to the analysis of vegetation and agrees with Gleason (1926) that plant distribution is primarily the result ,of past colonization events plus the fecundity and dispersal qualities of individual plants. In animal ecology this philosophical approach is represented by Andrewartha and Birch (1954). Based primarily on observations of insect populations in Australia, they viewed animal distribution as mainly determined by colonization events plus the rate of increase since the last disaster. With ordination methodology, we are not concerned with processes that affect adaptation, but rather with finding a biologially realistic empiric method of organizing data about the geographic distribution of species—one that does not make a priori assumptions about either the pattern of distribution or its cause. Can ordination research provide unbiased methods for expressing geographic population-level phenomena?

The ordination literature in plant ecology is very complex and not always consistent in terminology. Orloci (1975), Whittaker (1978), and Gauch (1982) use the term ordination only for the restricted problem of ordering species by samples (communities) or the reverse. Whittaker (1967, 1978) called arranging samples or species by environmental factors "direct gradient analysis," and he called arranging species by sites or sites by species "indirect ordination" or "indirect gradient analysis."

Cody's (1968) graphs of the distribution of several species of finches are the first examples of a multivariate approach to direct gradient analysis in ornithology. He used discriminant function analysis to define a composite axis expressing variation in structural features of grassland habitats that maximally separated several species of finches by their habitat affinities. This approach is an empiric model for the display of the distribution of birds along a multivariate resource axis. Cody considered the species to form a community and interpreted their location along the axis in terms of the Hutchinsonian (1965) model of the niche. According to the Hutchinsonian model, differences in the resource use of species are evidence of past or present interspecific competition. Subsequent work by Diamond (1979), Cody (1974, 1978), Dueser and Shugart (1979), Noon (1981), Sabo (1980), Sabo and Whittaker (1979), and others has invoked similar interpretations. This is a case in which a theoretic model has been used to interpret an observed pattern. Wiens and Rotenberry (1981) and Rotenberry and Wiens (1980) used similar methods, but concluded that patterns of species associations were probably not limited by interspecific relationships. Because none of these studies were either experimental or used principles of experimental design to try to sort out possible causes of the observed patterns, all the inferences about causes are unreliable. There is no need to invoke the theoretic implications of resource limitation in an analysis of patterns of bird distribution. James (1971), Smith (1977), Collins et al. (1982), Johnston (1977), Whitmore (1975), and others have used direct gradient analysis of vegetation structure to examine interspecific patterns of bird distribution without invoking the theoretical implications of resource limitation in their interpretations of the results. See Section 6.4 for further comments on multivariate niche analysis.

In phytosociological research, indirect ordination methods (= indirect gradient analysis) are used most often (Noy-Meir and Whittaker, 1977). Environmental data are not entered into the analysis at all. One of the first methods of indirect ordination was polar ordination (Bray and Curtis, 1957). Bond (1957) and Beals (1960) applied polar ordination to data for bird communities. Other indirect methods are multivariate methods that provide geometrically interpretable distances in graphic space (e.g., principal components; Orloci, 1966) or non-Euclidean methods that are suited to species-by-sites data (Hill, 1973, 1974; Gower, 1966; Noy-Meir and Whittaker, 1977; Gauch, 1982).

Reciprocal averaging (RA) is one of the non-Euclidean indirect ordination techniques that simultaneously orders species and sites. Sabo and Whittaker (1979), Holmes et al. (1979), Sabo (1980), and Collins et al. (1982) have applied RA to ordination studies of bird distribution. They used data for vegetation structure rather than the species-by-sites data for which RA was designed. Although most of these ordinations have provided intuitively pleasing results, and in these cases the results of RA were similar to those of direct gradient analysis based on principal components analysis of data for vegetatation structure, the biological relationships are necessarily vague with all indirect methods. The meanings of axes are difficult to interpret, and although the first axis in reciprocal averaging can be shown to be related to the first principal component for the same data (Hill, 1973), subsequent axes have no geometric interpretation. In fact, interpretation of higher axes with RA is often fuzzy or worse, misleading. This problem is not completely resolved in variants of RA such as detrended correspondence analysis (Gauch, 1982). Reciprocal averaging does not use Euclidean distances, but instead uses chi-square-like distances. This procedure may be appropriate for data that are counts, but whether there is statistical justification for applications to data consisting of percentages or continous measurements has not been investigated with continuous measurements. Euclidean distance often has clear interpretations. Also, some of the trappings of principal components analysis have been carried over to reciprocal averaging as if they were still valid. Some papers talk of "percent of the variance explained." Because variance arises from the use of the square of the Euclidean distance, this phrase has little meaning when chi-square distances are used. In fact, we think that one should only be using RA when variance is *not* a reasonable quantity to be discussing (in indirect ordination).

Direct ordination techniques have several advantages over indirect techniques, and they are potentially a very valuable tool for ecologists. Their main advantage over indirect techniques is that the researcher's notions about what environmental factors underlie the pattern of distribution can be built into the anlaysis. (Psychologists who use reciprocal averaging, under the name of optimal scaling, often do not have this same type of preliminary information.) In plant ecology, indirect methods are used for a very restricted class of problems, ones in which the data are merely the occurrence of species in associations. The indirect methods do not have the potentially broad ecological applicability that the direct methods have.

Whittaker's (1978) advocacy of direct gradient analysis was only within the Hutchinsonian model of community structure. This is an unnecessary restriction because in fact one of the main reasons that direct gradient analysis is such a good descriptive tool is that one can view distributional relationships among variables independently of any a priori theoretical construct about causes. Direct ordinations can be used to order species, associations of species, or attributes of species according to environmental factors. If an ordination that reflects foraging methods of birds and vegetation structure is desired, a canonical correlation analysis between two matrices of data for one set of species would probably be preferable to a principal components analysis in which the two data sets are pooled. Both principal components analysis and canonical correlation analysis are useful multivariate methods for constructing direct ordinations (James and Porter, 1979; see also Gibson, 1984, and the response by James and Porter, 1984).

In ornithology a clear proponent of the gradient philosophy was Joseph Grinnell. Although Grinnell did not attempt to reconstruct patterns of multispecies distributions, he emphasized the study of continuously varying adaptations of single species to resources throughout their geographic ranges (James et al., 1984). Note that Grinnell's concept of the niche (1917) was subsumed by community ecology and that ever since Elton's paper in 1927, the niche has been defined in terms of within-community parameters (Whittaker *et al.*, 1973).

6.2. Regression, Correlation, and Principal Components Analysis

Regression and correlation encompass a body of techniques used to investigate relationships among variables. The difference between regression and correlation methods is that in regression a variable is designated the dependent (or response) variable. It is predicted from or modeled as a function of the other variables, which are called independent (or predictor or carrier) variables. In correlation no such distinction is made.

The distinction between regression and correlation can be seen from a computational point of view if one looks at the calculations inherent in the two techniques. In regression, one finds the regression line by picking the line that makes the sum of squared distances in the direction of the dependent variable as small as possible. Thus, with two variables X and Y, the regression slope will be different if Y is regressed on X than if X is regressed on Y because two different distance measures are used. In the first case, distances parallel to the Y axis are used, and in the second case, distances parallel to the X axis are used. In correlation, the order of the two variables is unimportant and the same correlation is obtained no matter which is used as the first variable. Of course, no line is obtained from a correlation analysis. If a line is desired and the two variables are to be treated equally, the perpendicular distances from the line to the points may be the most appropriate distance measure to use. In this case, the line that minimizes the sum of squares of the perpendicular distances is the first principal component.

These concepts are illustrated in Fig. 4, which displays data from Mosimann and James (1979). In Fig. 4A, the regression of Y on X is plotted (where Y is the log of the mean bill length minus the log of the mean bill depth and X is the log of the mean wing length) along with the distances whose squared sum is minimized. Figure 4B shows a scatterplot of Y versus X along with the correlation, and Fig. 4C shows the first principal component along with the distances minimized. If interest is in predicting one variable from the other, then a regression analysis would be appropriate. On the other hand, in some cases, for example, allometric studies, no variable should be designated the dependent variable because no distinction between the variables is desired. Yet correlation analysis is not appropriate because the aim is to estimate a linear relationship between the variables if one exists. Hills

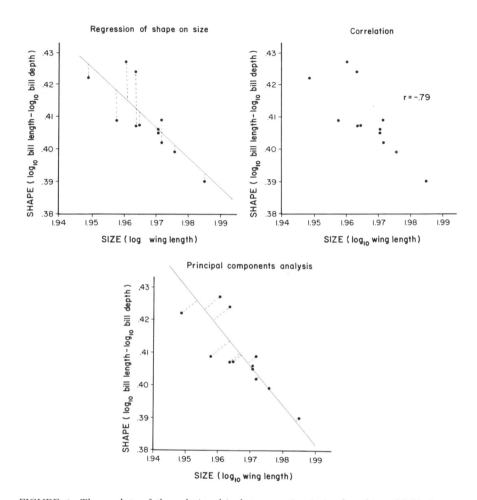

FIGURE 4. Three plots of the relationship between size (wing length) and bill shape (ratio of bill length to bill depth) in female Red-winged Blackbirds breeding in Florida. Data are averages for 12 one-degree latitude blocks (Mosimann and James, 1979) plotted as logarithms. Bill length/bill depth = log bill length minus log bill depth. (A) A is a linear regression of Y on X ($Y = 2.21 - .92X$). (B) gives the correlation ($-.79$). (C) Gives the first principal component ($Y = 2.8 - 1.2X$).

(1982) recommends fitting the first principal component of the sample covariance matrix (he calls it the major axis).

Another way one can use regression, besides prediction, is to try to interpret the coefficients in the regression equation. Often the goal is to try to establish association or causation between the dependent variable and one or more of the independent variables. This may be a difficult, if not impossible, task in observational studies. Of course, the mere fact that a regression analysis is being performed and the results of the analysis of the association of the variables is statistically significant is no basis for inferring causation.

One of the usual assumptions in using regression is that the independent variables are fixed or can be regarded as fixed at their values by the experimenter. Often this is not the case. In observational studies, the independent variables vary together and their joint values are merely observed during the course of the experiment or data gathering. In this circumstance, useful predictions can often still be made, as long as the relationships between variables are reasonably stable. However, interpretations of the coefficients of the sample regression model are fraught with difficulties. In particular, the usual interpretation of a multiple regression coefficient as the change in Y with a change in X, while all the other independent variables are kept fixed, is often nonsensical. If all the independent variables vary together, it may not make sense to consider fixing the values of all the independent variables except one. Power (1969) noticed these types of problems in a multiple regression analysis relating bill length of Red-winged Blackbirds to temperature and latitude. He had a difficult time interpreting the coefficient for latitude and came up with explanations which were, in his own words, "highly speculative." For a discussion of these difficulties, see Mosteller and Tukey (1977, Chapters 12 and especially 13) and Box et al. (1978, Section 14.7).

In the rush to test statistical hypotheses, investigators often forget that regression analysis can be a powerful descriptive statistical tool. By estimating a few parameters in a regression model, one may be able to summarize complicated relationships among many variables succinctly. Regression analysis can be used to remove the effect of, or "control for," a variable that is not of immediate interest, but that obscures the relationship between more interesting variables. New diagnostic statistics for regression analysis (such as influential case analysis or high leverage point analysis; see Weisberg, 1980) can aid in discovering observations that do not conform to the general patterns and may merit further investigation. Of course, an investigation of a data set not gathered to answer specific questions, with the fitting of several trial models, the removal of questionable points, and possibly transformation of some of the variables, largely invalidates formal inferential procedures. With a large enough data set, an investigator can use half of the data to investigate and form hypotheses and the other half to test hypotheses formally (Selvin and Stuart, 1966).

6.3. Size and Shape Analysis

In order to study morphological variation among animals, it is important to conceive of the problem in the framework of allometry—how shape varies with size. If size and shape variables can be defined precisely, then their association can be described. Mosimann (1970, 1975a,b), Mosimann and James (1979), Mosimann et al. (1978), and Mosimann (1984) give very simple direct methods for this problem that can be used in a multivariate framework. The techniques are best suited to the study of character variation and covariation at the population level. First, one or several biologically reasonable size variables are assigned. These may be the lengths of easily measured characters such as wing length or bill length or even a composite of variables such as the geometric mean of several measurements. Shape variables are either ratios or proportions. If the original data follow a multivariate lognormal distribution, their logs will jointly follow a multivariate normal distribution, and parametric statistical methods can be used to study the relationship of size and shape. The goal is not to produce uncorrelated size and shape variables, but rather to define size and shape precisely and then to analyze how they are related. Mosimann and James (1979), using frequency ellipses, showed that the assumption of a lognormal distribution was reasonable for their data. To see how bill shape is related to geographic size variation in Red-winged Blackbirds breeding in Florida, they plotted the log of the wing length (size) on a graph in which the other axis was the log of the bill length minus the log of the bill depth (= bill length/bill depth = shape). Average scores for samples for females from twelve one-degree latitude-longitude blocks have a correlation coefficient of $r = -0.79$. This number expresses the association of size and shape across localities (see Section 6.2 and Fig. 3).

Leisler (1980), James (1982), and Niemi (1983) show the value of performing a morphometric analysis solely on shape variables. By either dividing by a size statistic before conducting the analysis (Leisler 1980) or subtracting the log of a size statistic from the log of another measurement, they essentially created shape variables. Principal components analyses of these transformed data were then used to display the shape relationships of the species, and these in turn had striking ecological correlates. Such close linkage between morphology and ecology is less evident when measurements or their logs are entered directly into a principal components analysis or a discriminant analysis. This fact may explain why Wiens and Rotenberry (1980) did not see a relationship between external morphology and the ecology of the associations of birds that they studied.

Jolicoeur (1963) proposed that the first principal component of the log measurements be used as the size variable. The first principal component can usually be interpreted as a size variable, but some care must be exercised. All that principal components analysis guarantees is to pick as the first principal component that linear combination of the variables that has the maximal variance. There is no notion of size built into the analysis, and there is no guarantee that the first component will turn out to be a suitable size measure. If some of the coefficients are negative or they are all positive but not close to being equal, then it may be best to discard it in favor of a specified size measure, as opposed to a data-derived size measure. For a discussion, see Mosimann and Malley (1979).

Even if principal components analysis gives a first component that can be satisfactorily interpreted as a size variable, it may not adequately separate the effects of size and shape. Burnaby (1966) proposed a method that gives only components that are invariant to size. His method also works for discriminant function analysis. A feature of his method is that the size measure must be specified rather than derived from the data. Another facet of his method is that, as with principal components analysis, which chooses PC1 as the size variable, it gives components that are uncorrelated with the size variable. If the objective is to use the resultant components to classify observations into populations, this may be an advantage. On the other hand, if the aim is to determine whether size and shape are related, then the above approaches will not be useful. An interesting new proposal by Mosimann and Darroch (unpublished) derives scale-free shape components, which are not forced to be uncorrelated with size variables and hence can be used to analyze the relationships of size and shape.

Most of the methods discussed above use linear combinations of the logs of character measurements, which is essentially the same as looking at ratios of measurements. In our opinion some of the arguments of Atchley *et al.* (1976), who claim that ratios are necessarily bad, are misleading. For example, the commonly used *t* statistic is the ratio of a sample mean and the square root of the sample standard deviation divided by the sample size. Thus, there is nothing inherently wrong with ratios. Atchley *et al.* claim that "ratios of random normal variables are distributed as a Cauchy distribution" and quote some undesirable properties of that distribution. But the only case in which ratios of normal variables have a Cauchy distribution is when the means of both the numerator and the denominator are exactly zero. Surely this is an unreasonable model for strictly positive measurements such as character variables. The important point is that the resulting distribution of the ratios is known or can be approximated and worked with. Atchley *et al.* imply that there is something wrong with skewed distributions. Again it is only important that the specific characteristics of the distribution be known; for example, the widely used F distribution is a skewed distribution. Atchley *et al.* emphasize that they are only looking at the statistical properties of ratios, but they are missing an important point. There is nothing wrong with the fact that there is negative correlation between X/Y and Y when X and Y are uncorrelated. The only possible trouble is in the interpretation of the correlation as a meaningful biological relationship. Surely that is a biological problem, not a statistical one. Ratios, properly treated, are very important variables in biology.

6.4. Distance Measures and Multivariate Niche Analysis

Multivariate distance measures are widely used in many fields of biology (Atchley *et al.*, 1982) and are a key ingredient of the statistical analysis of niche size and niche overlap. Atchley *et al.* (1982) provide a very clear discussion of the use of various distance measures, including Manhattan distance, Cherry's distance, and Mahalanobis distance.

In the first paper on the use of multivariate methods in niche analysis, Green (1971) attempted to build a statistical model for testing niche overlap. It made some pioneering contributions by solving two problems in niche analysis:

1. Correlated variables were not analyzed because discriminant axes (canonical axes) are uncorrelated.
2. Dimensionality of the data was reduced in a sensible way.

On the other hand, we feel that it fails in one of its main goals—to derive a statistical model and statistical tests for niche overlap.

This is another case where inferential statistics, as they are applied, are inappropriate. First there are problems in translating niche theory into an inferential statistics problem. If we use Hutchinson's definition of a niche, then a niche can be represented mathematically as a hypervolume in N-dimensional space. The test in discriminant function analysis (DFA, which is the same as MANOVA) is that the *means* of the environmental variables where the organisms are found are the same. It does not really say anything about hypervolumes, unless one includes the *assumptions* of multivariate normality and equal variance-covariance matrices. Thus, the only case for which the DFA test can be regarded as a test of niche overlap is when the niches are assumed to be exactly the same size and distributed according to a multivariate normal shape (in which event, all that is left is the mean—where the hypervolume is located). In this case, major parts of the theory have been *assumed* rather than tested.

The other problem with the use of inferetial statistics is that it is hard to imagine *exact* equality of niches of two species. That is precisely what the DFA test is looking for—*exact* equality of the means of the environmental variables. Thus, the test degenerates into a test of whether or not the sample size is sufficient to detect a difference that almost assuredly exists.

Even though the inferential purpose of DFA is to separate means, using DFA to derive axes may be reasonable because DFA may succeed in separating the centers of the niches. Thus, the major contribution of Green's paper is on a descriptive level. It provides us with new ways to look at niches. In fact, much of the later part of the paper is just that—a use of the DFA to provide description. Williams (1983) reviews past applications of discriminant analysis in ecology and concludes, as we do, that authors who have interpreted confirmatory tests as evidence for processes should have confined their discussions to the description of patterns.

In addition to problems with statistical inference in multivariate niche analysis, other problems are discussed by Rotenberry and Wiens (1980), Carnes and Slade (1982), Van Horne and Ford (1982), Dueser and Shugart (1982), and James et al. (1984b). Two points are particularly important. First, grossly different sample sizes may affect distance calculations (Carnes and Slade, 1982; Van Horne and Ford, 1982). Second, when species are plotted in a graphic resource space defined by their values on the first two or three principal components, equivalent Euclidean distances between species means do not have the same probabilistic meaning in all directions. Differences depend not on distances alone but on the directions in which they differ (James et al., 1984b). When the graphic resource space is defined by discriminant function analysis, the distances have only probabilistic meaning. Population means may be judged far apart in cases in which the species are similar in all resource use except that they differ in the use of one resource that varies a small but statistically significant amount. Contrary to Carnes and Slade (1982), Euclidean distances in principal components space

and Mahalanobis distances in discriminant space should not be interpreted the same way.

7. CONCLUSIONS

We have tried to show that all scientists use models, whether they
realize it or not. The early ornithologists, whose excellent work has
made birds the best-known class of organisms, had very clear objectives
and worked with informal models. Bent (1919–1967) organized information about life histories, Rensch (1934, 1939) studied character variation, and Grinnell (1904, 1924) analyzed geographic distribution. In
many ways, they were no less scientific than modern ornithologists.
Now we have the tools of statistical inference and experimental design,
so the job is to use them where appropriate. If future ornithologists
place value judgments on periods in the development of biology, they
may well criticize our present practices for the misuse of methods that
were available. We are using inductive reasoning to justify the accep-
tance of untested theoretic models, and we err in the design of experiments. Our journals are full of tables decorated with stars and double
stars when in many cases the assumptions underlying the inferences
of statistical significance are not met. In many cases, we are using
statistical significance tests and confidence levels when they are not
only not appropriate but misleading.

There are parallels between the fields of statistics and views of
rationalist philosophers like Popper. Both advocate the formulation of
hypotheses and attempts to refute them. The difference is that Popper
says that research hypotheses (theoretic models) can be falsified by
nonsupport of critical predictions. Statistical hypothesis testing is simply about evaluating the predictions of both theoretic and empiric models
in those cases in which one has independent samples from the populations of interest. For testing the predictions of theoretical models,
both statisticians and rationalist philosophers advocate conducting critical experiments that are designed to discriminate among different models.

We have tried to make three more points from the field of statistics.
When the principles of experimental design that are used in manipulative experiments are incorporated into the anlaysis stage of strictly
observational studies ("natural experiments"), weak inferences about
causal relationships are possible; otherwise they are not. Second, although we agree with Popper's argument that one cannot prove statements by example (induction), statistical induction controls the chance
of making mistakes. Statistical procedures help to define the uncertainty

of inferences about populations from samples. Third, an important part of the scientific method is the early observational stage, before a formal model has been selected. In the stochastic biological world, exploratory data analysis should precede even empiric modeling, and empiric modeling should precede the formulation of theoretic models. More attention to these stages should permit scientists to formulate more realistic theoretic models.

APPENDIX I
Glossary

Topic	Description	References
Analysis of variance (ANOVA)	A numerical method for partitioning the total sum of squares of the data and its degrees of freedom. Often used to get F tests for differences between the means of different groups.	Sokal and Rohlf (1981, Chapters 8–13) Snedecor and Cochran (1980, Chapters 11–16, 20)
		See recent survey articles by Underwood (1981) and Hurlbert (1984).
Analysis of covariance (ANCOVA)	A special form of analysis of variance that allows adjustment for independent variables. Used to increase precision in analysis of variance by adjusting for possible bias introduced by independent variables and to estimate group means adjusted for the independent variables.	Sokal and Rohlf (1981, Section 14.9) Snedecor and Cochran (1980, Chapter 18)

(Continued)

APPENDIX I (Continued)

Topic	Description	References
Box plot	A graphical technique for displaying the median, range, and skewness of a data set.	Tukey (1977) McGill et al. (1978)
Canonical correlation	A multivariate statistical technique for calculating correlations between a linear function of one set of variables and a linear function of another set of variables. The objective is to find out whether there are significant correlations between the sets.	Tatsuoka (1971) Morrison (1976)
Chi-square tests	Several statistical methods for tests such as testing goodness of fit and testing independence of categorical variables.	Sokal and Rohlf (1981, Chapter 17) Snedecor and Cochran (1980, Chapter 11, Sections 20.7–9)
Cladistics	In systematics, a set of methods for constructing branching diagrams (cladograms) that express the relatedness of a set of units (taxa) based on variation in discrete traits. The methods contain criteria for determining branching points in the diagrams on the basis of shared derived characters.	Cracraft (1982) Hull (1979) Eldredge and Cracraft (1980) Patterson (1982)

APPENDIX I (Continued)

Topic	Description	References
Cluster analysis	Methods for constructing branching diagrams (dendrograms) that express the overall similarities (or differences) among a set of units.	Pielou (1977, Chapter 20) Hartigan (1975) Johnson and Wichern (1982, Chapter 11)
Correlation	Techniques to measure the closeness of linear relationships between variables. Multiple correlation measures the association between a single variable and a linear combination of a group of variables. Partial correlation measures the association between two variables while adjusting for other variables.	Sokal and Rohlf (1981, Chapter 15) Snedecor and Cochran (1980, Chapter 10)
Curve fitting	Methods of finding a curve, $Y = f(X)$ that passes closely through a cloud of data representing a plot of Y versus X. Common methods are least squares (linear and nonlinear regression) and splines.	Snedecor and Cochran (1980, Chapter 19)
Directional statistics	Techniques for analyzing data that consist of angles or directions. Commonly used in orientation studies.	Batschelet (1965, 1979)
Discriminant function analysis (DFA)	A multivariate statistical method to find a new composite variable or variables	Johnson and Wichern (1981, Chapter 10)

(Continued)

APPENDIX I (Continued)

Topic	Description	References
	(linear combinations of the original variables) that separate previously identified groups in the data.	Pielou (1977, Chapter 22)
Diversity indices	In ecology, descriptive statistics designed to measure simultaneously the species richness and the relative abundances of species in a community (or sample).	Pielou (1977, Chapter 19) Pielou (1975), Green (1979)
Factor analysis	A multivariate statistical method that attempts to find structure among the intercorrelations of an ensemble of variables.	Johnson and Wichern (1982, Chapter 9) Tatsuoka (1971)
Hypothesis	Hypotheses can be either research hypotheses or statistical hypotheses. Research hypotheses are theoretic models that make predictions. Statistical hypotheses are statements about population parameters. Usually a statistical hypothesis, called the alternative hypothesis, reflects the predictions of an empiric or theoretic model, and it has a logical complement, called the null hypothesis. Note that	

APPENDIX I (Continued)

Topic	Description	References
	the term, null hypothesis, is a statistical term and does not necessarily imply that the situation described by the null hypothesis is a null model.	
Jackknife	A nonparametric technique for finding a standard error for complicated statistics.	Sokal and Rohlf (1981, Section 18.4)
Multivariate statistical methods	A body of techniques for anlayzing data that consist of more than one variable. In addition to standard statistical objectives such as obtaining confidence interval estimates, common objectives are to separate data into groups and to find lower dimensional structure in the data.	Morrison (1976) Tatsuoka (1971) Johnson and Wichern (1982) Green (1980)
Multivariate analysis of variance (MANOVA)	Analysis of variance for more than one variable.	Johnson and Wichern (1982, Chapter 6) Tatsuoka (1971)
Nonparametric statistics	Statistical methods that do not assume that the data follow a particular distribution (such as the normal distribution).	Conover (1980)

(Continued)

APPENDIX I (*Continued*)

Topic	Description	References
Ordination	Techniques to order units (e.g., communities or species) along an axis or axes. In indirect ordination, the axes are determined by the co-occurrence patterns of species. In direct ordination, the axes are determined by environmental variables.	Pielou (1977, Chapter 21) Gauch (1982, Chapters 3, 4)
Parametric statistics	Statistical methods that depend on the assumption of a specific distribution for the data (e.g., normal).	Most of Sokal and Rohlf (1981) and Snedecor and Cochran (1980) deal with parametric statistics.
Phenetics	The study of the overall similarities of the phenotypes of groups of organisms; methods involve reducing the dimensionality of morphometric data or classifying "operational taxonomic units" (e.g., taxa) according to overall similarities.	Sneath and Sokal (1973)
Principal components analysis (PCA)	A multivariate statistical method designed to reduce the dimensionality of the original variables by creating new composite variables ordered by variance explained.	Johnson and Wichern (1982, Chapter 8) Tatsuoka (1971)

APPENDIX I (*Continued*)

Topic	Description	References
Probability models	Models that describe patterns of random variation. Common models: normal, binomial, Poisson, *t* distribution, chi square, and *F* distribution.	Sokal and Rohlf (1981, Chapters 5, 6) Snedecor and Cochran (1980, Chapters 4, 7)
Rarefaction	A statistical method that can be used to estimate the number of species expected in a random sample of individuals taken from a census.	Simberloff (1979) James and Rathbun (1981)
Reciprocal averaging	An indirect ordination technique.	Gauch (1982, Chapter 4) Hill (1973, 1974)
Regression	A technique for modeling one variable (the dependent variable) as a function of one or more other variables (the independent variables). Often used for predicting the values of the dependent variable.	Sokal and Rohlf (1981, Chapters 14, 16) Snedecor and Cochran (1980, Chapters 9, 17, 19)
Sampling	Methods of selecting units for study from a larger set of units (the population). Usually, the objective is to use the sample units to gain information about the population.	Snedecor and Cochran (1980, Chapters 1, 21)
Simulation	The use of random numbers, rather than real data, to find the properties of a model or a statistic.	Fishman (1978)

(Continued)

APPENDIX I (Continued)

Topic	Description	References
Spatial statistics	A body of statistics used for analyzing data on how populations distribute themselves in space.	Pielou (1977, Chapters 7–17) Ripley (1981)
t test	A test used to compare means of normal distributions. Also used to test contrasts in analysis of variance and regression coefficients in regression.	Sokal and Rohlf (1981, Chapter 7) Snedecor and Cochran (1980, Chapter 6)

ACKNOWLEDGEMENTS. We thank the graduate students in the ecology group at Florida State Univeristy who took part in early discussion of papers. D. Simberloff, S. Hurlbert, W. W. Anderson, J. E. Mosimann, J. Travis, J. Rhymer, G. Graves, D. Meeter, M. Keough, R. T. Engstrom, and S. Rathbun made helpful suggestions about drafts of the manuscript.

REFERENCES

Abbott, I., 1980, Theories dealing with the ecology of landbirds on islands, in: *Advances in Ecological Research*, Volume 11, (A. Macfadyen, ed.), Academic Press, London, pp. 329–371.

Abbott, I., Abbott, L. K., and Grant, P. R., 1977, Comparative ecology of Galápagos ground finches *(Geospiza* Gould): Evaluation of the importance of floristic diversity and interspecific competition, *Ecol. Monogr.* **47**:151–184.

Alatalo, R. V., 1982, Evidence for interspecific competition among European tits *Parus* spp.: A review, *Ann. Zool. Fenn.* **19**:309–317.

Allee, W. C., Emerson, A. E., Park, D., Park, T., and Schmidt, K. P., 1949, *Principles of Animal Ecology*, W. B. Saunders Co., Philadelphia.

Andersson, M., 1982, Female choice selects for extreme tail length in a widowbird, *Nature* **299**:818–820.

Andrewartha, H. G., and Birch, L. C., 1954, *The Distribution and Abundance of Animals* University of Chicago Press, Chicago.

Aschoff, J., and Pohl, H., 1970, Rhythmic variations in energy metabolism, *Fed. Proc, Fed. Amer. Soc. Exp. Biol.* **29**:1541–1552.

Atchley, W. R., Gaskins, C. T., and Anderson, D. T., 1976, Statistical properties of ratios. I. Empirical results, *Syst. Zool.* **25**:137–148.

Atchley, W. R., Nordheim, E. V., Gunsett, F. C., and Crump, P. L., 1982, Geometric and probabilistic aspects of statistical distance functions, Syst. Zool. 31:445–460.

Austin, G. T., Blake, E. R., Brodkorb, P., Browning, M. R., Godfrey, W. E., Hubbard, J. P., McCaskie, G., Marshall, J. T., Jr., Monson, G., Olson, S. L., Ouellet, H., Palmer, R. S., Phillips, A. R., Pulich, W. M., Ramos, M. A., Rea, A. M., and Zimmerman, D. A., 1981, Ornithology as a science, Auk 98:636–637.

Bakken, G. S., 1980, The use of standard operative temperature in the thermal energetics of birds, Physiol. Zool. 53:108–119.

Bart, J., and Robson, D. S., 1982, Estimating survivorship when the subjects are visited periodically, Ecology 63:1078–1090.

Batschelet, E., 1965, Statistical Methods for the Analysis of Problems in Animal Orientation and Certain Biological Rhythms, American Institute of Biological Sciences, Washington, D.C.

Batschelet, E., 1979, Introduction to mathematics for life scientists, 3rd ed., Springer-Verlag, New York.

Beals, E. W., 1960, Forest bird communities of the Apostle Islands of Wisconsin, Wilson Bull. 72:150–181.

Bent, A. C., 1919–1967, Life Histories of North American Birds, 20 vols., U. S. National Museum, Washington, D.C.

Bergmann, C., 1847, Ueber die Verhältnisse der Wärmeökonomie der Thiere zu ihrer Grösse, Gottinger studien, v. 3, pt. 1, p. 595–708.

Birch, L. C., 1979, The effect of species of animals which share common resources on one another's distribution and abundance, Fortschr. Zool. 25:197–221.

Bock, W. J., 1973, Philosophical foundations of classical evolutionary classification, Syst. Zool. 22:375–392.

Bond, R. R., 1957, Ecological distribution of breeding birds in the upland forests of southern Wisconsin, Ecol. Monogr. 27:351–384.

Box, G. E. P., Hunter, W. G., and Hunter, J. S., 1978, Statistics for Experimenters, John Wiley and Sons, New York.

Brady, R. H., 1979, Natural selection and the criteria by which a theory is judged, Syst. Zool. 28:600–621.

Bray, J. R., and Curtis, J. T., 1957, An ordination of the upland forest communities of southern Wisconsin, Ecol. Monogr. 27:325–349.

Brown, J. H., 1981, Two decades of homage to Santa Rosalia: Toward a general theory of diversity, Amer. Zool. 21:877–888.

Brown, W. H., and Wilson, E. O., 1954, The case against the trinomen, Syst. Zool. 3:174–176.

Brown, W. H., and Wilson, E. O., 1956, Character displacement, Syst. Zool. 5:48–64.

Brownie, C., Anderson, D. R., Burnham, K. P., and Robson, D. S., 1978, Statistical Inference From Band Recovery Data—A Handbook, U. S. Fish and Wildlife Service, Resource Publ. 131.

Burnaby, T., 1966, Growth invariant discriminant functions and generalized distances, Biometrics 22:96–110.

Cain, A. J., 1977, The efficacy of natural selection in wild populations, in: The Changing Scenes in the Natural Sciences (C. Goulden, ed.), Acad. Nat. Sci., Spec. Publ. 12, Philadelphia, pp. 111–133.

Calder, W. A., 1983, Ecological scaling in mammals and birds, Annu. Rev. Ecol. Syst. 14:213–230.

Calder, W. A., and King, J. R., 1974, Thermal and caloric relationships of birds, in: Avian Biology, Volume IV (D. S. Farner and J. R. King, eds.), Academic Press, New York, pp. 259–413.

Capen, D. E., ed., 1981, Proceedings of workshop: The use of multivariate statistics in studies of wildlife habitat (Burlington, Vermont, April 23–25, 1980), U. S. Forest Service Gen. Tech. Rep. RM-87, Ft. Collins, Colorado.

Carnes, B. A., and Slade, N. A., 1982, Some comments on niche analysis in canonical space, *Ecology* **63**:888–893.

Carpenter, F. L., and R. E. MacMillen, 1976, Threshold model of feeding territoriality and test with a Hawaiian honeycreeper, *Science* **194**:639–642.

Chambers, J. M., Cleveland, W. S., Kleiner, B., and Tukey, P. A., 1983, *Graphical Methods for Data Analysis*, Wadsworth, Belmont, California.

Clements, F. E., 1904. The development and structure of vegetation, *Bot. Surv. Nebraska* **7**:5–175.

Cody, M. L., 1968, On the methods of resource division in grassland bird communities, *Am. Natur.* **102**:107–147.

Cody, M. L., 1974, *Competition and the Structure of Bird Communities*, Princeton University Press, Princeton, New Jersey.

Cody, M. L., 1978, Habitat selection and interspecific territoriality among the sylviid warblers of England and Sweden, *Ecol. Monogr.* **48**:351–396.

Cody, M. L., and Mooney, H. A., 1978, Convergence versus nonconvergence in Mediterranean-climate ecosystems, *Annu. Rev. Ecol. Syst.* **9**:265–321.

Collins, S. L., James, F. C., and Risser, P. G., 1982, Habitat relationships of wood warblers (Parulidae) in northern central Minnesota, *Oikos* **39**:50–58.

Connell, J. H., 1975, Some mechanisms producing structure in natural communities: A model and evidence from field experiments, in: *Ecology and Evolution of Communities* (M. L. Cody and J. M. Diamond, eds.), Harvard University Press, Cambridge, Massachusetts, pp. 460–490.

Connell, J. H., 1980, Diversity and coevolution of competitors, or the ghost of competition past, *Oikos* **35**:131–138.

Connor, E. F., and Simberloff, D., 1979, The assembly of species communities: Chance or competition? *Ecology* **60**:1132–1140.

Connor, E. F., and Simberloff, D., 1983, Interspecific competition and species co-occurrence patterns on islands: Null models and the evaluation of evidence. *Oikos* **40**:455–465.

Connor, E. F., and Simberloff, D., 1984, Neutral models of species' co-occurrence patterns, in: *Ecological Communities: Conceptual Issues and the Evidence* (D. R. Strong, D. Simberloff, L. G. Abele, and A. B. Thistle, eds.), Princeton University Press, Princeton, New Jersey (in press).

Conover, W. J., 1980, *Practical Nonparametric Statistics*, 2nd ed., John Wiley, New York.

Cooley, W. W., and Lohnes, P. R., 1971, *Multivariate Data Analysis*, John Wiley, New York.

Cooper, R. J., 1981, Relative abundance of Georgia caprimulgids based on call counts, *Wilson Bull.* **93**:363–371.

Cowie, R. J., 1977, Optimal foraging in Great Tits *(Parus major)*, *Nature* (London) **268**:137–139.

Cracraft, J., 1974, Phylogenetic models and classification, *Syst. Zool.* **23**:71–90.

Cracraft, J., 1982, Geographic differentiation, cladistics, and vicariance biogeography: Reconstructing the tempo and mode of evolution, *Am. Zool.* **22**:411–424.

Cronmiller, J. R., and Thompson, C. F., 1980, Experimental manipulation of brood size in Red-winged Blackbirds, *Auk* **97**:559–565.

Crook, J. H., 1964, The evolution of social organization and visual communication in the weaver birds (Ploceinae), *Behav. Suppl.* **10**:1–178.

Dayton, P. K., 1979, Ecology: A science and a religion, in: *Ecological Processes in Coastal and Marine Systems* (R. J. Livingston, ed.), Plenum Press, New York, pp. 3–18.

Deevey, E. S., 1947, Life tables for natural populations of animals, *Qt. Rev. Biol.* **2**:283–314.

Diamond, J. M., 1975, Assembly of species communities, in: *Ecology and Evolution of Communities* (M. L. Cody and J. M. Diamond, eds.), Harvard University Press, Cambridge, pp. 342–444.

Diamond, J. M., 1978, Niche shifts and the rediscovery of interspecific competition, *Am. Sci.* **66**:322–331.

Diamond, J. M., 1979, Population dynamics and interspecific competition in bird communities, *Fortschr. Zool.* **25**:389–402.

Dobzhansky, T., 1975, Book review, *Evolution* **29**:376–378.

Dolby, G. R., 1982, The role of statistics in the methodology of the life sciences, *Biometrics* **38**:1069–1083.

Dow, D. D., 1978, A test of significance for Mayfield's method of calculating nest success, *Wilson Bull.* **90**:291–295.

Dueser, R. D., and Shugart, H. H., Jr., 1979, Niche pattern in a forest floor small mammal fauna, *Ecology* **60**:108–118.

Dueser, R. D., and Shugart, H. H., Jr., 1982, Reply to comments by Van Horne and Ford and by Carnes and Slade, *Ecology* **63**:1174–1185.

Eberhardt, L. L., 1972, Some problems in estimating survival from banding data, in: *Population Ecology of Migratory Birds—A Symposium*, Bureau of Sport Fishers and Wildlife, Wildl. Res. Rep. 2, pp. 153–177.

Eberhardt, L. L., 1976, Quantitative ecology and impact assessment, *J. Environ. Manage.* **4(1)**:27–70.

Eberhardt, L. L., 1978, Appraising variability in population studies, *J. Wildlife Manage.* **42**:207–238.

Ehrenberg, A. S. C., 1978, *Data Reduction: Analysing and Interpreting Statistical Data*, John Wiley, New York.

Ehrenberg, A. S. C., 1982, *A Primer in Data Reduction: An Introductory Statistical Textbook*, John Wiley, New York.

Eldredge, N., and Cracraft, J., 1980, *Phylogenetic Patterns and the Evolutionary Process*, Columbia University Press, New York.

Elton, C. S., 1927, *Animal Ecology*, Sedgwick and Jackson, London.

Emlen, S. T., 1981, The ornithological roots of sociobiology, *Auk* **98**:400–403.

Errington, P. L., 1945, Some contributions of a fifteen-year local study of the northern bobwhite to a knowledge of population phenomena, *Ecol. Monogr.* **15**:1–34.

Everett, I. B., 1978, *Graphical Techniques for Multivariate Data*, Heinemann, London.

Fishman, G. S., 1978, *Principles of Discrete Event Simulation*, John Wiley, New York.

Fretwell, S. D., 1975, The impact of Robert MacArthur on ecology, *Annu. Rev. Ecol. Syst.* **6**:1–13.

Gauch, H. G., Jr., 1982, *Multivariate Analysis in Community Ecology*, Cambridge University Press, Cambridge.

Ghiselin, M. T., 1969, *The Triumph of the Darwinian Method*, University of California Press, Berkeley, California.

Gibson, A. R., 1984, Reviews and comments, multivariate analysis of lizard thermal behavior, *Copeia* **1984**:267–272.

Gill, F. B., and Wolf, L. L., 1975, Foraging strategies and energetics of East African sunbirds at mistletoe flowers, *Am. Natur.* **109**:491–510.

Gleason, H. A., 1926, The individualistic concept of the plant association, *Bull. Torrey Bot. Club* **53**:7–26.

Gnanadesikan, R., 1977, *Methods for Statistical Data Analysis of Multivariate Observations*, John Wiley, New York.

Gould, S. J., and Johnston, R. F., 1972, Geographic variation, *Ann. Rev. Ecol. Syst.* **3**:457–498.

Gould, S. J., and Lewontin, R. C., 1979, The spandrels of San Marco and the panglossian paradigm: A critique of the adaptationist programme, *Proc. Roy. Soc. London, Ser. B* **205**:581–598.

Gower, J. C., 1966, Some distance properties of latent root and vector methods used in multivariate analysis, *Biometrika* **53**:325–338.

Grant, P. R., 1972, Convergent and divergent character displacement, *Biol. J. Linn. Soc.* **4**:39–68.

Grant, P. R., and Abbott, I., 1980, Interspecific competition, island biogeography and null hypotheses, *Evolution* **34**:322–341.

Grant, P. R., and Schluter, D., 1984, Interspecific competition inferred from patterns of guild structure, in: *Ecological Communities: Conceptual Issues and the Evidence* (D. R. Strong, L. G. Abele, D. Simberloff, and A. B. Thistle, eds.), in press, Princeton University Press, Princeton, New Jersey.

Green, R. F., 1977, Do more birds produce fewer young? A comment on Mayfield's measure of nest success, *Wilson Bull.* **89**:173–175.

Green, R. H., 1971, A multivariate statistical approach to the Hutchinsonian niche: Bivalve molluscs of central Canada, *Ecology* **52**:543–556.

Green, R. H., 1979, *Sampling Design and Statistical Methods for Environmental Biologists*, John Wiley, New York.

Green, R. H., 1980, Multivariate approaches in ecology: The assessment of ecological similarity, *Annu. Rev. Ecol. Syst.* **11**:1–14.

Grinnell, J., 1904, The origin and distribution of the Chestnut-backed Chickadee, *Auk* **21**:364–382.

Grinnell, J., 1917, The niche-relationships of the California Thrasher, *Auk* **34**:427–433.

Guttman, L., 1977, What is not what in statistics, *Statistician* **26**:81–107.

Haila, Y., 1983, Ecology of island colonization by northern land birds: A quantitative approach, Dept. Zool., Univ. Helsinki, Finland.

Hamilton, T. H., 1961, The adaptive significances of intraspecific trends of variation in wing length and body size among bird species, *Evolution* **15**:180–195.

Hanson, N. R., 1958, *Patterns of Discovery, An Inquiry into the Conceptual Foundations of Science*, Cambridge University Press, Cambridge.

Harper, J. L., 1977, The contributions of terrestrial plant studies to the development of the theory of ecology, in: *Changing Scenes in the Natural Sciences, 1776–1976* (C. E. Goulden, ed.), Special Publ. 12, Acad. Nat. Sci., Philadelphia, Pennsylvania, pp. 139–157.

Harper, J. L., 1982, After description, in: Spec. Publ. Brit. Ecol. Soc., No. 1 (E. I. Newman, ed.), Blackwell Sci. Publ., Oxford, pp. 11–26.

Hartigan, J. A., 1975, *Clustering Algorithms*, John Wiley, New York.

Harvey, P. H., Colwell, R. K., Silvertown, J. W., and May, R. M., 1983, Null models in ecology, *Annu. Rev. Ecol. Syst.* **14**:189–211.

Hayne, D. W., 1978, Experimental designs and statistical analyses in small mammal population studies, in: *Populations of Small Mammals under Natural Conditions* (D. P. Snyder, ed.), University of Pittsburgh Press, Pittsburgh, Pennsylvania, pp. 3–10.

Hennig, W., 1966, *Phylogenetic Systematics*, Unversity of Illinois Press, Urbana.

Hill, M. O., 1973, Reciprocal averaging: An eigenvector method of ordination, *J. Ecol.* **61**:237–249.

Hill, M. O., 1974, Correspondence analysis: A neglected multivariate method, *Appl. Statistics* **23**:340–354.

Hills, M., 1982, Allometry, in: *Encyclopedia of Statistical Sciences*, Volume 1, John Wiley, New York, pp. 48–54.

Holmes, R. T., Bonney, R. E., Jr., and Pacala, S. W., 1979, Guild structure of the Hubbard Brook bird community: A multivariate approach, *Ecology* **60**:512–520.

Horn, H., 1978, Optimal tactics of reproduction and life-history, in: *Behavioural Ecology* (J. R. Krebs and N. B. Davies, eds.), Blackwell Scientific Publications, Oxford, pp. 411–429.

Hull, D. L., 1979, The limits of cladism, *Syst. Zool.* **28**:416–440.

Hurlbert, S. H., 1984, Pseudoreplication and the design of ecological field experiments, *Ecol. Monogr.* **54**(2):187–211.

Hutchinson, G. E., 1958, Concluding remarks, *Cold Spring Harbor Symp. Quart. Biol.* **22**:415–427.

Hutchinson, G. E., 1965, *The Ecological Theater and the Evolutionary Play*, Yale University Press, New Haven, Connecticut.

Hutchinson, G. E., 1968, When are species necessary? in: *Population Biology and Evolution* (R. C. Lewontin, ed.), Syracuse University Press, Syracuse, New York, pp. 177–186.

James, F. C., 1970, Geographic size variation in birds and its relationship to climate, *Ecology* **51**:365–390.

James, F. C., 1971, Ordinations of habitat relationships among breeding birds, *Wilson. Bull.* **83**:215–236.

James, F. C., 1982, The ecological morphology of birds, *Ann. Zool. Fenn.* **19**:265–275.

James, F. C., 1983, Environmental component of morphological differentiation in birds, *Science* **221**:184–186.

James, F. C., and Porter, W. P., 1979, Behavior-microclimate relationships in the African rainbow lizard *Agama agama*, *Copeia* **1979**(4):585–593.

James, F. C., and Porter, W. P., 1984, Response to Gibson, *Copeia* **1984**(1):272–274.

James, F. C., and Rathbun, S., 1981, Rarefaction, relative abundance and diversity of avian communities, *Auk* **98**:785–800.

James, F. C., and Wamer, N. O., 1982, Relationships between temperate forest vegetation and avian community structure, *Ecology* **63**:159–171.

James, F. C., Engstrom, R. T., NeSmith, C., and Laybourne, R., 1984a, Inferences about population movements of Red-winged Blackbirds from morphological data, *Am. Midl. Nat.* **111**(2):319–331.

James, F. C., Johnston, R. F., Wamer, N. O., Niemi, G. J., and Boecklen, W. J., 1984b, The Grinnellian niche of the Wood Thrush, *Am. Natur.* **124**:17–30.

Janetos, A. C., and Cole, B. J., 1981, Imperfectly optimal animals, *Behav. Ecol. Sociobiol.* **9**:203–210.

Johnson, D. H., 1981, The use and misuse of statistics in wildlife habitat studies, in: *The Use of Multivariate Statistics in Studies of Wildlife Habitat* (D. E. Capen, ed.), Rocky Mountain Forest and Range Experiment Station, USDA Forest Service General Tech. Rep. RM-87, Ft. Collins, Colorado.

Johnson, D. H., Sargeant, H. B., and Allen, S. H., 1975, Fitting Richards' curves to data of diverse origins, *Growth* **39**:315–330.

Johnson, R. A., and Wichern, D. W., 1982, *Applied Multivariate Statistical Analysis*, Prentice-Hall, Englewood Cliffs, New Jersey.

Johnston, R. F., 1977, Composition of woodland bird communities in eastern Kansas, *Bull. Kansas Ornithol. Soc.* **28**:13–18.

Jolicoeur, P., 1963, The multivariate generalization of the allometry equation, *Biometrics* **19**:497–499.

Keough, M. J., and Butler, A. J., 1983, Temporal changes in species number in an assemblage of sessile marine invertebrates, *J. Biogeogr.* **10**:317–330.

Kitts, D. B., 1977, Karl Popper, verifiability, and systematic zoology, *Syst. Zool.* **26**:185–194.

Kitts, D. B., 1980, Theories and other scientific statements: A reply to Settle, *Syst. Zool.* **29**:190–192.

Klett, A. T., and Johnson, D. H., 1982, Variability in nest survival rates and implications to nesting studies, *Auk* **99**:77–87.

Kodric-Brown, A., and Brown, J. H., 1978, Influence of economics, interspecific competition and sexual dimorphism on territoriality of migrant Rufous hummingbirds, *Ecology* **59**:285–296.

Krebs, J. R., Kacelnik, A., and Taylor, P., 1978, Test of optimal sampling by foraging Great Tits, *Nature* (London) **275**:27–31.

Krebs, J. R., Stephens, D. W., and Sutherland, W. J., 1983, Perspectives in optimal foraging, in: *Perspectives in Ornithology* (A. H. Brush and G. A. Clark, eds.), Cambridge University Press, Cambridge, pp. 165–216.

Lack, D., 1969, The number of bird species on islands, *Bird Study* **16**:193–209.

Lieberson, J., 1982, The "truth" of Karl Popper, *N. Y. Rev. Books*, 18 November, pp. 67–69.

Leisler, B., 1980, Morphological aspects of ecological specialization in bird genera, *Ökol. Vögel* **2**:199–200.

Leisler, B., and Thaler, 1982, Differences in morphology and foraging behavior in the goldcrest, *Regulus regulus* and firecrest *R. ignicapillos*, *Ann. Zool. Fennici* **19**:277–284.

Lenington, S., 1983, Commentary, in: *Perspectives in Ornithology* (A. H. Brush and G. A. Clark, eds.), Cambridge University Press, Cambridge. pp. 85–91.

Levandowsky, M., 1977, A white queen speculation, *At. Rev. Biol.* **52**:383–386.

Lewontin, R. C., 1972, Testing the theory of natural selection, *Nature* **236**:181–182.

Lovejoy, T. E., 1980, Discontinuous wilderness: Minimum areas for conservation, *Parks* **5**:13–15.

Lynch, J. F., and Johnson, N. K., 1974, Turnover and equilibria in insular avifaunas, with special reference to the California Channel Islands, *Condor* **76**:370–384.

MacArthur, R. H., 1962, Growth and regulation of animal populations, *Ecology* **43**:579.

MacArthur, R. H., and Levins, R., 1967, The limiting similarity, convergence and divergence of coexisting species, *Am. Natur.* **101**:377–385.

MacArthur, R. H., and MacArthur, J. W., 1961, On bird species diversity, *Ecology* **42**:594–598.

MacArthur, R. H., and Pianka, E. R., 1966, On the optimal use of a patchy environment, *Am. Natur.* **100**:603–609.

MacArthur, R. H., and Wilson, E. O., 1967, *The Theory of Island Biogeography*, Princeton University Press, Princeton, New Jersey.

MacFadyen, A., 1975, Some thoughts on the behavior of ecologists, *J. Anim. Ecol.* **44**:351–363.

McGill, R., Tukey, J. W., and Larsen, W. A., 1978, Variations of box plots, *Am. Statist.* **32**:12–16.

McIntosh, R. P., 1980, The background and some current problems of theoretical ecology, *Synthèse* **43**:195–255.

Mahoney, S. A., and King, J. R., 1977, The use of equivalent black-body temperature in the thermal energetics of small birds, *J. Thermal Biol.* **2**:115–120.

Mares, M. A., 1976, Convergent evolution of desert rodents: Multivariate analysis and zoogeographic implications, *Paleobiology* **2**:39–63.

May, R. M., 1977, Optimal life-history strategies, Nature 267:394–395.

May, R. M., 1981a, The role of theory in ecology, Am. Zool. 21:903–910.

May, R. M., 1981b, Theoretical Ecology: Principles and Applications, 2nd ed., Saunders, Philadelphia.

Mayfield, H., 1961, Nesting success calculated from exposure, Wilson Bull. 73:255–261.

Mayfield, H., 1975, Suggestions for calculating nesting success, Wilson Bull. 87:456–466.

Maynard Smith, J., 1974, Models in Ecology, Cambridge University Press, Cambridge.

Maynard Smith, J., 1978, Optimization theory in evolution, Annu. Rev. Ecol. Syst. 9:31–56.

Mayr, E., 1963, Animal Species and Evolution, Harvard University Press, Cambridge.

Miller, H. W., and Johnson, D. H., 1978, Interpreting the results of nesting studies, J. Wildlife Manage. 42:471–476.

Moreau, R. E., 1957, Variation in the western Zosteropidae (Aves), Bull. Br. Mus. Nat. Hist. Zool. 4:311–433.

Morrison, D. F., 1976, Multivariate Statistical Methods, 2nd ed., McGraw-Hill, New York.

Mosimann, J. E., 1970, Size allometry: Size and shape variables with characterizations of the lognormal and generalized gamma distributions, J. Am. Stat. Assoc. 65:930–945.

Mosimann, J. E., 1975a, Statistical problems of size and shape. I. Biological applications and basic theorems, in: Statistical Distributions in Scientific Work, Volume 2 (G. P. Patil, S. Kotz, and K. Ord, eds.), D. Reidel Publ. Co., Dordrecht-Holland, pp. 187–217.

Mosimann, J. E., 1975b, Statistical problems of size and shape. II. Characterizations of the lognormal and gamma distributions, in: Statistical Distributions in Scientific Work, Volume 2 (G. P. Patil, S. Kotz, and K. Ord, eds.), D. Reidel Publ. Co., Dordrecht-Holland, pp. 219–239.

Mosimann, J. E., in press, Size and shape analysis, in: Encyclopedia of Statistical Sciences, John Wiley and Sons, New York.

Mosimann, J. E., and James, F. C., 1979, New statistical methods for allometry with application to Florida Red-winged Blackbirds, Evolution 33:444–459.

Mosimann, J. E., and Malley, J. D., 1979, Size and shape variables, in: Statistical Ecology, vol. 7, Multivariate Methods in Ecological Work (L. Orloci, C. R., Rao, and W. M. Stiteler, eds.), Internatinal Co-operative Publ. House, Fairland, Maryland, pp. 175–189.

Mosimann, J. E., Malley, J. D., Cheever, A. W., and Clark, C. B., 1978, Size and shape analysis of schistosome egg-counts in Egyptian autopsy data, Biometrics 34:341–356.

Mosteller, F., and Tukey, J. W., 1977, Data Analysis and Regression, A Second Course in Statistics, Addision-Wesley Publ. Co., Reading, Massachusetts.

Mugaas, J. N., and King, J. R., 1981, The annual variation in daily energy expenditure of the Black-billed Magpie: A study of thermal and behavioral energetics, Stud. Avian Biol. 5:1–78.

Myers, J. P., 1983, Commentary, in: Perspectives in Ornithology (A. H. Brush and G. A. Clark, eds.), Cambridge University Press, Cambridge, pp. 216–221.

Myers, J. P., Connors, P. G., and Pitelka, F. A., 1981, Optimal territory size and the Sanderling: Compromises in a variable environment, in: Foraging Behavior: Ecological, Ethological and Psychological Approaches (A. C. Kamil and T. D. Sargent, eds.), Garland, STPM Press, New York, pp. 135–158.

Nelson, G., 1978, Classification and prediction: A reply to Kitts, Syst. Zool. 27:216–218.

Newton-Smith, W. H., 1981, The Rationality of Science, Routledge and Kegan Paul, Boston, Massachusetts.

Niemi, G. J., 1983, Ecological morphology of breeding birds in Old World and New World peatlands, Ph.D. Dissertation, Florida State University, Tallahassee.

Noon, B. R., 1981, The distribution of an avian guild along a temperate elevational gradient: The importance and expression of competition, Ecol. Monogr. 51:105–124.

Noy-Meir, I., and Whittaker, R. H., 1977, Continuous multivariate methods in community analysis: Some problems and developments, *Vegetatio* **33**:79–98.

Oaten, A., 1977, Optimal foraging in patches: A case for stochasticity, *Theor. Popul. Biol.* **12**:263–285.

Olson, S. L., 1981, The museum tradition in ornithology—A response to Ricklefs, *Auk* **98**:193–195.

Orians, G. H., and Pearson, N. E., 1979, On the theory of central place foraging, in: *Analysis of Ecological Systems* (D. J. Horn, G. R. Stairs, and R. Mitchell, eds), pp. 155–177, Ohio State University Press, Columbus.

Orloci, L., 1966, Geometric models in ecology I. The theory and application of some ordination methods, *J. Ecol.* **54**:193–215.

Orloci, L., 1975, *Multivariate Analysis in Vegetation Research*, Junk, The Hague.

Parker, G. A., 1978, Searching for mates, in: *Behavioural Ecology, An Evolutionary Approach* (J. R. Krebs and N. B. Davies, eds.), Blackwell Scientific Publications, Oxford, pp. 214–244.

Patterson, C., 1982, Cladistics, in: *Evolution Now, A Century after Darwin* (J. Maynard Smith, ed.), pp. 110–120, W. H. Freeman Co., San Francisco.

Peters, R. H., 1976, Tautology in evolution and ecology, *Am. Natur.* **110**:1–7.

Peters, R. H., 1978, Predictable problems with tautology in evolution and ecology, *Am. Natur.* **112**:759–762.

Pielou, E. C., 1975, *Ecological Diversity*, Wiley, New York.

Pielou, E. C., 1977, *Mathematical Ecology*, 2nd ed., Wiley, New York.

Pielou, E. C., 1981, The usefulness of ecological models: A stock-taking. *Qt. Rev. Biol.* **56**:17–31.

Platnick, N. I., and Gaffney, E. S., 1977, Systematics: A Popperian perspective, *Syst. Zool.* **26**:360–365.

Platt, J. R., 1964, Strong inference, *Science* **146**:347–353.

Poole, R. W., 1978, Some relative characteristics of population fluctuations, in: *Time Series and Ecological Processes* (H. H. Shugart, Jr., ed.), SIAM, Philadelphia, Pennsylvania, pp. 18–33.

Popper, K. R., 1959, *The Logic of Scientific Discovery*, Basic Books, New York (revised in 1965).

Popper, K. R., 1965, Normal science and its dangers, in: *Criticism and the Growth of Knowledge* (I. Lakatos and A. Musgrave, eds.), Cambridge University Press, Cambridge, pp. 51–58.

Popper, K. R., 1968, *Conjectures and Refutations: The Growth of Scientific Knowledge*, 2nd ed., Harper & Row, New York.

Popper, K. R., 1972, *Objective knowledge: An Evolutionary Approach*, Clarendon Press, Oxford.

Popper, K. R., 1983, *Postscript to the Logic of Scientific Discovery*, Volume 1, *Realism and the Aim of Science*, Hutchinson/Rowan & Littlefield, London.

Popper, K., and Miller, D., 1983, A proof of the impossibility of inductive probability, *Nature* **302**:687–688.

Porter, W. M., and Gates, D. M., 1969, Thermal equilibria of animals with environment, *Ecol. Monogr.* **39**:227–244.

Power, D., 1969 Evolutionary implications of wing and size variation in the Red-winged Blackbird in relation to geographic and climatic factors: A multiple regression analysis, *Syst. Zool.* **18**:363–373.

Pyke, G. H., Pulliam, H. R., and Charnov, E. L., 1977, Optimal foraging theory: A selective review of theory and tests, *Q. Rev. Biol.* **52**:137–154.

Quinn, J. F., and Dunham, A. E., 1983, On hypothesis testing in ecology and evolution, *Am. Natur.* **122**:602–617.

Ramensky, L. G., 1930, Zur Methodik der vergleichenden Bearbeitung und Ordnung von Pflanzenlisten und andera Objecten, die durch mehrere verschiedenartig wirkende Factoren bestimmt werden, *Beitr. Biol. Pflanz.* **18**:269–304.

Rensch, B., 1934, Einwirkung des Klimas bei der Ausprägung von Vogelrassen mit besonderer Berücksichtigung der Flügelform und der Eizahl, *Proc. Eighth Internat. Congr.* (Oxford) p, 285–311.

Rensch, B., 1939, Klimatische auslese von Grössenvarianten, *Arch. für Naturg.* **NF 8**:89–129.

Richards, F. J., 1959, A flexible growth function for empirical use, *J. Exp. Bot.* **10**:290–300.

Ricklefs, R. E., 1967, A graphical method of fitting equations to growth curves, *Ecology* **48**:978–983.

Ricklefs, R. E., 1968, Patterns of growth in birds, *Ibis* **110**:419–451.

Ricklefs, R. E., 1969, Natural selection and the development of mortality rates in young birds, *Nature* **223**:922–925.

Ricklefs, R. E., 1973, Patterns of growth in birds II. Growth rate and mode of development, *Ibis* **115**:177–201.

Ricklefs, R. E., 1979, Patterns of growth in birds V. A comparative study of development in the Starling, Common Tern, and Japanese Quail, *Auk* **96**:10–30.

Ricklefs, R. E., 1983, Comparative avian demography, in: *Current Ornithology* (R. F. Johnston, ed.), Plenum Press, New York, London, pp. 1–32.

Ripley, B. D., 1981, *Spatial Statistics*, John Wiley, New York.

Robertson, R., 1972, Optimal niche space of the Red-winged Blackbird (*Agelaius phoeniceus*). I. Nesting success in marsh and upland habitat, *Can. J. Zool.* **50**:247–263.

Romesburg, 1981, Wildlife science: Gaining reliable knowledge, *J. Wildlife Manage.* **45**:293–313.

Rosenzweig, M. L., 1968, The strategy of body size in mammalian carnivores, *Am. Midl. Natur.* **80**:299–315.

Rotenberry, J. T., and Wiens, J. A., 1980, Habitat structure, patchiness, and avian communities in North American steppe vegetation: A multivarate analysis, *Ecology* **61**:1228–1250.

Roughgarden, J., 1983, Competition theory in community ecology, *Am. Natur.* **122**:583–601.

Sabo, S. R., 1980, Niche and habitat relations in subalpine bird communities of the White Mountains of New Hampshire, *Ecol. Monogr.* **50**:241–259.

Sabo, S. R., and Whittaker, R. H., 1979, Bird niches in subalpine forest: An indirect ordination, *Proc. Natl. Acad. Sci. U.S.A.* **76**:1338–1342.

Schluter, D., 1981, Does the theory of optimal diets apply in complex environments, *Am. Natur.* **118**:139–147.

Schoener, T. W., 1972, Mathematical ecology and its place among the sciences. Review of R. H. MacArthur, *Geographical Ecology, Science* **178**:389–391.

Schoener, T. W., 1979, Generality of the size-distance relation in models of optimal foraging, *Am. Natur.* **114**:902–914.

Schoener, T. W., 1982, The controversy over interspecific competition, *Am. Sci.* **70**:586–595.

Selvin, H. C., and Stuart, A., 1966, Data-dredging procedures in survey analysis, *Am. Statist.* **20**:20–23.

Settle, T., 1979, Popper on "When is a science not a science?" *Syst. Zool.* **28**:521–529.

Simberloff, D., 1979, Rarefaction as a distribution-free method of expressing and estimating diversity, in: *Ecological Diversity in Theory and Practice* (J. F. Grassle, G. P. Patil, W. K. Smith, and C. Taillie, eds.), *Statistical Ecology Series*, Volume 6, International Cooperative Publishing House, Fairland, Maryland, pp. 159–176.

Simberloff, D., 1980, A succession of paradigms in ecology: Essentialism to materialism and probabilism, *Synthèse* **43**:3–39.

Simberloff, D., 1982, The status of competition theory in ecology, *Ann. Zool. Fenn.* **19**:241–253.

Simberloff, D., 1983a, Biogeographic models, species' distributions and community organization, in: *Evolution, Time and Space: The Emergency of the Biosphere* (R. W. Sims, J. H. Price, and P. E. S. Whalley, eds.), Systematics Association, Special, Volume No. 23, Academic Press, New York, pp. 57–83.

Simberloff, D., 1983b, Biogeography: The unification and maturation of a science, in: *Perspectives in Ornithology*, Cambridge University Press, Cambridge, pp. 411–473.

Simberloff, D., 1983c, When is an island community in equilibrium? *Science* **220**:1275–1276.

Simberloff, D., 1983d, Competition theory, hypothesis-testing and other community ecology buzzwords, *Am. Natur.* **122**:626–631.

Smith, K. G., 1977, Distribution of summer birds along a forest moisture gradient in an Ozark watershed, *Ecology* **58**:810–819.

Sneath, P. H. A., and Sokal, R. R., 1973, *Numerical Taxonomy*, Freeman, San Francisco, California.

Snedecor, G. W., and Cochran, W. G., 1980, *Statistical Methods*, 7th ed., Iowa State University Press, Ames, Iowa.

Snow, D. W., 1954, Trends in geographical variation in the Palearctic members of the genus *Parus*, *Evolution* **8**:19–28.

Sobolev, L. N., and Utekhin, V. D., 1973, Russian (Ramensky) approaches to community systematization, in: *Handbook in Vegetative Science*, Volume 5, pp. 77–105, (R. H. Whittaker, ed.). Junk, The Hague.

Sokal, R. R., and Rohlf, F. J., 1981, *Biometry*, 2nd ed., Freeman, San Francisco, California.

Southwood, T. R. E., 1973, Continuing in the MacArthur tradition, *Science* **192**:670–672.

Stearns, S. C., 1976, Life-history tactics: A review of the ideas, *Q. Rev. Biol.* **51**:3–47.

Stearns, S. C., 1977, The evolution of life history traits: A critique of the theory and a review of the data, *Annu. Rev. Ecol. Syst.* **8**:145–171.

Stearns, S. C., 1980, A new view of life-history evolution, *Oikos* **35**:266–281.

Stephens, D. W., and Charnov, E. L., 1982, Optimal foraging: Some simple stochastic models, *Behav. Ecol. Sociobiol.* **10**:251–263.

Strong, D. R., 1980, Null hypotheses in ecology, *Synthèse* **43**:271–286.

Strong, D. R., Szyska, L. A., and Simberloff, D. S., 1979, Tests of community-wide character displacement against null hypotheses, *Evolution* **33**:897–913.

Strong, D. R., Simberloff, D. S., Abele, L. G., and Thistle, A. B. (eds), 1984, *Ecological Communities: Conceptual Issues and the Evidence*. Princeton University Press, Princeton, New Jersey.

Tatsuoka, M. M., 1971, *Multivariate analysis*, John Wiley, New York.

Tufte, E. R., 1983, *The Visual Display of Quantitative Information*, Graphics Press, Cheshire, Connecticut.

Tukey, J. W., 1977, *Exploratory Data Analysis*, Addison-Wesley Publishing Company, Inc., Reading, Massachusetts.

Underwood, A. J., 1981, Techniques of analysis of variance in experimental marine biology and ecology, *Annu. Rev. Oceanography Marine Biol.* **19**:513–605.

Van Horne, B., and Ford, R. G., 1982, Niche breadth calculation based on discriminant analysis, *Ecology* **63**:1172–1174.

Velleman, P. F., and Hoaglin, D. C., 1981, *Applications, Basics, and Computing of Exploratory Data Analysis*, Duxbury Press, Boston, Massachusetts.

Wagner, F. H., 1969, Ecosystem concepts in fish and game management, in: *The Ecosystem Concept in Natural Resource Management* (G. M. Van Dyne, ed.), Academic Press, New York, pp. 259–307.

Walsberg, G. E., 1983, Ecological energetics: What are the questions?, in: *Perspectives in Ornithology* (A. H. Brush and G. A. Clark, Jr., eds.), Cambridge University Press, New York, pp. 135–158.

Walsberg, G. E., and King, J. R., 1978a, The heat budget of incubating Mountain White-crowned Sparrows *(Zonotrichia leucophrys oriantha)* in Oregon, *Physiol. Zool.* **51**:92–103.

Walsberg, G. E., and King, J. R., 1978b, The energetic consequences of incubation for two passerine species, *Auk* **95**:644–655.

Weisberg, S., 1980, *Applied Linear Regression,* Wiley-Interscience, New York.

White, G. C., and Brisbin, I. L., Jr., 1980, Estimation and comparision of parameters in stochastic growth models for Barn Owls, *Growth* **44**:97–111.

Whitmore, R. C., 1975, Habitat ordination of breeding passerine birds of the Virgin River Valley, southwestern Utah, *Wilson Bull.* **87**:65–74.

Whittaker, R. H., 1967, Gradient analysis of vegetation, *Biol. Rev.* **42**:207–264.

Whittaker, R. H., 1978, Direct gradient analysis, in: *Ordination of Plant Communities* (R. H. Whittaker, ed.), Junk, the Hague, pp. 7–50.

Whittaker, R. H., Levin, S. A., and Root, R. B., 1973, Niche, habitat, and ecotope, *Am. Natur.* **107**:321–338.

Wiens, J. A., 1973, Pattern and process in grassland bird communities, *Ecol. Monogr.* **43**:237–270.

Wiens, J. A., 1980, Theory and observation in modern ornithology: A forum, *Auk* **97**:409.

Wiens, J. A., 1983, Avian community ecology: An iconoclastic view, in: *Perspectives in Ornithology* (A. H. Brush and G. A. Clark, eds.), Cambridge University Press, Cambridge, pp. 355–403.

Wiens, J. A., and Rotenberry, J. T., 1980, Patterns of morphology and ecology in grassland and shrubsteppe bird populations, *Ecol. Monogr.* **50**:287–308.

Wiens, J. A., and Rotenberry, J. T., 1981, Habitat associations and community structure of birds in shrubsteppe environments, *Ecol. Monogr.* **51**:21–41.

Wiley, E. O., 1975, Karl R. Popper, systematics and classification: A reply to Walter Bock and other evolutionary taxonomists, *Syst. Zool.* **24**:233–243.

Williams, B. K., 1983, Some observations on the use of discriminant analysis in ecology, *Ecology* **64**:1283–1291.

Williams, W. T., 1976, *Pattern Analysis in Agricultural Science,* Elsevier, New York.

Williamson, M., 1981, *Island Populations,* Oxford University Press, Oxford.

Wilson, E. O., 1975, *Sociobiology: The New Synthesis,* Belknap Press, Cambridge.

CHAPTER 2

THE EVOLUTION OF REVERSED SEXUAL DIMORPHISM IN SIZE
A Comparative Analysis of the Falconiformes of the Western Palearctic

HELMUT C. MUELLER and KENNETH MEYER

1. INTRODUCTION

Females are larger than males in most species of Falconiformes. This dimorphism can be great: the average weight of a female can be almost twice that of a male (Mueller et al., 1979; Newton and Marquiss, 1982). The first hypothesis on the functions of reversed sexual dimorphism (RSD) was probably advanced by Frederick II in the thirteenth century (Wood and Fyfe, 1943). Many hypotheses have been proposed in recent years; most have been supported only by theory or selected anecdotal evidence. As noted by Andersson and Norberg (1981), most hypotheses on RSD are extremely difficult (if not impossible) to test experimentally. The comparative method (Lack, 1968) appears to be simultaneously the best and easiest method of testing these hypotheses. In this paper we examine all of the reasonably viable hypotheses on RSD, utilizing the

HELMUT C. MUELLER and KENNETH MEYER • Department of Biology, University of North Carolina at Chapel Hill, Chapel Hill, North Carolina 27514.

comparative method and all of the data available on the more than 40 species of Falconiformes found in the Western Palearctic. The main premise of our approach is that if a given trait were important in the evolution of RSD, it should be exhibited to a greater degree in highly dimorphic species than in less dimorphic ones.

Most recent hypotheses on RSD rely on a synthesis of previous hypotheses (e.g., Snyder and Wiley, 1976; Newton, 1979; Walter, 1979; Andersson and Norberg, 1981; Cade, 1982). However, we will test each hypothesis separately, and it is thus useful to provide a list. Our list of references for each hypothesis is not intended to be exhaustive; only principal references are given. The hypotheses fit reasonably well into three categories.

1.1. Ecological Hypotheses

There are several ecological hypotheses, but basically they all hold that RSD has evolved to allow the sexes to take different sizes of prey, reducing competition between the sexes and allowing the pair to exploit a wider range of sizes of prey (Brüll, 1937; Hagen, 1942; Storer, 1966; Reynolds, 1972; Snyder and Wiley, 1976; Newton, 1979). If any of these hypotheses are valid, we would predict greater RSD in trophic appendages (those associated with the capture of prey) than in body size or nontrophic appendages. Selander (1972) states this more forcefully: "Only when the trophic structures alone are modified can we conclude that the dimorphism results primarily or wholly from selection for differential niche utilization."

Unfortunately, there is some controversy over what are the primary trophic structures in birds of prey. For example, Andersson and Norberg (1981) and Schantz and Nilsson (1981) hypothesize that the differences in flight performance between small males and large females strongly influence the ability to capture and transport prey and are key factors in the evolution of RSD. These hypotheses imply that weight, wing area, wing loading, and other factors, are important "trophic structures" and raise the possibility that all available measurements are of trophic entities. We propose the following empirical approach to this problem: the size of a primary trophic structure (relative to body size) for a species (sexes combined) should be correlated with diet. If the relative size of a structure is not correlated with diet, then it is not a primary trophic structure. This approach simply argues that diet–structure correlations are as important between species as they are between sexes.

1.2. Sex Role Differentiation Hypotheses

1.2.1. Egg Size

Large females can lay larger eggs than small females (Reynolds, 1972; Selander, 1972; Schantz and Nilsson, 1981; Cade, 1982). If RSD has evolved for the production of large eggs, we would predict a relationship between relative egg size and RSD.

1.2.2. Follicle Protection

Large females provide a better cushion for developing eggs than smaller ones (Walter, 1979). A relationship between RSD and egg size or clutch size could be considered evidence for this hypothesis.

1.2.3. Incubation

Large females incubate more efficiently than smaller individuals (Snyder and Wiley, 1976; Cade, 1982). In species showing high RSD, the female should perform a greater share of incubation than in species with low RSD.

1.2.4. Nest Protection

Large females, particularly in predatory birds, are more effective in deterring heterospecific predators than smaller individuals (Storer, 1966; Reynolds, 1972; Snyder and Wiley, 1976; Andersson and Norberg, 1981; Cade, 1982). If nest protection has been important in the evolution of RSD, females of species with high RSD should perform a greater share of nest defense than in species with low RSD.

1.2.5. Territorial Defense

Small, agile males are better in the defense of territory against conspecifics than larger, less agile individuals (Schmidt-Bey, 1913; Nelson, 1977). Males should perform a greater proportion of territorial defense in species with high RSD than in species with low RSD.

1.2.6. Feeding Efficiency

Large females are better able to tear prey into bits for feeding small young (Andersson and Norberg, 1981). Females should perform a greater proportion of direct, bill-to-bill, feeding in species with high RSD than in species with low RSD.

1.2.7. Foraging Interference

It is more efficient for only one member of the pair to hunt alert and agile prey than to have two birds flying about alerting and frightening prey and thus interfering with each other's efforts (Andersson and Norberg, 1981). This hypothesis does not predict which member of the pair should do the hunting, but see 1.2.8, below. If foraging interference has been important in the evolution of RSD, males should perform a greater share of the provisioning of the female and the young in species with high RSD than in species with low RSD.

1.2.8. Energy Conservation

Small males have lower energy consumption than large females and thus presumably consume less energy while foraging and perhaps can provide food for the young more efficiently than large females (Reynolds, 1972; Mosher and Matray, 1974; Balgooyen, 1976). If this presumed foraging efficiency was an important factor in the evolution of RSD, we would expect the males to perform a greater share of the provisioning of the female and young in species with high RSD than in species with low RSD.

1.2.9. Pyramid of Numbers of Prey Sizes

Small prey are more abundant than large prey, and small males should do most of the hunting until the food demands of the young become sufficiently great so that the female can help meet the demand for food by capturing and delivering large prey (Storer, 1966; Reynolds, 1972; Schantz and Nilsson, 1981). This hypothesis leads to the prediction that provisioning by males should be greater in species with high RSD than in species with low RSD.

1.3. Behavioral Hypotheses

RSD leads to female dominance of the male.

1.3.1. Anticannibalism

Large females prevent small males from eating the young (Hagen, 1942; Amadon, 1959). Amadon has since retracted his support for this hypothesis. If this hypothesis were valid, males should perform less brooding of young in species with high RSD than in species with low RSD.

1.3.2. Pair Bonding

Formation and maintenance of pair bonds are facilitated by female dominance and hence by a large female (Perdeck, 1960; Amadon, 1975; Ratcliffe, 1980; Cade, 1982; Smith, 1982). This hypothesis predicts that females should be more dominant in species with high RSD than in species with low RSD.

1.3.3. Role Partitioning

Female dominance forces the male into the role of provider of food (Cade, 1960, 1982; Monneret, 1974). This hypothesis predicts that males in species with high RSD should provision the female and young to a greater extent than in species with low RSD.

1.3.4. Sociability

Sociability may inhibit the evolution of RSD (Walter, 1979). Colonial species may be less dimorphic than noncolonial ones because communal defense against predators reduces the need for larger females (Snyder and Wiley, 1976). Either hypothesis predicts lower RSD in colonial species than in noncolonial ones.

2. METHODS

2.1. Sources of Data and Statistical Methods

The information used in this paper was taken almost entirely from Cramp and Simmons (1980); Glutz *et al.* (1971) was referred to occasionally in an attempt to clarify equivocal or insufficient statements. We used a secondary source, rather than primary ones, because Cramp and Simmons (1980) provide an excellent and thorough summary compiled without any bias relevant to the hypotheses concerned with RSD. The Spearman rank correlation coefficient, r_s (Siegel, 1956), was used to find correlations between RSD and other traits and to determine which structures are trophic. Nonnumerical information was placed in as many reasonably discrete categories as possible (6 to 20) and ranked with the lowest and highest ranks representing the extremes. Not all information was available for the 43 species for which mensural data were given; species were included in an analysis only if sufficient information was given to permit categorization. Details of the methods of categorization and ranking are presented for each trait.

The contingency coefficient (Siegel, 1956) was used in one case where the trait could be characterized only as present or absent. The sign test was used to compare the dimorphism ratios of various measurements.

2.2. Descriptions of Quantitative Measures

Dimorphism in weight. We agree with Amadon (1977) and Cade (1982) that weight is the best measure of the size of a bird. A problem with weight as a measure of size is that it is variable within the individual, depending on the season, condition of the bird, and other factors. However, the greatest problem is the unavailability of sufficient samples of weights of birds of known sex for many species of birds. In this paper, only one weight was available for one or both sexes in five species: *Elanus caeruleus, Gypaetus barbatus, Accipiter badius, Aquila verreauxii,* and *Falco eleonorae.* Only two to four weights for one or both sexes were available for seven additional species: *Circus macrourus, Melierax metabates, Accipiter brevipes, Aquila clanga, Hieraaetus fasciatus, Falco peregrinus,* and *F. pelegrinoides.* An additional 14 species were represented by five to nine weights for one or both sexes, and yet another nine species had 10 to 19 weights for one or both sexes.

Weights used were for adults, except where weights for adults were not segregated *(Milvus migrans, Aquila pomarina, A. rapax, A. chrysaetos, Hieraaetus pennatus, Buteo rufinus).* The mean weight used are those of Cramp and Simmons (1980), except for *Falco biarmicus* and *F. cherrug* where the mean is the average of the extremes, and for *F. naumanni* and *F. subbuteo,* which were taken from Glutz et al. (1971).

Dimorphism in weight is expressed as the cube root of mean male weight divided by the cube root of mean female weight multiplied by 100; or basically, male weight as a percentage of female weight (Table I). The cube root was used to permit comparisions with linear measurements. In calculating Spearman rank correlation coefficients, the most dimorphic species was ranked 1 and the least dimorphic was given highest rank.

Dimorphism in wing. Wing length is the distance from the carpal joint to the tip of the longest primary, with the wing pressed against a rule and stretched fully. Dimorphism in wing, and all other linear measurements, is expressed as the mean for males divided by the mean for females and multiplied by 100, or simply, male measurement as a percentage of female measurement (Table I). Wing length is possibly the second best measure of dimorphism in body size and has the advantage of the availability of larger samples. Only four species are represented by samples smaller than five measurements for one sex, and none with less than two: Aegypius monachus, Aquila verreauxii, Hieraaetus fasciatus, and Falco pelegrinoides. Only five to nine measurements for one sex were available for an additional nine species and yet another 16 species are represented by 10 to 19 measurements for one or both sexes. Only three species are represented by samples of less than five for both weight and wing measurement: Aquila verreauxii, Hieraaetus fasciatus, and Falco pelegrinoides.

There is a high correlation between weight and wing measurements: Spearman rank correlation coefficient r_s = 0.8027 t = 8.51 (a t value of 4.3 is significant at $P < 0.0001$). The 13 species for which there is a sample of less than five for a sex for either weight or wing measurement constitute 31% of the sample of 42 species, yet they contribute only 19% to the sum of squared differences between the ranks for weight and wing for the entire sample. It thus appears unlikely that additional weights or wing measurements would drastically alter our estimates of dimorphism.

Dimorphism in bill. Bill length is the chord of the culmen, from the edge of the feathers to the tip of the bill, including the cere. The cere was not included in the bill measurements of the vultures in Cramp and Simmons (1980), and we have not included them in Table I.

Dimorphism in toe. Cramp and Simmons (1980) give the length of the middle toe without claw. Claw is included in the toe measurement of vultures and we have not included these in Table I.

Dimorphism in claw. The measurement given in Cramp and Simmons (1980) is length of hind claw. No claw measurements were given for vultures.

Dimorphism in tarsus. The measurement is the length of the tarsometatarsus.

Dimorphism in tail. The measurement is from the skin surface between the central rectrices to the tip of the longest rectrix.

Clutch size. Mean clutch size was not expressed exactly for all species. Where the usual clutch size was given as, for example, three to four, we estimated the mean as 3.5. The species with the largest mean clutch sizes were ranked 1 and those with the smallest means were ranked last.

Egg weight. Most of the egg weights given in Cramp and Simons

TABLE I

Sexual Dimorphism in Size and Aspects of the Breeding Biology of 43 Species of Western Palearctic Raptors

Species	Wing[a]	$\sqrt[3]{wt}$[a]	Tail[a]	Tarsus[a]	Toe[a]	Claw[a]	Bill[a]	Clutch[b] Size	Egg[c] Weight	Clutch[d] Weight
Pernis apivorus	97.3	100.5	96.9	96.8	96.8	93.7	96.2	2	7.83	15.65
Elanus caeruleus	99.3	99.3	100.0	100.3	97.8	103.0	99.4	3.5	9.01	31.55
Milvus migrans	96.3	98.3	95.5	98.7	94.0	92.3	93.0	2.5	6.27	15.68
M. milvus	97.4	94.5	95.3	100.8	99.3	101.4	98.9	2.13	6.20	13.21
Haliaeetus albicilla	90.7	89.6	91.8	97.9	98.6	93.1	89.8	2.1	2.57	5.39
Gypaetus barbatus	97.8	99.4	100.6	100.6	—	—	97.8	1.5	3.43	5.14
Neophron percnopterus	99.2	99.9	95.7	100.1	—	—	100.6	2	4.48	8.95
Gyps fulvus	96.4	98.7	98.2	98.2	—	—	99.1	1	2.72	2.72
Torgos tracheliotus	98.5	—	100.6	104.3	—	—	99.0	1	—	—
Aegypius monachus	96.4	97.4	95.8	97.7	—	—	101.3	1	2.52	2.52
Circaetus gallicus	97.8	98.6	100.0	98.6	97.3	102.8	96.7	1	8.00	8.00
Circus aeruginosus	95.2	90.8	93.7	95.8	90.2	87.2	85.7	4.66	6.83	31.86
C. cyaneus	89.9	86.9	87.7	93.0	87.1	84.5	86.1	4.46	7.09	31.64
C. macrourus	91.6	88.8	88.1	92.6	87.2	85.6	84.9	4.5	7.53	33.87
C. pygargus	98.1	89.0	96.8	96.0	93.4	92.5	92.7	4.2	7.91	33.23
Melierax metabates	97.8	92.5	95.9	100.0	98.2	93.6	95.5	1	6.59	6.59
Accipiter gentilis	88.4	85.5	87.1	88.6	85.9	84.6	88.0	3.51	5.62	19.71
A. nisus	84.6	81.7	82.8	87.5	82.8	83.2	79.2	4.9	11.27	55.25
A. badius	88.4	85.3	83.1	91.9	87.5	85.3	85.3	3.5	10.47	36.65
A. brevipes	93.6	89.8	92.8	98.4	95.3	91.8	88.6	4	9.95	39.82
Buteo buteo	97.2	94.1	96.7	98.0	95.4	95.3	93.9	2.73	6.82	18.61

B. rufinus	93.8	92.4	93.4	98.3	97.4	87.9	93.9	3.5	6.21	21.74
B. lagopus	94.3	94.8	94.2	92.1	91.8	87.4	91.5	3.8	6.43	24.45
Aquila pomarina	96.8	90.4	96.7	98.9	95.9	92.6	90.2	1.78	5.16	9.19
A. clanga	94.5	93.9	94.1	96.2	94.6	90.3	92.4	1.37	5.63	7.72
A. rapax	94.0	93.7	92.9	94.7	89.7	85.2	92.4	2	4.22	8.45
A. heliaca	93.8	87.6	92.7	92.7	88.5	87.5	92.7	2.5	4.05	10.12
A. chrysaetos	89.4	88.3	91.2	94.5	94.2	90.3	91.4	1.91	3.31	6.32
A. verreauxii	94.3	96.4	—	—	—	—	93.1	2	4.34	8.68
Hieraaetus pennatus	91.1	89.9	95.1	98.4	91.8	83.8	89.0	2	7.36	14.73
H. fasciatus	96.2	91.8	—	—	97.2	95.6	96.0	1.96	5.88	11.52
Pandion haliaetus	94.7	95.7	93.1	100.0	97.8	98.0	91.6	2.63	4.71	12.39
Falco naumanni	99.6	94.1	97.3	98.7	97.8	100.0	100.0	4	10.53	42.11
F. tinnunculus	96.1	94.5	95.3	100.0	98.1	95.8	92.7	4.72	8.58	40.52
F. vespertinus	99.2	96.5	97.0	101.0	99.2	96.6	95.4	3.48	10.90	37.92
F. columbarius	91.7	91.4	92.2	95.0	92.3	91.5	89.9	3.96	11.50	45.33
F. subbuteo	95.5	95.6	96.3	95.4	95.7	91.2	90.0	2.93	11.85	34.72
F. eleonorae	95.4	96.6	94.9	96.4	96.8	94.6	90.7	2.5	7.05	17.62
F. biarmicus	88.7	88.3	85.1	95.0	91.5	88.0	88.5	3.6	7.11	25.60
F. cherrug	90.7	90.5	88.8	94.1	93.4	88.7	89.6	4	5.36	21.46
F. rusticolus	90.9	86.3	88.2	94.3	91.7	88.9	87.5	3.53	4.91	17.34
F. peregrinus	86.8	87.9	83.2	87.7	87.9	85.6	83.3	3.4	4.68	15.91
F. pelegrinoides	86.4	81.5	83.1	85.0	88.5	87.1	85.8	3	8.61	25.83

[a]Dimorphism is expressed as (x̄ ♂♂/x̄ ♀♀)100.

[b] x̄ clutch size.

[c] (egg weight/x̄ ♂♂ weight + x̄ ♀♀ weight/2))100.

[d] ((egg weight)(x̄ clutch size)/(x̄ ♂♂ weight + x̄ ♀♀ weight/2))100.

(1980) are not actual weights, but are estimates based on the calculations of Schönwetter (1967). For our purposes, we have expressed egg weight as the percentage of mean body weight of the species (mean for both males and females). The species with the relatively largest egg was ranked 1.

2.3. Description, Categorization, and Ranking of Qualitative Measures

Incubation behavior. The verbal accounts of the relative roles of the sexes in incubation were placed into eight categories (Table II). Species in which all incubation is performed by the female were given the lowest rank.

Brooding. The verbal accounts of the relative roles of the sexes in the brooding of young were placed into six categories (Table III). Species in which all brooding is performed by the female were given the lowest rank.

Direct feeding of young. The participation of the sexes in direct, bill-to-bill, feeding of young was placed into one of 7 categories (Table IV). Those species in which only the female feeds young directly were given the lowest rank.

Provisioning. The degree and duration of provisioning of the female and young solely or primarily by the male was placed into one of seven categories (Table V). Those species in which the male was the sole provider of food through most of the breeding cycle were given the lowest rank.

TABLE II
Relative Roles of the Sexes in Incubation

Rank	Categorization
1	♀ only
2	Probably ♀ only; ♂ rarely, in absence of ♀
3	♀ all or most; ♂ may take a small part, ♂ often cover briefly in absence of ♀
4	♀ mainly or most, ♂ while ♀ feeds
5	♀ larger share, ♂ appreciable share ♂ assists slightly diurnally
6	Shared, but ♀ larger share
7	Approximately equal shares diurnally, ♀ all or most nocturnally
8	Shared equally

TABLE III
Relative Roles of the Sexes in the Brooding of Nestlings

Rank	Categorization
1	♀ only
2	♀ all or most, normally exclusively; ♂ rarely, exceptionally, seldom, or never
3	♀ most; ♂ occasionally and briefly when ♀ absent
4	♂ occasionally, sometimes
5	♀ principally, but ♂ also takes part; ♂ often when ♀ absent
6	Both sexes; approximately equal

Diet. The verbal accounts of the composition of diet were placed into 20 categories (Table VI). The categorization was based largely on the introductory statements under "food" in Cramp and Simmons (1980), but the entire account was read carefully to help us interpret the adjectives and adverbs used in the introductory statements. Our ranking method is based on the analysis of Newton (1979), who found that dimorphism was greatest in bird feeders (given the lowest ranking in our analysis) and least in consumers of carrion, and intermediate, in ascending ranking, in those that fed largely on fish, mammals, reptiles, insects, and snails.

Territorial defense. The relative roles of the sexes in defense against conspecifics were placed into 11 categories (Table VII). The species for which all recorded defense is by the male was ranked 1 and the species in which all defense is by the female was given the highest rank. Suf-

TABLE IV
Relative Roles of the Sexes in Direct, Bill to Bill, Feeding of Nestlings

Rank	Categorization
1	If ♀ dies, ♂ cannot feed small young; if ♀ dies, no direct contact between ♂ and young
2	♀ only
3	♂ seldom, exceptionally, not normally
4	♂ occasionally some ♂♂, when ♀ absent
5	♀ mostly; ♂ usually gives food to ♀ but may feed young directly even when ♀ there
6	♀ early, both sexes later
7	Both sexes

TABLE V

Duration of Period during which the Male is Primary or Sole Provisioner of
Food for the Female and Young

Rank	Categorization
1	Through much of nestling period, most, all or most, until young are well grown, main role in older young
2	At least through brooding (in some cases throughout cycle), until chicks feathered, through brooding or until young fledge
3	At least through brooding, especially in early stages, through guarding stage
4	Through brooding, early nestling stage, early first stage
5	Through at least part of brooding, variable into brooding
6	During brooding only (not during incubation)
7	None, apparently none

ficient information to ascertain the relative roles of the sexes in territorial defense was available for only 21 species.

Nest defense. The relative roles of the sexes in the defense of the nest against heterospecific predators were placed into seven categories (Table VIII). Nest defense by the female only was given the lowest ranking. Sufficient information was available for only 14 species.

Coloniality. The incidence of coloniality was placed into eight categories (Table IX). Those species that never nest colonially were given the highest ranking.

Mortality. In a few species, more than one set of mortality rates was available. In these cases, the most "reasonable" set was used, particularly the one in which the effects of shooting or other mortality caused by man were eliminated or reduced. For our purposes, "natural" mortality rates, independent of recent changes wrought by man, are the ones of interest. Mortality rates for the first year of life are available for only nine species, and this sample is reduced to eight for mortality for the second year (Table X).

Female dominance. Verbal accounts of behavior indicate dominance of the female over the male in 14 species: *Pernis apivorus, Circus aeruginosus, C. cyaneus, Accipiter gentilis, A. nisus, B. buteo, Aquila pomarina, A. heliaca, A. verreauxii, Falco eleonorae, F. biarmicus, F. cherrug, F. rusticolus,* and *F. peregrinus.* It is virtually impossible to place these accounts into categories; the available information simply indicates the presence of female dominance. The lack of statements of female dominance in other species does not necessarily indicate that

TABLE VI
Categories of Diet[a]

Rank	Categorization
1	Almost entirely birds
2	Chiefly birds
3	Mainly birds and mammals
4	Chiefly mammals and birds
5	Mammals, birds, lizards
6	Primarily fish
7	Predator, scavenger, kleptoparasite; fish, birds
8	Predominately mammals, birds also important
9	Chiefly mammals
10	Wide diversity, principally mammals
11	Opportunist predator and scavenger, mammals main prey
12	Predator and scavenger, wide range of species taken
13	Mammals, reptiles, birds, and insects
14	Chiefly birds and insects
15	Chiefly insects and birds
16	Chiefly reptiles
17	Chiefly reptiles and large insects
18	Adults: insects; young fed small vertebrates
19	Mainly insects
20	Carrion

[a]The order in which prey are given in each category indicates their relative importance in the diet. The ranks are arranged in an order based on the average elusiveness and alertness of prey.

TABLE VII
Relative Roles of the Sexes in Territorial Defense against Conspecifics

Rank	Categorization
1	♂ only
2	♂ mainly or usually
3	♀ vicinity of nest, ♂ rest of territory
4	Both sexes, mainly by ♂ when nesting, when ♀ defends immediate area of nest
5	♂ more than ♀
6	Both sexes, ♂ pursues intruders for a greater distance than ♀
7	♂ early, ♀ after incubation
8	Each sex defends against its own sex, ♂–♂ encounters more frequent
9	Both sexes, or each defends against its own sex, or considerable discussion with no indication of sex differences
10	♀ mainly
11	♀ only

TABLE VIII
Relative Roles of the Sexes in Defense of the Nest
against Heterospecific Predators

Rank	Categorization
1	♀ only
2	♀ primarily
3	♀ more than ♂
4	Both sexes, no indication of sex differences
5	♂ more than ♀
6	♂ primarily
7	♂ mostly

it is lacking, but dominance is more likely to be noted and reported in species where it is clearly expressed. Because we have only two categories for female dominance (presence or possible absence), we will utilize the contingency coefficient (Siegel, 1956) to compare the 14 species for which female dominance has been recorded with those for which it has not (N = 29, wing; N = 28, weight).

3. PREDICTIONS AND RESULTS

3.1. Trophic Appendages

3.1.1. Identification of Trophic Structures

Trophic structures were identified as those that showed a correlation with diet. We have limited these anlayses to the 36 species for which all measurements were available.

TABLE IX
Occurrence of Colonial Breeding

Rank	Categorization
1	Never
2	Rarely
3	Occasionally
5	Often
6	Highly
7	Nearly always
8	Always

We calculated the relative size of bill, claw, toe, tarsus, and tail by taking the mean wing measurements for the species (sexes combined) and dividing by the species mean for each of the structures. Relative size of bill, tarsus, and tail are not correlated with diet (our categorization of diet is presented in a later section): Bill, Spearman $r_s = 0.1954$, $t = 1.27$, $P > 0.20$; tarsus, $r_s = 0.1776$, $t = 1.14$, $P > 0.25$; tail, $r_s = -0.0610$, $t = 0.37$, $P > 0.60$. Relative size of claw and toe are correlated with diet: Claw, $r_s = 0.3846$, $t = 2.46$, $P < 0.02$; toe, $r_s = 0.3662$, $t = 2.33$, $P < 0.03$. Probably the most important relationship between wing and weight is wing loading; it has profound effects on flight performance. An index of wing loading for a species, crude but totally adequate for ranking species, can be derived by dividing mean species weight by the square of mean species wing measurement. This index shows no correlation with diet: $r_s = -0.2732$, $t = 1.68$, $P > 0.10$.

3.1.2. Comparison of RSD of Trophic and Nontrophic Structures

The sign test was used to compare dimorphism ratios of the various measurements. The null hypothesis in each case is that half of the species should be more dimorphic in one measurement than in the other (ignoring zero-value ties). The analysis is intended to test the hypothesis that trophic structures, those associated with obtaining food, should show greater dimorphism than body size or structures not associated with taking food.

The dimorphism ratios are listed in Table I and the results of the analysis are given in Table XI. Bill length shows the greatest dimorphism but is only scarcely, and not significantly, greater than body weight. Both are more dimorphic in significantly more species than claw, tail, wing, toe, or tarsus. Hind claw is essentially as dimorphic as bill and body weight, differing primarily in that the number of species that show greater dimorphism in length of hind claw is not significantly greater than those showing greater dimorphism in tail length. Tail, wing, toe, and tarsus (in that order) appear to be considerably less dimorphic than bill, weight, or claw. Significantly more species are more dimorphic in tail length than in either wing length or tarsus. Only tarsus length has significantly more species showing less dimorphism than wing length or toe length, and only tarsus has significantly more species showing less dimorphism than all other six measurements.

3.2. Correlation of RSD and Various Traits

The Spearman rank correlation coefficient was used to seek correlations between RSD and various traits of species. Each account be-

TABLE X
Aspects of the Behavior and Ecology of Falconiformes

Species	Inc[a]	Brood[b]	Feed[c]	Prov[d]	Diet[e]	Def[f]	Def[g]	Col[h]	1Y	>1Y
Pernis apivorus	5	5	5	3	19	—	—	8	—	—
Elanus caeruleus	5	1	3	3	13	—	—	8	—	—
Milvus mirgans	2	1	2	3	12	6	—	6	—	—
M. milvus	4	1	4	3	12	2	—	8	—	—
Haliaeetus albicilla	5	4	4	4	7	9	—	8	—	—
Gypaetus barbatus	5	4	6	—	20	—	—	8	—	—
Neophron percnopterus	8	6	7	4	20	—	—	8	—	—
Gyps fulvus	8	6	7	7	20	—	—	7	—	—
Torgos tracheliotus	2	6	7	—	20	—	—	4	—	—
Aegypius monachus	4	6	7	7	20	—	—	—	—	—
Circaetus gallicus	2	1	2	3	16	5	—	6	—	—
Circus aeruginosus	2	1	2	3	4	9	5	8	48	39
C. cyaneus	2	1	2	3	3	9	4	5	61.6	27.6
C. macrourus	2	1	2	3	4	—	—	5	—	—
C. pygargus	2	1	2	3	5	7	—	6	—	—
Melierax metabates	2	1	4	4	16	—	—	5	—	—
Accipiter gentilis	3	1	1	2	3	11	1	8	63	20
A. nisus	1	1	1	1	1	10	1	8	64.8	44.2
A. badius	3	1	2	4	17	—	—	8	—	—
A. brevipes	1	1	3	4	17	—	—	8	—	—
Buteo buteo	4	1	2	4	10	9	—	8	46.4	25.3
B. rufinus	—	—	—	—	13	—	—	8	—	—
B. lagopus	3	1	2	4	9	—	6	8	—	—

Species	a	b	c	d	e	f	g	h	1Y	>1Y
Aquila pomarina	4	3	2	1	9	3	—	8	—	—
A. clanga	1	—	2	1	11	—	—	8	—	—
A. rapax	1	—	3	1	12	—	—	8	—	—
A. heliaca	5	5	5	6	9	9	—	8	—	—
A. chrysaetos	3	2	3	6	4	—	4	8	—	—
A. verreauxii	4	4	3	6	9	9	—	8	—	—
Hieraatus pennatus	2	2	—	3	5	4	2	8	—	—
H. fasciatus	3	4	4	5	4	1	2	7	57.3	18.5
Pandion haliaetus	3	2	2	1	6	9	4	3	—	—
Falco naumanni	5	1	7	5	19	9	4	6	51	41
F. tinnunculus	3	1	2	2	9	9	4	2	—	—
F. vespertinus	7	1	3	4	18	—	4	8	—	—
F. columbarius	5	1	4	1	2	5	3	8	55	—
F. subbuteo	2	2	3	1	14	2	—	8	—	—
F. eleonorae	3	1	3	1	15	—	—	1	—	—
F. biarmicus	5	1	2	2	2	—	—	8	—	—
F. cherrug	5	1	2	2	8	—	7	8	—	—
F. rusticolus	5	2	4	3	2	9	4	8	—	—
F. peregrinus	5	3	4	3	2	8	5	8	59	32
F. pelegrinoides	—	—	—	—	1	—	—	8	—	—

[a] Relative roles of the sexes in incubation (see Table I).

[b] Relative roles of the sexes in brooding of young (see Table II).

[c] Relative roles of the sexes in the direct feeding of nestlings (see Table III).

[d] Extent of provisioning of food to ♀ and young by the ♂ (see Table IV).

[e] Food habits (see Table V).

[f] Relative roles of the sexes in territorial, conspecific interactions (see Table VI).

[g] Relative roles of the sexes in defense of the nest, eggs, and young against heterospecific predators.

[h] Incidence and degree of colonial breeding.

[i] Mortality, expressed in %. 1Y = mortality in first year of life. >1Y = annual mortality in subsequent years.

TABLE XI
Comparisons of Dimorphism of Trophic and Other Structures[a]

	$\sqrt[3]{\text{Weight}}$	Claw	Tail	Wing	Toe	Tarsus
Bill	22	15	26	31	31	33
	>0.18	>0.31	<0.008	<0.0001	<0.0001	<0.0001
$\sqrt[3]{\text{Weight}}$		18	25	27	26	31
		1.00	<0.02	<0.003	<0.008	<0.0001
Claw			23	28	31	31
			>0.09	<0.001	<0.0001	<0.0001
Tail				25	29	30
				<0.02	>0.75	<0.0001
Wing					18	29
					1.00	<0.0004
Toe						31
						<0.0001

[a]The number of species showing greater dimorphism in the measurement listed on the left than in the measurement listed above and the *probability* that one of the measurements is more dimorphic more frequently than the other (sign test, two-tailed).

gins with a statement of the two traits being compared, followed by the hypothesis that justifies the use of a one-tailed test, the results of the test, and a brief interpretative statement. Unless otherwise indicated, the hypotheses are derived from those listed in the introduction. For correlations of RSD with other traits, the results of comparisions with dimorphism in wing length are listed first, followed by dimorphism in weight. All 43 species listed in Table IX show RSD in wing length, and 32 species show a significant difference between the sexes in this measurement. Only one species fails to show RSD in weight (Cramp and Simmons present no statistics for weight). We are thus dealing with differences in the degree of RSD, not its presence or absence.

3.2.1. RSD and Clutch Size

Hypothesis: More dimorphic species lay larger clutches than less dimorphic ones. Wing: Spearman rank correlation coefficient, $r_s = 0.3729$, $n = 43$, $t = 2.57$, $P < 0.01$. Weight: $r_s = 0.4300$, $n = 42$, $t = 3.01$, $P < 0.002$. Interpretation: More dimorphic species lay significantly larger clutches than less dimorphic species.

3.2.2. RSD and Egg Weight

Hypothesis: More dimorphic species lay relatively larger eggs than less dimorphic ones. Wing: $r_s = 0.0455$, $n = 43$, $t = 0.30$, $P > 0.75$. Weight: $r_s = 0.1240$, $n = 42$, $t = 0.80$, $P > 0.45$. Interpretation: There is no correlation between dimorphism and egg weight.

3.2.3. RSD and Incubation Behavior

Hypothesis: The female should perform more of the incubation in more dimorphic species than in less dimorphic ones. Wing: $r_s = 0.2042$, $n = 41$, $t = 1.30$, $P > 0.10$. Weight: $r_s = 0.2361$, $n = 40$, $t = 1.50$, $P > 0.07$. Interpretation: there is no correlation between dimorphism and the relative roles of the sexes in incubation. In most (38) species females perform at least a larger share of incubation than the males and the lack of correlation with RSD is not surprising.

3.2.4. RSD and Brooding Behavior

Hypothesis: The female should do more of the brooding of young in more dimorphic species than in less dimorphic ones. Wing: $r_s = 0.0371$, $n = 39$, $t = 0.23$, $P > 0.35$. Weight: $r_s = 0.1431$, $n = 38$, $t = 0.87$, $P > 0.20$. Interpretation: there is no correlation between dimorphism and the relative roles of the sexes and the brooding of young. The female is at least the principal brooder of young in most (35) species and the lack of correlation with RSD is not surprising.

3.2.5. RSD and the Direct Feeding of Young

Hypothesis: The female should perform more of the direct, bill-to-bill feeding of young than the male in more dimorphic species than in less dimorphic ones. Wing: $r_s = 0.4100$, $n = 40$, $t = 2.77$, $P < 0.005$. Weight: $r_s = 0.3963$, $n = 39$, $t = 2.66$, $P < 0.01$. Interpretation: the female performs significantly more of the direct feeding of young in more dimorphic species than in less dimorphic species.

3.2.6. RSD and the Provisioning of Food by the Male

Hypothesis: The male should do more of the hunting and provisioning of the female and young in more dimorphic species than in less dimorphic ones. Wing: $r_s = 0.2020$, $n = 39$, $t = 1.64$, $P > 0.05$.

Weight: $r_s = 0.1083$, $n = 38$, $t = 0.66$, $P > 0.25$. Interpretation: There is no correlation between dimorphism and the extent of provisioning of food by the male. Males provide all or most of the food for the female and young through the brooding stage in most (32) of the species, and the lack of correlation between RSD and brooding behavior is not surprising.

3.2.7. RSD and Diet

Hypothesis: Highly dimorphic species prey mainly on birds, species with little dimorphism feed on carrion, and species feeding primarily on fish, mammals, reptiles, or insects are intermediate in dimorphism. Wing $r_s = 0.7033$, $n = 43$, $t = 6.33$, $P << 0.0001$. Weight: $r_s = 0.7501$, $n = 42$, $t = 7.17$, $P << 0.0001$. Interpretation: Dimorphism is highly correlated with diet.

3.2.8. RSD and Territorial Defense

Hypothesis: The male plays a greater role in the defense of territory against conspecifics in more dimorphic species than less dimorphic ones. Wing: $r_s = -0.4567$, $n = 21$, $t = 2.29$, $P < 0.02$. Weight: $r_s = -0.6429$, $n = 21$, $t = 3.36$, $P < 0.002$. Interpretation: The male plays a lesser role and the female a greater role in territorial defense in species with greater dimorphism than in species with less dimorphism. Note that the correlation is opposite to that predicted by the hypothesis.

3.2.9. RSD and Coloniality

Hypothesis: Colonially nesting species are less dimorphic than noncolonial ones. Wing: $r_s = 0.2841$, $n = 42$, $t = 1.87$, $P < 0.05$. Weight: $r_s = 0.3013$, $n = 42$, $t = 2.00$, $P < 0.04$. Interpretation: Colonially nesting species are less dimorphic than non-colonial ones.

3.3. Other Correlations

3.3.1. Coloniality and Diet

Hypothesis: Colonially nesting species feed on less active and alert prey than noncolonial ones; $r_s = 0.2822$, $n = 42$, $t = 1.86$, $P < 0.05$. Interpretation: Colonial nesting is correlated with feeding on less alert and agile prey.

3.3.2. RSD and Defense of the Nest Against Predators

Hypothesis: The female plays a greater role in the defense of the nest and young against predators in species with greater dimorphism than in less dimorphic ones. Wing: $r_s = 0.1078$, $n = 14$, $t = 0.38$, $P > 0.35$. Weight: $r_s = 0.0413$, $n = 14$, $t = 0.14$, $P > 0.14$. Interpretation: Dimorphism is not correlated with defense against predators.

3.3.3. Diet and Clutch Size

Hypothesis: Clutch size is larger in those species that feed on birds and other prey that is difficult to capture than in those that feed on insects and other food that is easy to capture (the rationales for this and the remainder of the hypotheses presented in this section are given in the discussion); $r_s = 0.4941$, $n = 43$, $t = 3.64$, $P < 0.0005$. Interpretation: species feeding on birds and other prey that is difficult to capture have significantly larger clutch sizes than species feeding on easily captured prey.

3.3.4. Diet and Mortality Rates in the First Year of Life

Hypothesis: Species that feed on birds and other prey that is difficult to capture have higher postfledgling mortality rates in their first year of life than species feeding on prey that is easier to capture; $r_s = 0.7782$, $P < 0.05$ (for $P = 0.05$, $r_s = 0.600$; for $P = 0.01$, $r_s = 0.783$). Interpretation: species that feed on birds and other prey that is difficult to capture have significantly higher mortality rates in their first year of life than species feeding on prey that is easier to capture.

3.3.5. Diet and Mortality Rates for Subsequent Years

Species feeding on birds and other prey difficult to capture have higher adult mortality rates than species feeding on prey that is easier to capture; $r_s = 0.1437$, n.s. (for $P = 0.05$, $r_s = 0.643$). Interpretation: Diet is not correlated with mortality rates beyond the first year of life.

3.4. RSD and Female Dominance

The contingency coefficient was used to determine whether the presence or apparent absence of a behavioral trait was correlated with RSD. Hypothesis: females are more dominant in species with high RSD than in those with low RSD. Wing: $c = 0.29$, $df = 1$, $P < 0.05$. Weight: $c = 0.37$, $df = 1$, $P < 0.02$. Interpretation: Females are significantly more dominant in species with high RSD than in those with low RSD.

4. DISCUSSION

We propose that a lack of correlation between the degree of RSD found in species and the degree to which a given trait is expressed in these species is strong evidence that the particular trait was not important in the evolution of RSD. On the other hand, the presence of a correlation between RSD and a trait does not indicate that the trait is a "cause" of RSD; it could just as readily be an "effect." Most hypotheses concerned with RSD are eminently logical; we would expect, for example, that larger females would take larger prey than smaller males.

4.1. Trophic Appendages

If the separation of foraging niches has been important in the evolution of RSD, we would predict that the dimorphism of trophic structures (those appendages used in foraging) should be greater than dimorphism in body size or than in appendages not directly used in foraging. Our empirical analysis suggests that of the available measurements, only toe and claw are trophic structures; the remaining four measurements are of nontrophic structures. No nontrophic measure shows greater dimorphism in significantly more species than claw (Table XI), and claw is more dimorphic than two nontrophic measures (wing, tarsus). However, two nontrophic measures (bill, weight) are also more dimorphic than wing and tarsus and, in addition, are also more dimorphic than tail. The other trophic structure, toe, is less dimorphic than two nontrophic measures (bill, weight) and more dimorphic than only one (tarsus). If claw and toe are the primary trophic structures in Falconiformes, then there is no evidence that trophic structures are more dimorphic than nontrophic entities. This strongly suggests that dimorphism did not evolve primarily for differential niche utilization (Selander, 1972).

The only way to erect an argument that trophic structures are more dimorphic than nontrophic measures would be to insist that bill is a trophic structure, toe is not, and that the use of cube root of weight in comparisons with linear measurements is invalid. Cube root of weight has been used extensively in studies of RSD and need not be defended here. Bill length is usually regarded as a measurement of a trophic appendage, but in birds of prey its importance is questionable since prey is seized by the feet and the beak is used merely to tear morsels from the prey item (Accipitridae) or at best to deliver a killing bite (Falconidae). In raptors, food-handling time is probably an insignificant fraction of hunting time.

Since the feet are used to seize prey in most species of Falconiformes, it is difficult to imagine how toe could be dismissed as a trophic appendage. One might argue that claw is more important than toe because the long, needle-sharp claws of raptors can inflict serious injury and have the potential of rapidly incapacitating prey. However, this is not true of the three North American species of *Accipiter*, which specialize on birds and have both long toes and claws. We have trapped thousands of *Accipiter striatus*, *A. Cooperii*, and *A. gentilis* using tethered lure birds. Relatively few lure birds are injured in the few seconds that elapse between the hawk seizing the lure and the springing of the net. We have also observed several cases in each of the three species where the raptor began to pluck or even tear flesh from avian prey that appeared to be otherwise uninjured (see also Newton and Marquiss, 1982). We suggest that the size of the grasp of the foot, toe plus claw, is more important than claw length alone. Unfortunately, no such measurement is given in Cramp and Simmons (1980) and we must thus deal with length of middle toe without claw and length of hind claw separately.

4.2. Flight Characteristics, Foraging, and Transport of Food

Although we have dismissed wing loading and other factors influencing flight performance as important factors in the evolution of RSD, some may not agree with the assumptions made in our analysis; further discussion of the hypotheses of Andersson and Norberg (1981) and Schantz and Nilsson (1981) is warranted. The approach of Andersson and Norberg is theoretical: they derive a series of equations defining flight performance that show, basically, that for optimal pursuit of prey an avian predator on birds should be slightly smaller than its prey. They note that the subduing and transport of prey should limit the size of prey taken below that optimal for pursuit. However, raptors can capture, subdue, and transport prey weighing more than they do (see, e.g., Cade, 1982; Newton and Marquiss, 1982; Ratcliffe, 1980). In spite of these capabilities, most prey taken by raptors that specialize on avian prey weigh only a fraction of the raptor's weight. In *Accipiter nisus*, the average weight of prey delivered to nests is about 10% of the weight of the adult hawk (derived from Newton, 1978; Geer, 1981). The prey

taken by *Falco peregrinus* averages 33% to 45% of the body weight of
the falcons (references in Cramp and Simmons, 1980). Cramp and Simmons present data indicating that the prey of *Accipiter gentilis* averages
40% of the hawk's body weight. The data are from Opdam (1975) and
are for winter (March); Opdam presents the breeding season, declining
to 29% of hawk body weight in June. However, Opdam's data are based
on questionable assumptions about the species and sex of the predator
that left the remains constituting the data, and further, Opdam fails to
consider adequately the possible biases in collecting data from prey
remains (see Newton and Marquiss, 1982, for a discussion of the problem). Cramp and Simmons (1980) do not present mean prey weights
for other species of the Falconiformes and what information is given
is insufficient for an accurage estimate. However, our evaluation of the
information leads us to the tentative conclusion that *F. peregrinus* is
at or near the high extreme, and that in most species of raptors that
specialize on birds the mean size of prey taken is considerably closer
to the 10% of hawk body weight shown by *A. nisus* than the 33% to
45% shown by *F. peregrinus*.

We believe that Andersson and Norberg place too much emphasis
on pursuit of prey. In our experience, most successful captures on avian
prey involve surprise and little or no pursuit. Pursuit prolonged for
more than a few seconds rarely results in a capture, particularly in
Accipiter (but see Page and Whitacre, 1975, for similar observations on
Falco columbarius). Indeed, raptors often appear to abandon pursuit
of a flying bird or even an attack on a sitting bird once the prey is aware
of the raptor's approach (see also Page and Whitacre, 1975; Schipper,
1977). However, we have seen many more pursuits of avian prey than
captures: why do raptors frequently engage in unsuccessful pursuits?
We agree with Ratcliffe (1980), Treleaven (1980), and Brown (1976):
Raptors often pursue prey that they have no intention of capturing.
"Intent" is impossible to measure in the field and the judgments of the
few highly experienced observers on intent have been questioned. Mueller
and Berger (1970) have shown that migrating *Accipiter striatus* frequently initiate attacks on tethered lure birds without actually striking
the lure. More than 50% of the *A. striatus* that initiated attacks on
tethered *Passer domesticus* resulted in the hawk aborting the attack
within 2 m of the prey. *Passer domesticus* is well within the size range
of prey taken by both male and female *A. striatus* and it is difficult to
explain this relatively low "success rate" with tethered prey without
considering intent. Mueller and Berger also used tethered *Columba livia*
as lures; attacks were frequently initiated on these large birds, but none
was actually struck by *A. striatus*. Although some of these aborted

attacks possibly can be attributed to faulty initial assessment of size by the hawk, it seems likely that many attacks were initiated with no intent of capture. There are at least a few possible explanations as to why raptors pursue prey without intent to capture; it is unnecessary to entertain these speculations in this paper.

The only variable in the equations of Andersson and Norberg (1981) is total body mass (weight). Important factors influencing maneuverability and other flight characteristics such as wing loading and pectoral muscle mass are assumed to be proportional to total body mass. Cade (1982) has noted that this is a gross simplification, and he presents considerable data to show that such assumed "geometric similarity" does not exist within the genus *Falco*. Differences between raptors and their prey are certainly greater. The calculations of Andersson and Norberg indicate that we should give greater attention to the effect of body size on flight performance, but the data available strongly suggest that flight performance was not an importance factor in the evolution of RSD.

The major, novel contribution of the hypothesis of Schantz and Nilsson (1981) is their emphasis on the transport of prey. Their paper is similar to ours in that they use empirical analyses (in an attempt to prove two parts of their hypothesis).

In one of the analyses they use the data from Cramp and Simmons (1980). Their prediction is: "Species that take large prey in relation to their own body size should be more dimorphic than those that take small prey." This prediction is generally consonant with the correlation we found between RSD and diet for which there are several explanations other than the ability to transport prey. Schantz and Nilsson (1981) confirm their prediction with a linear regression performed on the data in their Fig. 1, which is a scatter diagram with RSD (wing) on the ordinate and estimated relative weight of prey (estimated mean weight of prey/raptor species mean weight) on the abscissa. The 33 data points (not n = 34, as indicated on the figure) are unlabeled, and we have been unable to determine the identity of the species represented for more than half of the data points and unable to ascertain how estimated weight of prey was derived for most of these. The information given in Cramp and Simmons (1980) scarcely permits more than a wild guess of mean prey weight for most species, yet Schantz and Nilsson give Cramp and Simmons as the sole source for their apparently accurate estimates, some of which appear to differ by only 1%. We were unable to verify, let alone replicate, the analysis of Schantz and Nilsson and thus must conclude that a relationship between RSD and mean weight of prey remains to be demonstrated, and will probably have to await the collection of quantitative data to permit determination of mean prey weight in a greater number of species.

Another prediction of Schantz and Nilsson (1981) is: "Species that transport their prey by the crop should be less dimorphic. In this case crop size is the factor limiting the amount of food that can be transported." They offer as proof for this hypothesis a comparison of three species of snake eagles (Circaetus) and 22 unspecified species of Accipiter. It appears that Falconiform birds that transport food in their crops exhibit relatively low dimorphism, but this is correlated with diet. Crop size hardly seems to be limiting: the species of Circaetus transport very large snakes in their crops and stomachs (Newton, 1979), and vultures frequently engorge themselves so greatly that taking flight is difficult or even impossible (see, e.g., Cramp and Simmons, 1980).

In summary, the arguments of Andersson and Norberg (1981) and of Schantz and Nilsson (1981) for the primacy of body size and the associated flight performance as an influence on the presumed differences of the sexes to capture and transport prey are not convincing. It seems more likely that the primary trophic appendages are those directly associated with the capture of prey and not virtually every aspect of the anatomy of a bird. Our analysis strongly suggests that all hypotheses that invoke sexual differences in any aspect of foraging behavior as an important factor in the evolution of RSD are of dubious validity.

4.3. Sex Role Differentiation

Our comparative analysis provides little evidence that sex role differentiation played an important part in the evolution of RSD. Clutch size is directly correlated with RSD, but not egg weight. Although "bigger mothers" (Ralls, 1976) might be expected to lay larger clutches, it is more plausible that they would lay larger eggs. The larger clutch, but not larger egg, could be taken as evidence for Walter's (1979) hypothesis, that larger females are better able to cushion developing eggs, and are thus able to better avoid damage to the eggs during impact or struggles with prey. The correlation of larger clutch size and dimorphism can be interpreted as evidence for Walter's (1979) hypothesis because the longer the period of egg development and laying, the greater the possibility of damage. Also, a female might be able to acquire sufficient energy reserves to produce one or perhaps two eggs, but a female laying a large clutch would have to obtain food in the relatively brief interval between eggs, and thus must have food while eggs are developing. The problem with this hypothesis is that the available evidence indicates that females do not hunt during egg production in almost all species of Falconiformes; the males provision the females. Females become very inactive during this period (Newton, 1979), and there seems to be less need to protect eggs from damage due to activity of the female than in most nonraptorial birds.

Although Walter's (1979) hypothesis cannot be dismissed outright by our data, there is a possible alternative explanation: RSD is highly correlated with diet, clutch size is significantly correlated with diet, and diet is significantly correlated with mortality in the first year of life. This high mortality in the first year is probably the result of the difficulties juvenile hawks encounter in learning to capture birds, the most alert and elusive prey. Adult hawks feeding on birds can lay large clutches and rear large broods because nestlings and fledglings of prey species are easily captured and are predictably abundant when the hawks are breeding (see, e.g., Geer, 1981; Newton and Marquiss, 1982).

The correlation between high RSD and the predominance of the female in direct feeding of the young could be taken as evidence for the hypothesis that the larger female is more efficient in tearing prey into bits for feeding the young (Andersson and Norberg, 1981). However, it is doubtful that the very small increase in feeding efficiency that might be gained could be a driving force in the evolution of RSD. It is more likely that this correlation is an artifact of female dominance; a larger, more dominant female simply takes the food more frequently from a smaller, more subordinate male in species with high RSD than in those with low RSD. There is another plausible explanation for the high incidence of feeding by the female in species with high RSD: the adult female may be less intimidated by her daughters than the male might be, and this would lead to better apportionment of food among the young. Newton (1978) has shown that nestling *Accipiter nisus* show the best and most even growth when a female is diligent in feeding young, rather than allowing the young to feed themselves. It seems doubtful that a small male could perform such a task once his daughters are larger than he. Our observations of *A. striatus* in Canada suggest that the male avoids the proximity of his young as much as is possible (Mueller *et al.*, 1981; Meyer, unpublished data).

The only other significant correlation found between RSD and an aspect of sex role differentiation is in the defense of territory against conspecific intruders. Females show increased involvement in territorial defense in more dimorphic species than in less dimorphic ones. This is exactly opposite to that predicted by the hypothesis of Nelson

(1977). We suggest that only females can effectively deal with determined female intruders in highly dimorphic species and that this is merely a consequence of the female dominance over males associated with RSD.

All other hypotheses concerning aspects of sex role differentiation as important factors in the evolution of RSD are unsupported by our analysis and appear unlikely. Females of more dimorphic species do not incubate more than less dimorphic ones. Large females may be better incubators and brooders than small males but this appears to be irrelevant. Females simply do most of the incubating and brooding, regardless of the degree of sexual dimorphism.

There is no correlation between RSD and the female's defense of the nest, eggs, and young from potential heterospecific predators. Nest defense by females has been the hypothesis most frequently used as the major reason why females rather than males are the larger sex in raptors (Storer, 1966; Snyder and Wiley, 1976; Andersson and Norberg, 1981). The lack of correlation of this and other sex role separation hypotheses that have been used to explain why females are the larger sex in the various ecological hypotheses casts further doubt on the validity of the latter.

Andersson and Norberg (1981) suggested that it was adaptive for only one member of a pair to hunt alert vertebrate prey because if both hunted they would interfere with each others' success. High RSD is correlated with alert and elusive prey in the diet, but there is no correlation between RSD and the extent of provisioning by the male. Thus, males do not hunt more in species that prey on alert and elusive prey, and females less, than in species hunting less elusive and alert prey. There is no empirical evidence for the hypothesis of Andersson and Norberg.

Mosher and Matray (1974) have shown that small males use less energy per unit time than large females and have hypothesized that it is more efficient for males to hunt, leaving females to other parental duties. There is no correlation between RSD and the extent of provisioning by the male. In 32 of the 39 species treated in our analysis, the male provided most or all of the food for the female through the period of brooding of the young, essentially regardless of the degree of RSD. The experimental results of Mosher and Matray (1974) appear to have no relevance to the evolution of RSD.

The correlation between RSD and coloniality could be taken as evidence for Walter's (1979) hypothesis that sociability selects for reduced RSD. Walter does not provide a rationale for how coloniality might reduce RSD and our finding of a correlation between coloniality and diet suggests that diet is the primary factor in reducing RSD in colonial species.

4.4. Female Dominance

The comparative method does provide suggestive evidence on the validity of the behavioral hypotheses of RSD. There is a highly significant correlation between RSD and direct, bill-to-bill feeding of the young by the female and a reduction or absence of this behavior by the male in highly dimorphic species. This could be regarded as evidence for the anticannibalism hypothesis. However, there is no correlation between RSD and the roles of sexes in brooding. Although one might argue that a male is more likely to eat his young while feeding than when brooding them, we regard this argument as absurd. We have suggested earlier that the high incidence of direct feeding of the young by the female is a result of female dominance. This could be taken as evidence that female dominance leads to role partitioning of the sexes. However, we found no other correlations between RSD and the sex role differentiation of other behaviors, and it is unlikely that sex role partitioning of behaviors was important in the evolution of RSD.

We have no direct evidence commenting on the hypothesis that female dominance facilitates the formation and maintenance of the pair bond. We did find a significant correlation between RSD and female dominance. It is obvious that straightforward sexual selection will lead to much more rapid evolution of sexual dimorphism than selection involving any other indirect factors, although this has been overlooked by most work on RSD. Amadon (1975) has pointed out that dimorphism in size in animals, regardless of which sex is the larger, is nearly always related to mating behavior. No one appears to have seriously questioned that it is sexual selection that has produced larger males in the many species of birds exhibiting this form of dimorphism, and no one seems to have questioned the role of sexual selection in producing larger females in such birds as jacanas and phalaropes in which sex roles are largely reversed. The current reluctance to accept sexual selection as a possibility in the evolution of RSD may be a result of the concentration of modern studies on intrasexual selection, with virtually no work or thought devoted to epigamic selection (terminology of Huxley, 1938). Selander (1972) has shown that it is extremely difficult to differentiate between the effects of the two types of selection. It is obvious that the female does the choosing in most species of birds, but it is not nearly as obvious whether she chooses the male or the territory he defends against other males. The structures, behaviors, and other characteristics, including body size, that appear to function intrasexually also appear to function epigamically. It is difficult to understand the cavalier dismissal of the possibility that epigamic selection might function in the evolution of RSD in raptors.

We have suggested that if there is no correlation between the degree of RSD and the degree to which a trait is expressed in the various species of Falconiformes, then that trait has not been important in the evolution of RSD. Using this comparative method and the data available for the species of the Western Palearctic, we are left with remarkably few viable hypotheses on the evolution of RSD: (1) it is possible that the correlation between clutch size and RSD is evidence for the hypothesis that dimorphism evolved because larger females provide a better "cushion" for developing eggs, although we have offered a tentative, alternative hypothesis that is supported by the available data; (2) all other correlations involve female dominance. We conclude, tentatively, that female dominance is the key to understanding the evolution of RSD and this factor should be strongly considered in designing further studies of RSD in the birds of prey.

If one insists that damage to developing eggs was a factor in the evolution of RSD, it is possible that female dominance aids the protection of developing follicles: (1) female dominance would inhibit possible damage by an overly motivated male during copulation, and (2) female dominance might facilitate courtship feeding, reducing or eliminating the need for foraging by the female (see below). It is possible that this would lead to greater female dominance in species with large clutch sizes, but we believe that the correlations with diet and first-year mortality offer a more plausible explanation for the correlation between dimorphism and clutch size.

How might female dominance be the important factor in the evolution of RSD in Falconiformes? Smith (1980) lists more than 40 species of more than 20 families of birds in which the female is dominant over the male during the breeding season. All these species are monogamous. Data are lacking for other species, but it appears likely that female dominance during the breeding season occurs in virtually all monogamous species. Most species of raptors are monogamous.

We know very little about the courtship behavior of raptors; indeed, in most species it is unknown whether many displays and intraspecific interactions represent courtship or agonistic behavior, or both (Cramp and Simmons, 1980; Brown and Amadon, 1968). Courtship behavior has been studied in detail in a variety of other monogamous birds, and the early phases invariably involve at least some aggression on the part of the male (Armstrong, 1965; Hinde, 1973). In many birds the male is fully as hostile to an arriving female as to an intruding male. Smith (1982), noting that raptors possess potentially lethal weapons, suggests that RSD has evolved to facilitate female dominance, thus reducing the risk of injury to the female. In most other species of birds, males are dominant and some aggression occurs before dominance is reversed. It should be unnecessary to repeat that females choose the male (or the territory occupied by the male) in most species of birds, and not vice versa. It is easy to visualize why female birds of prey would choose males that are less aggressive in the initial stages of pair formation. We suggest that this is probably the key factor in the evolution of RSD.

Maintenance of the pair bond and the efficient partitioning of sex roles in parental care may also be facilitated by female dominance. Monneret (1974) and Cade (1982) have suggested that female falcons can control or stimulate the hunting efforts of their mates using vocalizations or other means. One of us (H.C.M.) has observed an incubating female *Falco columbarius* vocalize while on the nest, which was followed by the immediate departure of the male from a perch near the nest, and his return in a few minutes with prey. This behavior was observed several times in watches of only a few hours at two nests and is in full agreement with the suggestions of Cade and Monneret. Kenneth Meyer has similar observations for each of seven intensively studied pairs of *Accipiter striatus*. The behavior of the females ranged from vocalizations to extensive chases of the male, often including physical contact. In all but a few cases, the male promptly left the immediate area of the nest and was not observed again until he returned with food. Whether females have such proximate control of the behavior of their mates is not necessary for an ultimate, or evolutionary, argument. A male deprived of the opportunity to, for example, incubate or brood young can increase his inclusive fitness by provisioning the female and has little other choice in species that are monogamous and in which the total time required to breed occupies essentially all of the season suitable for breeding. Newton et al. (1983) have shown that female *Accipiter nisus* fail to breed successfully if provisioning by the male is inadequate during the prelaying and laying period.

4.5. Evidence from Other Birds

The differentiation of parental roles between the sexes has been an important component of most recent hypotheses concerned with RSD in raptors. Partitioning of parental roles is far from unique in raptors. Females perform all incubation in about 25% of the avian families for which sufficient information is available (Van Tyne and Berger,

1976), and although we were unable to find a quantitative statement, it appears that females usually do considerably more incubating than males in most of the species in which incubation is shared by the sexes. Incubation by the female only is found in a considerable number of species that are monogamous and have altricial young: for example, the Tyrannidae and many passerines (Van Tyne and Berger, 1976). Skutch (1976) notes that the roles of the sexes in incubation are generally continued into brooding.

Provisioning of the female by the male during incubation also occurs in a great variety of birds (Johnston, 1962). Provisioning to an extent so that the female's absences from the eggs are infrequent or short, or both, occurs in a diversity of birds: parrots (Psittacidae), some jays (Garrulinae), Cedar Waxwing (Bombycilla cedrorum), goldfinches and siskins (Carduelis), and crossbills (Loxia) (Skutch, 1976). The extreme in the partitioning of parental roles between the sexes is found in hornbills (Bucerotidae), where the female is imprisoned in the nest until the young are ready to fledge (in some species) and the male provides all food for his family (Skutch, 1976). Yet hornbills show extreme normal size dimorphism, with the minimum wing chord of males greater than the maximum of females in many species (McLachlan and Liversidge, 1978).

Any hypothesis concerning RSD that relies on sex role partitioning should at least consider its occurrence in a diversity of birds that do not exhibit RSD. Further, RSD is found in boobies (Sulidae) and frigate birds (Fregatidae) without any sexual partitioning of parental roles. Preliminary analysis of data from these groups suggests that female dominance is also important (Mueller and Nelson, unpublished data).

4.6. Role of Diet

A trait that is a consequence rather than cause of RSD may yet contribute to further evolutionary development of RSD. Sex difference in diet is a prime example of such a trait (Andersson and Norberg, 1981, have previously suggested that sex differences in diet are a consequence of RSD). There appears to be universal agreement that sex differences in the size of prey taken functions to reduce competition between the sexes and this reduction in competition has been a selective force in the evolution of RSD. Newton (1979) has provided the most comprehensive and incisive analysis of the possible role of competition in the evolution of RSD. Newton points out that RSD in raptors is correlated with the speed and agility of prey. The most dimorphic species specialize on birds; fish specialists are less dimorphic than bird specialists, but slightly more dimorphic than mammal specialists, which in turn are more dimorphic than reptile and insect specialists. Vultures, which feed entirely on carrion, are the least dimorphic. Newton further points out that predators specializing on rapidly moving and agile prey have fewer competitors than those that take slower, less elusive prey. He then argues that this scarcity of interspecific competition permits the reduction of intraspecific competition and the development of RSD.

As we have noted above, a hypothesis that invokes the reduction of intraspecific competition as the selective force in the evolution of RSD leads to the prediction that trophic appendages should be more dimorphic than body size or nontrophic appendages. The data falsify this prediction and alternative hypotheses concerning the correlation between RSD and diet should be considered. We propose that RSD is limited by the range of sizes of prey available to a specialist on a given type of prey (e.g., birds). Since the evolution of RSD was probably a gradual process, a continuous range of prey sizes of the preferred type must be available. The evolution of RSD in a species would halt when a marked discontinuity in prey sizes was encountered. Thus, specialization on a type of prey that is available in a wide and continuous range of prey sizes merely permits but does not in any way cause RSD. We suggest that RSD is relatively low in species that are generalist predators because the constraints prohibiting them from becoming specialists on a given type of prey would also constrain them from becoming specialists on differing sizes of prey. The monomorphism in size of vultures might be explained by the prevalence of intraspecific competition at large carcasses and other concentrated food sources: small males would be at a severe disadvantage in the presence of large females, thus preventing the evolution of RSD.

As prey items, birds are reasonably unique in possessing the following combination of characteristics: (1) most are active diurnally, (2) they are relatively abundant in most habitats where raptors occur, and (3) they usually occur in a virtual continuum of size classes within the range of sizes that might be taken by raptors. Mammals, in contrast, are mainly active nocturnally, and of the relatively few species readily available diurnally in a given habitat, there are usually rather large gaps between size classes (e.g., vole, ground squirrel, rabbit). Fish may tend to occur in a continuum of sizes, although there is some question about the relative availability of all the size classes to a surface-feeding predator. Because of the prolonged period of growth of most species of reptiles, these might also be available in a variety of sizes throughout much of the year. However, young reptiles often occupy different habitats and have habits quite different from adults (Bakker, 1982), and a specialist preying on adults would probably have difficulties in finding and capturing young. Although insects occur in an incredible variety of sizes, only the relatively large ones are profitable prey for raptors and the effective range of sizes is therefore small.

4.7. Conclusions: A New Working Hypothesis

In our hypothesis diet is a limiting but not selective force in the evolution of RSD. The hypothesis leads to no particular prediction about which appendages should show greater sexual dimorphism than others, and thus it is not in conflict with our finding that trophic appendages are not more dimorphic than body size. Since all but one of the correlations between RSD and other traits that we have found are related to female dominance, it might be worthwhile to speculate which physical aspects or appendages function in female dominance. We have insufficient evidence from most species, but most terrestrial aggressive displays include facing the opponent head-on and raising the feathers of the neck and back (see, e.g., the illustrations for *Buteo buteo* and *Falco peregrinus* in Cramp and Simmons, 1980). The raising of feathers exaggerates body size and a head-on display emphasizes the beak. Our analysis has shown that body weight and bill size are dimorphic in more species than the measurements of any other structure, offering some further support for our suggestion that female dominance is an important factor in RSD.

Much work remains to be done on the functions and evolution of RSD. We suggest that there is a surfeit of hypotheses and a dearth of pertinent data. Further studies should be designed to provide data that offer a possibility of proving or falsifying predictions based on hypotheses. For the present, we suggest that the most viable working hypothesis is that RSD has evolved to facilitate female dominance, which in turn facilitates formation and maintenance of the pair bond.

5. SUMMARY

There are 14 hypotheses that have been used, in various combinations, to explain the functions of reversed sexual dimorphism (RSD) in birds of prey. We generate 13 predictions from these hypotheses and test each using nonparametric statistics on the published data on the Falconiformes of the Western Palearctic. The most frequently cited hypothesis holds that RSD has evolved to reduce intraspecific competition for food. This hypothesis is falsified by our analysis, which shows that trophic appendages are not more dimorphic than nontrophic structures. The remaining analyses are based on the assumption that if a character were important in the evolution of RSD, the character should be exhibited to a greater degree in species with high RSD than in those with low RSD. Seven of the remaining 12 predictions show a correlation with RSD. High RSD is correlated with large clutch size and high clutch weight (but not egg weight), prevalence of the female in direct, bill-to-bill feeding of young, predation on alert and elusive prey, female dominance in intrapair interactions, and increased involvement of the female in territorial defense (the last of these correlations is opposite to that predicted by hypothesis). Low RSD is correlated with colonial nesting.

Further correlations suggest that some of the above correlations are artifacts: colonial nesting is correlated with diet, suggesting that diet is responsible for the low RSD in colonial species. Clutch size is also correlated with diet, and diet is correlated with high postfledgling mortality in the first year of life (but not in subsequent years). This suggests that large clutch sizes in species with high RSD are an adaptation to compensate for the high mortality suffered by young when they are learning to capture alert and elusive prey. The increased involvement of the female in direct feeding of the young and in territorial defense are probably the result of female dominance in intersexual interactions. We suggest that diet (primarily specializations in birds) permits but does not cause increased RSD. Female dominance appears to be the key factor in the evolution of RSD, probably by facilitating the formation and maintenance of the pair bond.

ACKNOWLEDGMENTS. We thank V. Tucker for consultations on avian aerodynamics, and P. Bencuya, J. A. Feduccia, N. S. Mueller and R. H. Wiley for comments on earlier versions of the manuscript.

REFERENCES

Amadon, D., 1959, The significance of sexual differences in size among birds, Proc. Amer. Phil. Soc. **103**:531–536.

Amadon, D., 1975, Why are female birds of prey larger than males? Raptor Res. **9**:1–11.

Amadon, D., 1977, Further comments on sexual size dimorphism in birds, Wilson Bull. **89**:619–620.

Andersson, M., and Norberg, R. A., 1981, Evolution of reversed sexual size dimorphism and role partitioning among raptors, with a size scaling of flight performance, Biol. J. Linnean Soc. **15**:105–130.

Armstrong, E. A., 1965, Bird Display and Behaviour, 2nd ed., Dover Publications, Inc., New York.

Bakker, R. T., 1982, Juvenile-adult habitat shift in Permian fossil reptiles and amphibians, *Science* **217**:53–55.

Balgooyen, T. G., 1976, Behavior and ecology of the American Kestrel *(Falco sparverius)* in the Sierra Nevada of California, *Univ. Cal. Publ. Zool.* **103**:1–88.

Brown, L., 1976, *British Birds of Prey*, Collins, London.

Brown, L., and Amadon, D., 1968, *Eagles, Hawks and Falcons of the World*, McGraw-Hill, New York.

Brüll, H., 1937, *Das Leben deutscher Greifvogel*, Gustav Fischer, Jena.

Cade, T. J., 1960, Ecology of the Peregrine and Gyrfalcon populations in Alaska, *Univ. Cal. Publ. Zool.* **63**:151–290.

Cade, T. J., 1982, *The Falcons of the World*, Cornell University Press, Ithaca, New York, pp. 36–39.

Cramp, S., and Simmons, K. E. L., 1980, *Handbook of the Birds of Europe, the Middle East, and North Africa*, Volume II, Oxford University Press, Oxford.

Geer, T., 1981, Factors affecting the delivery of prey to nestling Sparrowhawks *Accipiter nisus), J. Zool. Lond.* **195**:71–80.

Glutz, von Blotzheim, U. N., Bauer, K. M. and Bezzel, E., 1971, *Handbuch der Vögel Mitteleuropas*, Volume 4, Falconiformes, Akademisch Verlagsgesellschaft, Frankfurt.

Hagen, Y., 1942, Totalgewichts—Studien bei norwegischen Vogelarten, *Arch. Naturgeschichte* **11**:1–173.

Hinde, R. A., 1973, Behavior, in: *Avian Biology*, Volume III (D. S. Farner and J. R. KIng, eds.), Academic Press, New York, London, pp. 479–525.

Huxley, J. S., 1938, The present standing of the theory of sexual selection, in: *Evolution* (G. R. DeBeer, ed.), Clarendon Press, Oxford, pp. 11–42.

Johnston, R. F., 1962, A review of courtship feeding in birds, *Kansas Ornithol. Soc. Bull.* **13**:25–32.

Lack, D., 1968, *Ecological Adaptations for Breeding in Birds*, Methuen, London.

McLachlan, G. R., and Liversidge, R., 1978, *Roberts Birds of South Africa*, 4th ed., John Voelcker Bird Book Fund, Cape Town.

Monneret, R. J., 1974, Repertoire comportemental du faucon pelerin *(Falco peregrinus)*: Hypothese explicative des manifestations adversives, *Alauda* **42**:407–428.

Mosher, J. A., and Matray, P. F., 1974, Size dimorphism: A factor in energy savings for Broad-winged Hawks, *Auk* **91**:525–541.

Mueller, H. C., and Berger, D. D., 1970, Prey preferences in the Sharp-shinned Hawk: The roles of sex, experience, and motivation, *Auk* **87**:452–457.

Mueller, H. C., Berger, D. D., and Allez, G., 1979, Age and sex differences in size of Sharp-shinned Hawks, *Bird-Banding* **50**:34–44.

Mueller, H. C., Mueller, N. S., and Parker, P. G., 1981, Observation of a brood of Sharp-shinned Hawks in Ontario with comments on the functions of reversed sexual dimorphism, *Wilson Bull.* **93**:85–91.

Nelson, R. W., 1977, Behavioral ecology of coastal Peregrines *(Falco peregrinus pealei)*, Ph.D., Dissertation, University of Calgary, Canada.

Newton, I., 1978, Feeding and development of Sparrowhawk nestlings, *J. Zool.* **184**:465–487.

Newton, I., 1979, *Population Ecology of Raptors*, Buteo Books, Vermillon, South Dakota.

Newton, I., and Marquiss, M., 1982, Food, predation and breeding season in Sparrowhawks *(Accipiter nisus), J. Zool.* **197**:221–240.

Newton, I., Marquiss, M.,and Village, A., 1983, Weights, breeding, and survival in European Sparrowhawks, *Auk* **100**:344–354.

Opdam, P., 1975, Inter- and intraspecific differentiation with respect to feeding in two sympatric species of the genus *Accipiter*, *Ardea* **63**:30–54.

Page, G., and Whitacre, D. F., 1975, Raptor predation on wintering shorebirds, Condor **77**:73–83.

Perdeck, A. C., 1960, Observations on the reproductive behavior of the Great Skua or Bonxie, Stercorarius skua skua (Brünn) in Shetland, Ardea **48**:111–136.

Ralls, K., 1976, Mammals in which females are larger than males, Q. Rev. Biol. **51**:246–276.

Ratcliffe, D., 1980, The Peregrine Falcon, T. Carlton and D. Poyser, Ltd., Carlton, United Kingdom.

Reynolds, R. T., 1972, Sexual dimorphism in accipiter hawks: A new hypothesis, Condor **74**:191–197.

Schantz, T. von, and Nilsson, N. I., 1981. The reversed size dimorphism in birds of prey: A new hypothesis, Oikos **36**:129–132.

Schipper, W. J. A., 1977, Hunting in three European Harriers (Circus) during the breeding season, Ardea **65**:53–72.

Schmidt-Bey, W., 1913, Neckereien der Raubvögel nebst Gedanken uber die Entstechung ihrer secundoren Geschlechtsunterscheide, Ornithologische Monatsschrift **38**:400–414.

Schönwetter, M., 1967, Handbuch der Oologie I. Akademie Verlag, Berlin.

Selander, R. K., 1972, Sexual selection and dimorphism in birds, in: Sexual Selection and the Descent of Man 1871–1971 (B. Campbell, ed.), Aldine, Chicago, pp. 180–230.

Siegel, S., 1956, Nonparametric Statistics for the Behavioral Sciences, McGraw-Hill, New York.

Skutch, A. F., 1976, Parent Birds and Their Young, University of Texas Press, Austin.

Smith, S., 1980, Henpecked males: The general pattern in monogamy? J. Field Ornithol. **51**:55–63.

Smith, S., 1982, Raptor "reverse" dimorphism revisited: A new hypothesis, Oikos **39**:118–122.

Snyder, N. F. R., and Wiley, J. W., 1976, Sexual size dimorphism in hawks and owls of North America, Ornithol. Monogr. **20**:1–96.

Storer, R. W., 1966, Sexual dimorphism and food habits in three North American accipiters, Auk **83**:423–436.

Treleaven, R., 1980, High and low intensity hunting in raptors, Z. Tierpsychol. **54**:339–345.

Van Tyne, J., and Berger, A. J., 1976, Fundamentals of Ornithology, 2nd Ed., John Wiley and Sons, New York.

Walter, H., 1979, Eleonora's Falcon: Adaptations to Prey and Habitat in a Social Raptor, Chicago University Press, Chicago.

Wood, C. A., and Fyfe, F. M., (trans. and eds.), 1943, The Art of Falconry of Frederick II of Hohenstaufen, Stanford University Press, Stanford, California.

CHAPTER 3

VOCAL "DIALECTS" IN NUTTALL'S WHITE-CROWNED SPARROW

DONALD E. KROODSMA,
MYRON CHARLES BAKER, LUIS F. BAPTISTA,
and LEWIS PETRINOVICH

1. INTRODUCTION

The White-crowned Sparrow (*Zonotrichia leucophrys*) is a common breeding species on the west coast of North America, ranging from southern California to Alaska, although it is restricted to more northerly latitudes in the central and eastern part of the continent (Banks, 1964). Vocalizations of the several recognized subspecies (including *Z. l. pugetensis*, *Z. l. oriantha*, *Z. l. gambelli*, *Z. l. leucophrys*, and *Z. l. nuttalli*) have been examined, but it is the song of *Z. l. nuttalli* that has been the greatest focus of studies. Unlike the other subspecies, *Z. l. nuttalli* is nonmigratory and the characteristic song of the male is an ever-present trademark of the coastal chapparal in central California during the summer breeding season.

DONALD E. KROODSMA • Department of Zoology, University of Massachusetts, Amherst, Massachusetts 01003.   MYRON CHARLES BAKER • Department of Zoology/Entomology, Colorado State University, Fort Collins, Colorado 80523   LUIS F. BAPTISTA • Ornithology and Mammalogy, California Academy of Sciences, Golden Gate Park, San Francisco, California 94118.   LEWIS PETRINOVICH • Department of Psychology, University of California, Riverside, California 92521.   This article was written by the first author and amended according to the recommendations of the other three authors. The order of authorship of Baker, Baptista, and Petrinovich was determined at random.

It is this unique song that first attracted the attention of Peter Marler during the early 1960s. With colleagues, he described a system of song "dialects" in which neighboring males within a population shared the same song pattern, but populations as little as 2 miles (3.3 km) apart had strikingly different songs. He subsequently demonstrated that these adult songs were learned during a sensitive phase early in life, and he speculated that there may be a direct "relationship between song 'dialects' and the genetic constitution of populations . . . if young birds . . . are attracted to breed in areas where they hear the song type which they learned in their youth" (Marler and Tamura, 1962, p. 375; see also Marler and Tamura, 1964; Marler, 1970).

This possible direct link between genetic populations and dialects provides an exciting hypothesis for the function of an early song learning period: dialect differences could inhibit dispersal, thereby promoting population isolation and perhaps even speciation. Nottebohm (1969) elaborated on those possible relationships in his study of the Rufous-collared Sparrow (*Zonotrichia capensis*). Unfortunately, this hypothesis has proved exceedingly difficult to test. Early efforts did not reveal an unequivocal link between dialects and genetic populations in the Rufous-collared Sparrow (Nottebohm and Selander, 1972; Handford and Nottebohm, 1976), but sample sizes, methodologies, and study sites may have been inappropriate for demonstrating such a relationship. More recently, Baker (1975, 1983), Baptista (1975, 1977; Baptista and King, 1980; Baptista and Morton, 1982), and Petrinovich and associated colleagues (Petrinovich and Baptista, 1984; Petrinovich and Patterson, 1981, 1983; Petrinovich *et al.*, 1981) have launched research programs involving the study of dispersal, song development, mate choice, and detailed microgeographic variation of song in the White-crowned Sparrow. In addition, Baker has collected information on the genetics, as inferred from electrophoretic data, of free-living populations. All of these facets are crucial for testing Marler's hypothesis.

These studies have met with differing levels of success, varying interpretations, and diverse responses from the scientific community. In an attempt to sort facts and opinions, to clarify what has and what has not been accomplished, and to provide a summary of progress, we evaluate both published and unpublished data on the biology of the dialect system of the White-crowned Sparrow, primarily in the subspecies *Z. l. nuttalli*. After a brief description of the song, we review (1) its distribution over space both from human evaluation of sonagrams *and* avian reaction to song playbacks, (2) vocal development by the male and the female, (3) dispersal and "mate choice" data, and (4) the genetics of the dialect system. Last, we present our overall conclusions.

2. DESCRIPTION OF THE SONG

The songs of most Z. l. nuttalli males usually begin with a rather pure whistle or series of whistles; proceed, perhaps after a "vibrato" phrase, to a series of relatively complex syllables followed by a series of simpler syllables; and then sometimes conclude with a briefer and less noticeable whistled or frequency-modulated note (see Fig. 1). For discussion here, we will refer to the introduction (sometimes more than one part), the complex syllables, the simple syllables, and the conclusion (or terminal element).

Most adult males utter only a single song pattern, although in plastic song males may have two song types (Marler, 1970); Baptista (1975) has recorded more than one song type from 4.2% (17 of 402) of the birds in the San Francisco Bay area. This is undoubtedly a minimum value, for males may not always perform with all of their song types. Furthermore, matched countersinging with neighbors may lead to infrequent use of a second song type (Baptista, 1974), and sampling males in a transect would tend to uncover only the most commonly used songs. In Baptista's (1975) study, the additional song types in a repertoire were recognized primarily on the basis of different introductory phrases, whereas the remainder of the song often did not vary (e.g., Fig. 6 in Baptista, 1975). Thus, according to methods used by most investigators, the songs within a repertoire would usually be of the same dialect.

3. MICROGEOGRAPHIC VARIATION (I.E., DIALECTS) IN THE SONG

3.1. Classification of Recorded Songs by Humans

Thanks to careful and painstaking scrutiny of a vast array of sonagrams by several authors, several basic conclusions may be made.

1. Singing males at one locality typically share song patterns that are generally unlike those patterns shared by other groups of males at other localities. The greater the distance between localities, the easier is the distinction between song forms. Thus, Marler and Tamura (1962), when working with three populations separated by 3 to 160 km, had no difficulty in identifying three different "dialects." On a more local scale, these same trends persist since neighborhoods can often be identified where males have especially similar songs. Baptista (1975, p. 35) referred to "hybrid swarms" of song patterns in which a group of males

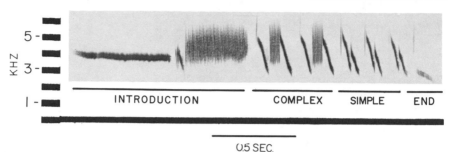

FIGURE 1. A representative song pattern for a male White-crowned Sparrow of the subspecies *Z. l. nuttalli*. The introduction, complex syllables, simple syllables, and ending are the center of focus in this review. The sonagram was made with the wide-band setting (analyzing filter bandwidth, 300 Hz) on a Kay Elemetrics Sona-Graph.

had songs intermediate between two larger dialect areas, and he documented "bilingual birds," which often occurred at the boundary between two larger dialect areas. Furthermore, these bilinguals matched song themes with a tape recorder or with appropriate neighbors, suggesting that "individuals may distinguish between subdialects as well as dialects" (p. 45). Baker *et al.* (1982c) referred to "subdialect clusters," and the distribution of different song components clearly illustrated many clusters of males with shared song characteristics (Baker and Thompson, unpublished data). Trainer (1983) examined this phenomenon quantitatively by applying multivariate statistical analysis to 21 variables measured from songs over approximately 2.5 km in the Strawberry Canyon area of Berkeley; in one of two years she found that songs from the same localities often fell close together on a scatter diagram of the first versus the third principal component, and similarly "songs from the same localities often clustered together" in her cluster analysis. Trainer's spatial autocorrelation analyses did not show that neighboring males consistently converge in any one variable, a result perhaps of the mosaic distribution of some song characters, convergence in different song characters by different groups of neighboring males, or inappropriateness of selected variables.

2. *When geographic areas larger than the local neighborhood are examined, it is clear that not all components of the song change at the same geographic points. A "dialect area" then depends to some extent on which portion(s) of the song is (are) chosen for the dialect marker(s).* Marler and Tamura (1962, p. 374), for example, found that the simple "syllables in the latter part of the trill are the most reliable index and permit a complete separation" of all songs in their samples from three

widely separated areas. Likewise, Baptista (1975) used the terminal portions of the songs, and the "feature that best distinguishes the San Francisco theme from Presidio . . motifs [so thoroughly studied by Baptista and Petrinovich since 1975—see below] is the terminal trill," or the simple syllables. On the other hand, the complex syllables have been used in other studies: "the form of the complex syllables provides the clearest regional differentiation of themes" in Z. l. oriantha (Baptista and King, 1980; p. 279), and has been used with Z. l. pugetensis as well (Baptista, 1977). Baker (e.g., 1975) has consistently used the complex syllable as the dialect marker in his Z. l. nuttalli populations, primarily because it was the most distinguishable feature audible in the field.

It is important to recognize these differences in "taxonomic" approaches, for a delineation of two dialect areas by one approach could conceivably be considered one dialect by a second approach, and vice versa. For example, if complex syllables are used as dialect markers, the Presidio and San Francisco dialects merge into one; if simple syllables are used as dialect markers, the dispersal study of Baker and Mewaldt (1978) does not span a dialect boundary.

The study that best illustrates these complications is one by Baker and Thompson (unpublished data) in which an analysis of the distribution of song components on the Point Reyes National Seashore was presented. By considering the song introduction only, they described two "superdialects"; the northern half of the study area usually contained songs with a single whistle, whereas the southern half typically had two introductory phrases, either two whistles or a whistle and a buzz. Study of the complex syllables prompted the separation of the northern and southern areas into a total of six "dialects", three with each type of introduction. To complicate matters, the four northern dialects typically had a buzzy ending, whereas the two southern dialects ended with a single descending whistle. Furthermore, these dialects, when defined by the complex syllable, "usually share a simple syllable in the region of the dialect border" such that the "boundaries of regions with common simple syllables do not coincide with any of the introduction, complex syllable, or ending boundaries" (Baker and Thompson, unpublished data).

Details from this extensive survey are worth examining, for they nicely illustrate some of the problems and frustrations in delineating dialect areas (see Table I). The vocal homogeneity of dialect areas, as delineated by Baker and Thompson (unpublished data), might be considered from two different approaches. First is the number of males

TABLE I

The Distribution of Complex Syllables among Six Dialect Regions of
White-crowned Sparrows in the Point Reyes National Seashore[a]

Types of complex syllables	Clear	Buzzy	Limantour	Drake	McClure	Barries Bay	Σ
1	42[b]	1	0	0	0	0	43[b]
2	1	34	0	2	1	20	58
3	1	4	50	4	1	0	60
4	0	0	1	33	1	0	35
5	0	0	1	33	3	4	41
6	0	0	0	7	41	0	48
7	0	0	1	0	0	34	35
8	0	0	0	0	0	10	10
9	1	0	0	5	0	7	13
Σ	45[c]	39	53	84	47	75	343
Rare[d]	2	1	0	14	15	5	37
Σ	47[c]	40	53	98	62	80	380

[a]From Baker and Thompson, unpublished.
[b]The number of males singing a particular syllable type.
[c]Row and column totals were used to calculate percentages discussed in text.
[d]Rare, by definition occurring in fewer than six males in the sample.

within the dialect that sang the particular complex syllable used as a dialect marker. If all males sampled in each area are included, the percentages of males with the particular dialect marker for Clear, Buzzy, Limantour, Drake, McClure, and Barries Bay dialects were 89%, 85%, 94%, 68% (grouping syllable types 4 and 5), 66%, and 74% (grouping syllable types 7, 8, and 2, the latter also marking the Buzzy dialect). These percentages increased to 93%, 87%, 94%, 78%, 87%, and 85% if males singing rare syllable types (by definition, occurring in 5 or fewer males on the study site) are excluded from the analysis.

Alternatively, one might consider how concentrated were particular complex syllable types into designated dialect areas. For example, 98% of the males singing syllable type 1 were in the Clear dialect area. For types 2 through 9, respectively, percentages were 59% in Buzzy (with an additional 34% in Barries Bay, even though these two dialects were not adjacent to one another), 83% in Limantour, 94% and 80% in Drake, 85% in McClure, 97%, 100%, and 54% in Barries Bay.

With regard to simple syllables, very intriguing distributional patterns arise. For example, at the Clear/Buzzy dialect border were 15 males singing Buzzy complex syllables combined with the simple syllable typical of the Clear dialect; this simple syllable occurred nowhere else in the study area. The Limantour dialect, very homogeneous with respect to the complex syllable (see above), was dominated by two simple syllables, one which was the dominant simple syllable in both the Buzzy and Drake dialects, the other of which occurred to some extent, though never dominating, in all five other dialects.

The microgeographic variation depicted by these data are typical of other study areas as well, including those studied by Baptista (1975), Trainer (1983), and Petrinovich *et al.* (1981, p. 180), whose Presidio study area contained not only the birds singing the local Presidio dialect, but also contained "scattered throughout breeding birds singing the dialect of the adjacent San Francisco city." As discussed above, an even closer scrutiny of song characters among neighboring males reveals subtle convergences, that is, subdialects, introducing yet another level of analysis that should be considered when attempting a thorough characterization of the microgeographic variation in these White-crowned Sparrow songs.

3.2. Responses of Birds to Playback Songs

The ultimate question of how songs differ over space and even whether it is important to distinguish "dialect areas" must be asked of the birds themselves. If microgeographic variation in songs is not recognizable and meaningful to the birds, then the elaborate evaluations of sonagrams by humans become rather meaningless. All indications are, though, that the birds are very adept at distinguishing different songs both within and between localities.

3.2.1. Birds Can Distinguish among Different Songs from the Same "Dialect" Area

As has been demonstrated in several other species (e.g., the White-throated sparrow, *Zonotrichia albicollis*, Brooks and Falls, 1975), male White-crowned Sparrows can distinguish the songs of neighbors from those of "strangers" more than 1 km distant, even if the strangers are from within the same dialect (Baker *et al.*, 1981c). Two factors may be involved here. First, males are undoubtedly recognizing or habituating to the nuances of neighboring songs while maintaining a higher response level to the songs of the potential intruder that would upset the status quo. Second, Trainer (1983) has found that within a dialect an increasing number of song parameters begin to change beyond 1 km. It is possible that males may be responding more strongly to songs that are similar, that is, classified to the same dialect by complex syllable, yet slightly different from the familiar songs of the immediate neighbors (see more extensive discussion of this idea in Section 3.2.2).

Similarly, males of one dialect area may respond differently to two sample songs from an adjacent dialect area (Tomback et al., 1983), giving further evidence of a recognizable and significant heterogeneity of songs within defined dialect areas. Given the responses of the birds to this variability, however, it appears that these "sub-dialects" should in the future receive increased attention.

3.2.2. Birds Can Distinguish among Songs from the Home and Alien Dialects

The pattern of responses discovered in the playbacks of Milligan and Verner (1971), Baker et al. (1981d), Petrinovich and Patterson (1981), Tomback et al. (1983), and Baker et al. (unpublished data) is confusing at first glance. In some cases the birds seem to respond more strongly to the home dialect, in other cases more strongly to the neighboring dialect, although never most strongly to distant dialects of the same or other subspecies. In addition, response measures are difficult to evaluate. Plaguing playback studies on any species is the question of what indicates the strongest response to the song playback: the number of full songs, partial songs, flights, nonsong vocalizations (chinks, trills), wing flutters, attacks on the speaker, and so forth.

As suggested by Petrinovich and Patterson (1981) and elaborated on by Baker et al. (unpublished data), one plausible explanation may lie in the degree of difference between the songs of the dialects in question (see Fig. 2). For example, Petrinovich and Patterson (1981, pp. 10, 12), doing playbacks within the San Francisco dialect, found that "females responded more strongly to the adjacent Presidio dialect than to the local dialect . . . The males sang more in response to the local dialect, flew more to the local and adjacent dialects [than to more distant dialects], and trilled and [wing] fluttered more often to the adjacent dialect. . . . The Presidio song is *somewhat similar* (our italics), but not identical, to the San Francisco song," and differs primarily in the simple syllable of the trill.

Baker et al. (1981d) found that males from the Clear dialect responded less to songs of the home dialect than to songs of the adjacent Buzzy dialect, and even less to songs of a dialect 55 km distant. Tomback et al. (1983), on the other hand, found that Limantour males responded more to their home dialect than to either the adjacent Buzzy

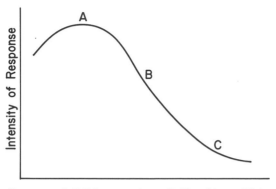

Degree of Difference from Self or Home Dialect

FIGURE 2. The response intensity of a male to song playback may depend on the degree to which that song is different from his own, which is an exemplar of the home dialect. The strongest response may be to songs that are slightly different from the familiar (A); these could be from distant males within the same dialect, or from an adjacent dialect, depending on how dialects are defined by the investigator. Weaker responses are found to songs of distant dialects (B) or other subspecies (C). (A female could use the same approach when pairing, comparing her natal song with that of a potential mate.)

or Drake dialects; as stressed by Baker *et al.* (unpublished data), though, songs from these adjacent dialects may have been especially different from those of Limantour. Because of its different introduction, Drake was classified in a different "superdialect" (Baker and Thompson, unpublished data); in addition it had the usual difference in complex and simple syllables. Furthermore, Limantour and Buzzy differed not only in the complex and simple syllables, but also in the nature of the song ending. Baker *et al.* (unpublished data) presented further evidence that the differences "in singing responses by the subjects are correlated with the degree to which the stimulus differed from the local dialect of the subjects."

Other factors could also contribute to the observed variability among playback experiments. These include possible prior experience with the songs of adjacent (or even more distant?) dialects, and the recency of contact or divergence between two dialects (Tomback *et al.*, 1983). As Petrinovich and Patterson (1981, p. 11) emphasize, carefully controlled playback experiments may require characterizing "different songs in terms of a similarity index to permit an unambiguous and objective scaling along a physical stimulus dimension." Baker *et al.* (unpublished data) have in effect controlled for distance in using songs of adjacent dialects that differ in degree, and an additional step might be to use computer-synthesized songs to quantify and vary systematically selected song parameters (e.g., Margoliash, 1983).

3.3. Discussion

Delineating "dialect areas" that are biologically meaningful is an extremely difficult problem. Mundinger (1982) advocated methods borrowed from linguistics by which boundaries are mapped for geographic changes in each song component; a dialect boundary may then be drawn at the locations where a number of these "isogloss" lines coincide. Trainer's (1983) multivariate approach may hold promise for *Z. l. nuttalli* dialects, although perhaps more meaningful variables could be measured than simple frequency and temporal characters.

It is crucial to realize how different investigators define dialects and important to realize there are disagreements about where boundaries should be drawn. As recognized in recent playback studies, new progress in understanding the nature and significance of microgeographic variation in *Z. l. nuttalli* songs will be achieved only when the birds themselves are asked. This requires a full appreciation of how songs vary spatially over a few territories to several kilometers, such as provided by the excellent surveys of Baptista (1975) and Baker and Thompson (unpublished data), and careful selection of playback stimuli (e.g., Petrinovich and Patterson, 1981; Baker *et al.*, unpublished data) to assess how the birds respond to existing variation in the field.

Furthermore, even though complex and simple syllables have been used primarily in describing dialects, it is not clear to what features of the song the birds are most responsive. Systematically varying the introduction, complex syllables, simple syllables, or ending, either singly or in combination, should yield interesting insights concerning how territorial males or, for that matter, females in the laboratory (see below) perceive these variations. It is even possible that males and females are responsive to different portions of the song, and that the songs of males in nature are a product of a conservative intersexual influence countered by a tendency for vocal elaboration as a result of intrasexual interactions (King and West, 1984).

All playback experiments, from Milligan and Verner (1971) to Baker *et al.* (unpublished data), used a single-speaker design. Petrinovich and Patterson (e.g., 1981) used a playback that consists of 16 two-minute periods, each containing ten songs, separated alternately by one and five minutes of silence. Baker's approach (e.g., Baker *et al.*, unpublished data) involves 20 repetitions of a song stimulus during five minutes followed by ten minutes of observation. Milligan and Verner (1971) alternated five minute periods of silence and playback, and over a 35-minute period presented two playbacks of one stimulus separated by the playback of a second stimulus; as with Baker's approach, 20 songs were presented during each five-minute playback session. Disadvantages of this one-speaker experimental design include possible habituation or sensitization if the same male is used in successive playbacks and considerable variability in response because of the time of day, stage of nesting cycle, or a male's previous or intervening experiences with neighbors or intruders. Statistical techniques may reduce these problems (Petrinovich and Patterson, 1979; Patterson and Petrinovich, 1979) and there is value in consistently using the same technique so that one's past and future data can be compared. In addition, a variety of response measures can be gathered for each playback stimulus.

Alternatively, some problems can be eliminated by using a two-speaker design in which two stimuli are presented alternately from two speakers 16 to 20 m apart (e.g., Lanyon, 1978; Searcy et al., 1981). The location of the male during the playback is the principal measure of his response. This design presents a male with an immediate choice, and the variability resulting from a comparison of successive playbacks to a single male or playbacks to different males is controlled. On the other hand, all behavioral measures are really a response to both playback stimuli. Song from the subject, which is relied on heavily in other studies (Falls, 1978), cannot be used reliably, and a bird's position could be a result of fleeing one speaker, attacking the other, or both. Although this two-speaker design does not solve all problems, it may be more sensitive and therefore preferable for testing some of the subtler microgeographic variation of song in future playbacks.

4. VOCAL DEVELOPMENT

As in the imprinting literature (Bateson, 1979) and throughout the song-learning literature (Kroodsma, 1982), experimental treatments and methods are rarely, if ever, replicated by other independent investigators or even the same investigator. Consequently, interpretations of data must be made cautiously and with full attention given to experimental design. Several key factors to be considered include (1) the age at which birds are taken from the field (as eggs, nestlings, fledglings, or adults, and how accurately ages can be estimated), (2) the photoperiod on which the birds are maintained, (3) the age of tutoring in the laboratory, (4) the treatment or experiences before and after tutoring (field experiences; caging conditions in the laboratory— whether caged singly or with another male or female, whether isolated from only conspecific or all avian sounds), (5) the presentation of the tutor sound (via near or distant loudspeakers, or via an adult singing male and his placement relative to the juveniles), (6) the amount of exposure to the tutor songs (e.g., number per day from loudspeaker or adult model), (7) sample size (birds in the same room may not be independent sample points because of group interactions during the developmental process), and (8) who judges the similarity of model songs and those developed by the young male (whether this person was blind relative to experimental treatment). Given the number of permutations and probable interactions of these possible treatments, it is no wonder that many issues remain unresolved in the vocal development of songbirds, including the White-crowned Sparrow. Nevertheless, a number of preliminary conclusions concerning song development in the male and female White-crowned Sparrow can be made.

4.1. Song Development by the Male

From those studies of vocal development by the male, we believe the following conclusions are warranted:

1. White-crowned Sparrows must hear and imitate the songs of conspecific adults in order to develop normal songs. Birds taken as nestlings and isolated individually or in a group from 5 to 8 days of age developed abnormal songs (Marler, 1970).

2. The ability to learn songs in the laboratory is greatest from approximately day 10 to day 50. One male from day 8 to 28 and a second male from day 35 to 56 (acoustically deprived until day 35) learned songs from loudspeakers in the laboratory, but one bird tutored from day 3 to 8 and two birds tutored from day 50 to 71 developed greatly abnormal songs; these birds were all individually isolated in sound attenuation chambers and most of the important variables discussed above were held constant (Marler, 1970). Additionally, nestlings collected at 5 to 8 days and later grouped in an aviary learned predominantly but not exclusively those songs heard from about day 11 to 51 rather than those from day 52 to 91 (Cunningham and Baker, 1983).

3. The sensitive period in the laboratory can be extended beyond 50 days by increasing the potency of the song stimulus with live tutors, lowering the learning threshold in the subjects by song-deprivation during early life, or both. When nestling males were acoustically isolated from adult songs until approximately day 50 (range, 30 to 54;

median, 50 days) and then tutored with a singing adult conspecific in an adjacent cage, most (13/17) juvenile males learned copies of the tutor songs; no males (n = 7) that were tape tutored after 50 days learned (Baptista and Petrinovich, 1984). Of six males tutored by either tape (birds A, B, C, D) or an adjacent live model (Birds E, F) from day 10 to 50 and then by live model from day 51 to 100, three (A, E, F) sang the second song, one (D) the earlier song, and two (B, C) sang hybrid songs (Petrinovich and Baptista, unpublished data). In another experiment, males grouped in an aviary were tutored via loudspeaker and simultaneously by a caged male placed in the middle of the aviary *both* before and after 50 days, and four of 15 showed some influence from the tutoring after 50 days (Cunningham and Baker, 1983) Additional hints that live tutors may enable some song learning during later life are provided by possible convergences in unique song characteristics among juveniles that are housed together (see illustrations in Marler, 1970; Cunningham and Baker, 1983).

4. It may be possible for a male in nature to change his song when a year old. Data are limited on this point, but one male observed by Baptista and Petrinovich (1984) sang an abnormal song as a yearling, but sang a more typical song the next year. Another changed to a new subdialect in a second year, and three Z. l. oriantha males changed dialects by changing both their complex and simple syllables in a second year (Baptista, unpublished data).

Most likely these birds actually learned the new songs as adults, but one other possibility exists. Marler and Peters (1981) found that song patterns present in late stages of song development (plastic song) may not appear in the final crystallized song during a Swamp Sparrow Melospiza georgiana) male's first season. Whether or not these apparently imitated songs can occur in later years remains to be determined.

4.2. Song Development by the Female

As with the males, several preliminary conclusions may be derived from laboratory studies:

1. Even though treated the same as the males, females develop relatively abnormal songs. The songs of eight out of nine females reared by Cunningham and Baker (1983) "deviated significantly from normal White-crowned Sparrow song patterns," and as a group were far more abnormal than were the songs of the males. Petrinovich and Baptista (unpublished data) found that under a variety of experimental conditions involving both tape and social tutors before and after 50 days

only two of 29 females imitated parts or all of the tutor song; the others developed either complex isolate song (n = 17) or simple isolate songs (n = 10). Wild-caught testosterone-treated females do not sing entirely typical male songs either, but their songs do appear more normal than those from laboratory-reared females (compare data in Konishi, 1965; Kern and King, 1972; Baptista, 1974; Baptista and Morton, 1982).

2. The sensitive period for song learning (i.e. imitation, not rec-ognition—see Section 4.3) may be comparable to that of the male. For songs of five females analyzed in some detail, Cunningham and Baker (1983) found one song that had been influenced by tutoring after 50 days of age. Of Petrinovich's and Baptista's (unpublished data) two females that learned songs, one did so before and one after 50 days of age. There are really insufficient data from imitating females to make a quantitative comparison with the data of males.

3. When in breeding condition, a laboratory-reared female displays selectively to songs heard during her early sensitive period. Nestling and fledgling females with normal experience in the field were taken to the laboratory and tutored over loudspeakers with songs from the same dialect until 60 to 100 days old. As adults in breeding condition, they displayed selectively to this dialect as opposed to an adjacent dialect or to one 25, 275, or 1900 km distant. In addition, females appeared to be more active and they gathered more nesting material (strings from dispenser) when hearing the familiar songs (Baker, 1983; Baker et al., 1981b, 1982b; Spitler-Nabors and Baker, 1983).

Building on this basic experimental approach, one could pose several types of questions leading to more naturalistic experiments with female response. Thus, now that we know females can discriminate between song types, future experiments can be refined to address how the experiences of a female in nature might influence her preference for different songs. It will also be interesting to determine whether data from estradiol-induced sexual displays and testosterone-induced singing are concordant; that is, whether a female displays preferentially to the song she can sing.

4.3. Discussion

There are some interesting differences in research strategies demonstrated by these studies. Marler (1970) was attempting to define the time period when learning occurred and provided an outline for further studies. Baptista and Petrinovich (1984) were interested in testing the flexibility of this sensitive phase by depriving birds during the first 50 days and then enriching the song stimulus with live tutors thereafter. Cunningham and Baker (1983), on the other hand, attempted to simulate field conditions more closely by providing identical tutoring regimes both before and after 50 days.

Although the White-crowned Sparrow may be an excellent "preparation" for studying the mechanisms of vocal development, learning, and memory in the laboratory, one must be careful when using these laboratory data to infer what processes or patterns of dispersal may be occurring in nature. Controlling all relevant variables in the laboratory and simulating interactions that a juvenile might have with conspecifics in nature are extremely difficult. Plasticity in the timing of learning phases as a consequence of stimulus quality and environmental (caging) conditions are well known in other systems (e.g., imprinting; Bateson, 1979) and to be expected in oscine song learning as well (e.g., Kroodsma, 1982). With the data of Cunningham and Baker (1983), for example, a small modification in experimental design, such as changing the nature of the interaction between the caged adult tutor and the free-flying juveniles, could have altered the patterns of learning by the juveniles. And in nature few juveniles are isolated from male song to day 50 (Baptista and Petrinovich, 1984). At the same time, these laboratory studies have greatly advanced our knowledge of *potential* explanations of what is going on in natural populations. We do not wish to discourage reductionistic approaches, only encourage improved experimental designs and suitable caution in extrapolations to field conditions.

5. DISPERSAL AND MATE SELECTION

Marler and Tamura (1962) outlined the significance of the dispersal problem. The *crucial* question is whether young birds that are hatched and raised to independence in the presence of one song pattern are sufficiently influenced by this early experience to be inhibited from dispersing and breeding beyond the geographic region where this particular song pattern predominates. In other words, do females preferentially pair and males preferentially establish territories in areas where song patterns are the same as the natal site (father and immediate neighbors), or do fewer birds disperse across dialect boundaries than would be expected by chance alone?

Documenting that some birds disperse into other dialect areas is insufficient evidence to negate the hypothesis of *reduced* dispersal between song populations. This is an important point, and one that has been frequently misunderstood. Only a comparison of observed dispersal to that expected from a random dispersal model is adequate to test this hypothesis. Additionally, whether a given level of dispersal is sufficient to *maintain* genetic differences between populations is a related but separate question (see Section 6).

Two indirect approaches to answering this dispersal question involve female androgen-induced singing and estradiol-induced copulatory responses to male song. Songs of wild-caught females should match, that is, be similar to, songs of their mates if there is "positive assortative mating," as predicted by the dispersal hypothesis outlined above. This approach assumes that both males and females learn the song at their natal site or at least within their natal dialect, and that observed pairs in the field are the result of a preference or choice, probably on the part of the female; each of these assumptions is risky. Alternatively, females may be given estradiol in the laboratory and their displays to various songs studied; this technique could be especially useful in the study of "mate choice" although we realize that real mate selection in the field is considerably more complex.

Data have been collected using these two approaches; each is discussed separately below.

5.1. Dispersal

There are dispersal data available for several populations (Baker and Mewaldt, 1978, 1979; Petrinovich et al., 1981; Baker et al., 1982a; Baptista and Morton, 1982; Petrinovich and Patterson, 1982). Several of these studies demonstrated that dispersal occurs across "dialect" boundaries, but only Baker and Mewaldt (1978) have had sufficient data to test the dispersal hypothesis statistically.

The possibility that vocal learning during an early sensitive period might inhibit dispersal, promote genetic structuring of populations, and lead to more rapid speciation is fascinating. It is an exciting hypothesis, one that if accepted would be a truly novel and important contribution to understanding the interplay between genetic and cultural evolution. Because this subject has profound implications for the evolution and consequences of vocal learning in Z. l. nuttalli, perhaps other songbirds, and maybe even other animals that learn to "speak," dispersal data have been scrutinized very carefully (e.g., Petrinovich et al., 1981; Payne, 1981).

From 1966 through 1970, Mewaldt ran a banding program in the Point Reyes National Seashore making roughly 2000 captures (including recaptures) per year. Baker continued this effort from 1971 through 1974, capturing about 350 birds per year during 1971 to 1972 and 450 per year during 1973 to 1974. Mewaldt's efforts included by chance the boundary between Clear and Buzzy complex syllables (but not simple syllables) and Baker concentrated his efforts there during 1973 to 1974. Their Fig. 6 indicates that the overall capture efforts were divided with about 85% in the Clear and 15% in the Buzzy dialect. Of the 1768 juveniles first captured (i.e., banded) in their hatching year (May through December), 371 were selected for closer study because they met two conditions: they were (1) banded during the breeding season in May, June, or July and therefore likely to be nearer the natal site than those banded later and (2) recaptured in the next or any subsequent breeding season.

Of the 371 birds, only five (or 15, depending on whether ten birds that were first caught or subsequently found breeding near the dialect border are included) crossed the dialect boundary marked by the change in complex syllables. Either value (5 or 15) is different statistically from the 26 expected as calculated from a theoretical dispersal distribution. Undesirable features of this study that could be improved in future studies include (1) nonuniform distribution of banding success throughout the study area, (2) a trapping system not equally distributed on both sides of the dialect transition, and (3) possible local movements of juveniles prior to banding (Petrinovich et al., 1981; Baker and Mewaldt, 1981; Hafner and Petersen, 1984).

Banding nestlings, however, introduces an element of risk. Even though several field studies have failed to show increased nest mortality owing to investigators visiting the nest, many field biologists suspect that the chances of predation are increased when a nest is visited (e.g., Bart, 1977; Bart and Robson, 1982). Such predation could depress the reproductive success of an area, increase immigration, and also increase emigration of unsuccessful adults, especially females (Harvey et al., 1979). Susceptibility to predation depends to a large extent on the bird species (ground nesters may be especially vulnerable) and the types of predators at particular study sites. The Presidio and Twin Peaks populations studied by Petrinovich and Patterson (1983) are probably "subject to more intensive snake predation than is usual for small passerines." Overall, about 2.5 fledglings are produced by each pair in a season in these two study areas (see Table II), but whether visiting nests increases nestling mortality is a debated issue (see Baker and Mewaldt, 1981; Petrinovich et al., 1981; Petrinovich and Patterson, 1983).

There seem to be several important differences in the demographic data available from the Presidio and Twin Peaks areas combined (those of Petrinovich and Patterson, 1982, 1983) and the Point Reyes National Seashore (the study site of Baker and Mewaldt, 1978, 1981). In the

Presidio and Twin Peaks sites, average adult mortality, especially of females, appears higher (see Table II, footnote C, for a comment on estimates of mortality) and reproductive success seems low. In order to maintain a stationary population, not including a sizeable nonbreeding population, juvenile survival from fledging to breeding age would have to be about the same as adult survival from year to year (about 0.45). This seems unlikely, for juvenile survival is more typically about one quarter of adult survival (Ricklefs, 1977). These data suggest the need for a net immigration if population levels are to be maintained.

TABLE II

Demography of Z. l. nuttalli White-crowned Sparrows from Selected Study Sites

	Presidio and Twin Peaks[a]	Point Reyes National Seashore[b]
Adult mortality[c]		
Male	0.50	0.47
Female	0.65	0.46
x̄ (of male and female)	0.57	0.46
Reproductive success (fledglings/pair per year)	2.50	——
Minimum juvenile mortality, from fledging at 1.7 weeks to first breeding, required for stationary population	0.54	——
Estimated mortality of juveniles from banding at about 4 weeks to first breeding	——	0.70[d]
Estimated number of young/pair per season required at banding (about 4 weeks) to maintain stationary population	——	3.57

[a]From Petrinovich and Patterson (1982, 1983).

[b] From Baker et al. (1981a).

[c]Assumption that no birds disperse from the study areas is violated to an unknown degree; given short dispersal distances for adults, we will assume that all breeding adults are located. In Point Reyes, dispersal from year to year is very similar in the two sexes (154 m and 138 m, n = 67 and n = 97 for females and males, respectively; see Fig. 5 in Baker and Mewaldt, 1978). Mortality for Point Reyes birds is based on recaptures of breeding birds at feeding stations. Mortality for Presidio and Twin Peaks is based on direct observation and trapping of breeding birds in successive years. Although actual mortality may be somewhat lower if some adults become nonbreeders (i.e., floaters— see Petrinovich et al., 1984), the return rate of breeding adults (and hence the adult "mortality") presented in this table is the datum needed for comparison with the minimum juvenile mortality required to maintain a stationary breeding population.

[d]Perhaps 5% of the 70% represents dispersal beyond the study site, but presumed high mortality from fledging at 1.7 weeks to banding at about 4 weeks probably makes 0.70 a conservative estimate of mortality from fledging to breeding (Baker et al., 1981a).

Whether these San Francisco sites are marginal habitats or whether atypically severe weather or investigator influences during the study reduced reproductive success is unclear. If these populations are not stationary as the demographic data suggest, then some caution should be exercised before making evolutionary statements about data collected in the Twin Peaks and Presidio study sites (Lack, 1965). On the other hand, if the dialect system itself develops in small unstable populations in isolated habitat patches, the Presidio and Twin Peaks sites may be especially relevant to understanding historic events that shape dialect structures. Burning of the White-crowned Sparrow habitat in the Point Reyes area followed by careful study of the recolonization process would at this point be most instructive, yet this is unlikely to happen with modern practices of fire control.

Because of differences in banding strategies, comparable data on reproductive success are not available for Point Reyes. Burn patterns are undoubtedly different today than in the past, but as argued by Baker and Mewaldt (1981), the relatively undisturbed and continuous nature of the Point Reyes area would seem to make it prime habitat for White-crowned Sparrows.

Dispersal into or from a population is to some extent a consequence of reproductive success within that population. With the possibility, however likely or remote, that nest visitation does affect reproductive success, a good compromise in studies of dispersal may be the technique now used by Baker et al. (1982a). Tail feather growth of juveniles, as determined in the laboratory, is used to estimate the ages of birds first captured in nature. Provided (1) the laboratory data are calibrated with growth rates of known-age individuals in the field (feather growth rates of laboratory juveniles and wild molting adults appear similar— see Baker et al., 1982a), and (2) the earliest age of dispersal is conservatively estimated (about 30 days of age), this approach should provide reliable data on an individual's natal territory, although not necessarily on reproductive success.

In retrospect, it is unfortunate that the dialect border between the Clear and Buzzy complex syllables did not lie in the middle of Mewaldt's banding area and that juveniles were not aged more accurately. Furthermore, it is now realized that almost all of the sampling effort in the Buzzy dialect area was in a small pocket of birds that sang the simple syllable characteristic of the Clear dialect (Baker and Thompson, unpublished data). The change to the typical Buzzy simple syllable was roughly on the edge of the study area. Hence, songs in these two areas differed in only one portion of the song, the complex syllable. If, as song playbacks seem to indicate, birds are sensitive to the degree of difference between songs, an especially good location to test dispersal in the future would be at a dialect boundary where two, three, or perhaps even all four portions of the songs change.

5.2. Mate Selection as Inferred from Female Song and Displays

Given that we know little of the dynamics of song learning and possible song changing in the field, data from androgen-induced singing in females must be interpreted cautiously. Nevertheless, there are some striking differences between the studies from Point Reyes (Tomback and Baker, 1984) and San Francisco (Petrinovich and Baptista, 1984). In the Point Reyes National Seashore, Tomback and Baker (1984) found evidence for "positive assortative mating" near three dialect transition zones, with songs of mates matching far more often than expected by chance. On the other hand, Petrinovich and Baptista (1984) found considerable mismatch in the songs of mated pairs at the contact zone between their Presidio and San Francisco dialect regions. Whereas most males sang the Presidio dialect, females tended to sing the neighboring San Francisco dialect. As discussed above, annual disappearance of females from these two areas appeared higher (0.65) than for local males (0.50) or Point Reyes males (0.47) or females (0.46). If these Presidio and Twin Peaks sites are to maintain a stationary "female population," survival of juvenile females must be as great as or greater than that of adult females (0.52 versus approximately 0.35—see Table II for accuracy of these values). A sizeable number of females must immigrate into these two sites in order to maintain the population (although Petrinovich and Patterson, 1983, have argued differently). Thus, between the *Z. l. nuttalli* study sites there are not only probable differences in mortality (or, at least disappearance) of females and reproductive success of pairs, but also apparent differences in dispersal and the degree to which females pair assortatively with males. Quite likely these factors are all interrelated.

In a population of the subspecies *Z. l. oriantha* in the Sierra Nevada of California, Baptista and Morton (1982) also found considerable mismatch in songs of females and their mates. One female sang her natal dialect yet paired to a male in a foreign dialect. Similarly, there are examples of *Z. l. nuttalli* males singing *Z. l. pugetensis* songs but pairing with females singing the local dialect (Baptista, 1974). Thus, it is clear that males and females that sing different songs *can* pair and breed successfully, indicating that song types are not *perfect* ethological isolating mechanisms. The significance of these population differences and examples of matching and mismatching of songs in mated pairs is unclear, but they certainly warrant a close examination of population dispersal patterns, the relative abundance of the different song forms in the populations, and the role that male song actually plays in mate selection in nature, where the female may also be attentive to territory quality, male plumage patterns, or other male attributes. The importance of several of these factors has already been demonstrated in studies of certain closely related species pairs, including the buntings (*Passerina cyanea* and *P. amoena*; Emlen et al., 1975), meadowlarks (*Sturnella magna* and *S. neglecta*; Lanyon, 1957), towhees (*Pipilo erythrophthalmus* and *P. ochroa*; Sibley and Sibley, 1964), and warblers (*Vermivora chrysoptera* and *V. Pinus*; Ficken and Ficken, 1968; Murray and Gill, 1976).

Although some of these White-crowned Sparrow data do indicate pairing patterns, caution must be used before actually inferring mate choice, as stressed by Baker (1983). Given two males and two females, and female selection of a mate, only the first female has a choice: the other female must take the male that remains. The sequence of pairing in nature and the options available to a female should really be known before using these data to invoke mate "choice."

Another interesting possibility is that a female selects a male that has a song slightly different from those under which she was hatched and reared (Bateson, 1978; McGregor and Krebs, 1982). Although a human classification of sonagrams might provide some index of similarity or difference, the female will be the final judge, and human-drawn dialect boundaries could be irrelevant. Laboratory experiments that tutor females with one song pattern and then test her (under the influence of estradiol; see Baker et al., 1981b) on an array of systematically varying songs could test this hypothesis.

These estradiol-induced sexual displays by the female are an excellent technique to use in studying the significance of various song parameters to a female. Baker et al. (1981b) demonstrated that females are selectively responsive to songs experienced early in life or, in the case of wild-caught adult females, to the dialect from which they were removed. Further refinements in this approach and the questions asked could provide exciting advances in appreciating how the female responds to and perhaps even "chooses" males with different vocal behaviors.

6. CORRELATIONS BETWEEN SONG DIALECT MARKERS AND ELECTROPHORETIC TRAITS

There are really two separate but related questions here. The first concerns the origin of possible genetic differences between dialect areas, and the second involves the maintenance (or decay) of genetic differentiation over time. A likely origin of genetic differences between adjacent dialects is the "founder effect," and Nei *et al.* (1975) and Lande (1980) offer some insights as to how founder effects (or bottlenecks) can have genetic consequences. Baker (1975) elaborated the hypothesis that dialect areas may be a consequence of the chapparal fire climax community, and that dialect boundaries are secondary contact zones between formerly isolated populations that descended by chance from genetically different ancestors. Genetic drift might also play a role in such small isolated populations. After secondary contact, possible genetic differentiation would be maintained only to the extent that dispersal across the boundary is inhibited. Thus, even if there are demonstrable genetic differences between dialects, nothing can be inferred about dispersal unless one knows the history of the populations; genetic differentiation might be preserved because of reduced dispersal between populations, but if "sufficient" dispersal occurs between populations genetic differences might be observable only during a brief time window following secondary contact. Our primary interest is in whether the birds actively maintain a genetic divergence at a dialect boundary, and it is important to remember that from genetic data alone this is impossible to determine.

Baker *et al.* (1982c) have gathered the largest and most detailed data base concerning the interaction of song dialects and genetic structure of populations. As with the dispersal study of Baker and Mewaldt (1978), the questions are of such importance and have such far-reaching consequences that the data have been and will continue to be analyzed repeatedly by critics with diverse backgrounds and expertise in different statistical approaches. As with two other topics above, we will take a stepwise approach in addressing this issue.

6.1. Genetic Facts(?)

6.1.1. There Is Genetic Heterogeniety within Dialect Areas

Baker *et al.* (1982c; p. 1023) demonstrated that in the Point Reyes National Seashore two of four dialects, Clear and Buzzy, representing geographic distributions of complex syllables, were "significantly heterogeneous over all loci." These data, and especially some former similar data (Baker, 1975) have been used by critics (e.g., Petrinovich *et al.*, 1981; Payne, 1981; Zink and Barrowclough, 1984) to point out that genetic structuring can occur in the absence of dialect boundaries, and therefore any concordance between dialects and genetic differences may be only coincidental. However, Baker *et al.* (1982c) have been concerned with whether there is an *additive* effect of the dialect boundary. The value of quantifying the genetic variation within dialects is then useful primarily as a yardstick for comparing between-dialect differences, and the test is whether there are greater between- than within-dialect differences, regardless of the magnitude of genetic structuring within dialects.

6.1.2. The Genetic Distance (A Measure of the Degree of Genetic Difference) and Geographic Distance between Populations Are Positively Correlated

When all pairwise comparisons of the Baker *et al.* (1982c) samples are plotted, "genetic distance increases with geographic distance (Fig. 1; $r = 0.66$, $p < 0.05$)" (Zink and Barrowclough, 1984). As a result of this distance effect, song dialects would be expected to be genetically different. The question of whether there is an added effect attributable to dialect boundaries, over and above mere distance effects, becomes more sticky.

6.1.3. There Does Appear to be an Added Genetic Difference Related to Song Dialects. Both May be a Consequence of Founder Effects

Baker *et al.* (1982c) calculated an average F_{ST} of 4.2% among dialects, whereas Zink and Barrowclough (1984), using an "hierarchical *F*-statistic analysis," calculated a total F_{ST} of 5.7%, of which 3.3% can be attributed to a between-dialect effect, over and above any distance effect. F_{ST} values of this magnitude are comparable to those seen in subdivided populations of other species (e.g. humans; Neel and Ward, 1972).

After controlling for distance effects, Baker *et al.* (1982c, 1984) found less genetic variation within dialects than between, whereas Zink and Barrowclough (1984) argued that there is insufficient evidence for that conclusion. Evan Balaban (unpublished data) explored the G statistic as a substitute for the D statistic (used by Zink and Barrowclough, 1984) and found that the data are consistent (at $P = 0.05$) with the claim of Baker *et al.* (1982c, 1984). In yet another analysis of the data, Hafner and Petersen (1984) found "two zones of major shifts in gene frequency" roughly concordant with the two southernmost dialect transition zones, and they therefore concluded that the data support "the existence of discrete genetic units" among the White-crowned Sparrows in the Point Reyes study sites.

6.2. Discussion

"If dialect populations originate by colonization, these founder events may be the source of the original genetic divergences" (Baker *et al.*, 1984). If secondary contact does not promote immediate panmixia, genetic effects of dialects are to be *expected*. The magnitude of the genetic differences will depend on the original stock of a founder population, selection and/or drift occurring in isolation, recency of secondary contact, and finally the nature of the dispersal across the boundary after contact. Furthermore, it must be recognized that dispersal rates at dialect boundaries cannot be inferred from any genetic data.

There are no direct estimates of gene flow in these Point Reyes populations. Some individuals dispersed across dialect boundaries (Baker and Mewaldt, 1978), but careful study of both dispersal *and* subsequent reproductive success of individuals with different genetic backgrounds is needed to clarify this issue. Also, possible levels of selection in these different areas are simply unknown. It is therefore impossible to say at this time whether or not the observed genetic shifts associated with some dialect transitions can be maintained by the behavior of the birds.

The statistical arguments of Zink and Barrowclough (1984), Baker *et al.* (1984), and Hafner and Peterson (1984) are based on dialect boundaries drawn by Baker *et al.* (1982c). Just as the genetic distance among samples is quantitatively estimated, though, so too can the "vocal distance" of the song populations be estimated. Dialects are probably not an "all-or-none" phenomenon, although it has been convenient to treat them so. Adjacent dialects may differ in one to four portions of the song, and the degree of difference (not easily quantified, admittedly) within each part may vary. The approach of Baker *et al.* (1982c) was a good first approximation, but with new information now available on song variation it will be worth submitting the entire data set to reanalysis. For example, their DP site was classified in the Buzzy dialect, but the simple syllable there is characteristic of the Clear dialect (Baker and Thompson, unpublished data); it is intriguing that this site is also "at a zone of gene frequency transition, which renders clustering of this sample with neighboring samples arbitrary" (Hafner and Petersen, 1984). Thus, in *both* genic and vocal characters this site is intermediate. Furthermore, the Drake dialect belongs to a different "superdialect" (Baker and Thompson, unpublished data), and the two Drake samples, together with a nearby sample in the adjacent Limantour dialect, are very divergent from the other six samples (see Fig. 2 in both Zink and Barrowclough, 1984, and Hafner and Petersen, 1984). These relationships are tantalizing and encourage further study of a possible relationship between dialects and genetics.

7. CONCLUSIONS

1. We have repeatedly used the term "dialect" throughout this review, even though it is becoming increasingly obvious that there is little agreement on an operational definition. Discrete boundaries drawn after inspection of sonagrams are proving inadequate to characterize the complex patterns of microgeographic variation found in nature.

There is no simple solution to this problem. We suggest an approach that first determines the variations to which the birds, both males and females, are most responsive, followed by a multivariate, quantitative approach, weighting variables according to the responses of the birds; we are sensitive to the problems that this approach encounters in systematics, but think it is a good first step. Computer-synthesized songs will be essential to this approach. Perhaps less attention should be paid to the boundary between two dialects than to some quantitatively estimated "phenetic distance" or index of dissimilarity between songs from two or more males or localities. Subdialects, dialects, and superdialects may be unnecessary terms that actually bias investigations of relevant variation in the field.

2. One must be cautious when inferring dispersal rates between song populations from (1) the matching of song types by mates in the field, (2) timing of sensitive periods for male or female vocal learning in the laboratory, (3) microgeographic variation of songs in nature, (4) copulatory displays of wild-caught or laboratory-reared females, or (5) any technique other than careful study of dispersal itself. To determine whether the microgeographic distribution of vocal patterns can influence dispersal patterns, it would be useful to undertake an intensive banding program in a habitat as homogeneous as possible where changes in vocal behavior are most pronounced (e.g., in Limantour-Drake areas of Baker and Thompson, unpublished data). Care would have to be taken to insure that (1) juveniles were accurately aged (for this particular question it might be advisable to band fledged juveniles, not nestlings, just to insure against possibly increased mortality in the nest), and that (2) juveniles dispersing in all directions have an equal probability of being relocated the next breeding season. Until more and better dispersal studies are completed, the role of dispersal in maintaining either observed distributions of songs or genetic structure of populations in continuous stretches of natural habitat will remain unclear.

Furthermore, careful study of breeding adults banded as juveniles will provide critical information on the vocal phenotype "chosen" by the female, whether males or females are most likely to breed in areas dissimilar to their natal area, and whether the reproductive success of birds exploiting different options varies.

3. There are real differences in the demographic data obtained from the different *Z. l. nuttalli* study sites and we believe these differences in study sites contribute in large part to the differences in data and opinion about what is "typical" behavior for Nuttall's White-crowned Sparrow. Reproductive success, together with characteristically high juvenile mortality, appears insufficient to maintain a stationary population at the two San Francisco study sites of Baptista, Patterson, and Petrinovich. A net immigration of individuals into these sites is therefore likely and could contribute substantially to their findings there. Although the characteristics of the Point Reyes site (large size, continuous habitat, relatively undisturbed) would suggest that it is closer to optimal breeding habitat, data on reproductive success to confirm this point are desirable.

4. The research questions and strategies of Baptista, Patterson, and Petrinovich and those of Baker and colleagues often differ. The former have consistently demonstrated the inherent flexibility of the song-learning process, showing that under various conditions the timing of vocal learning in both the laboratory and field can be greatly extended beyond what was originally thought to be the case. Their field studies at Twin Peaks and the Presidio are also valuable in revealing the interrelationships among reproductive success, mortality, dispersal, and pairing of birds with different song types. Furthermore, estimates of reproductive success and unequivocal identification of the parents and natal territory can only be achieved by banding nestlings. On the other hand, Baker has sacrificed data on reproductive success and precise natal sites to insure against the possibility of increasing the mortality of nestlings by leading predators to the nest. He has focused on quantifying dispersal and genetic changes over distance in the relatively continuous and natural habitat of the Point Reyes seashore. It is important to realize the differences between study sites, methods, and primary focus of the studies, for only then does a coherent picture of the biology of Nuttall's White-crowned Sparrow dialects begin to emerge.

5. Genetic differences between adjacent areas that differ in vocal behavior might be expected if, as postulated, these groups represent descendents from small founder populations. Analysis and reanalysis of the Baker *et al.* (1982c) data tend to confirm some concordance between dialect transitions and shifts in gene frequencies. Genetic differences, of course, need not imply anything about dispersal rates, but

any genetic differentiation would be maintained over time across the secondary contact zone in direct porportion to the degree that dispersal and gene flow across that zone is inhibited. To refer to "dialect effects" on genetic structuring implies that there is an effect of song dialect in promoting or maintaining genetic differentiation, but this should be regarded as a research hypothesis, not a conclusion.

6. Readers must be careful not to make unwarranted inferences. The genetic data of Baker *et al.* (1982c) do not prove directly that females are mating assortatively in nature, that dispersal is limited across zones of song change, and so forth. As a second example, the terms natal and alien can have two meanings: a natal song can be the song of a female's hatching area in nature, or it might specify the song a laboratory female heard during the first 30 days or so of life in a laboratory experiment. If the second use is intended, stating that "females are more responsive to their natal than an alien song" need not imply strong support for the "assortative mating" hypothesis. This is an important distinction, for if the two uses are confounded, a naive reader may be misled in some discussions of mate choice.

7. The relevance of the *Z. l. nuttalli* studies for those of other subspecies or species remains unclear. Biologists seek generalities and given that sensitive periods, microgeographic vocal variation, and so forth are also found in other songbirds, we hope that the Nuttall's White-crowned Sparrow is not unique in these characteristics. Until studies of comparable sophistication are conducted on other species, though, generalizations may not be forthcoming.

ACKNOWLEDGMENTS. We thank Joyce Britt, Debra Boudreau, and Cindy Staicer for their help in preparing the text. Evan Balaban, Bob Bowman, Tod Highsmith, Rich Lenski, Peter Marler, David Spector, Cindy Staicer, and Bob Zink all made constructive comments on earlier drafts of this paper. In addition we thank N.S.F. and N.I.H. for financial support of studies on White-crowned Sparrows (DEB-78-22657, BNS-82-14008 to M.C.B.; HD04343, BNS-7914126, BNS-8004540 to L.P.; DEB7712980 to L.F.B.; MH38782 to L.F.B. and L.P.) and other aspects of vocal behavior in birds (BNS-8201085 to D.E.K.).

REFERENCES

Baker, M. C., 1975, Song dialects and genetic differences in White-crowned Sparrows (*Zonotrichia leucophrys*), *Evolution* **29**:226–241.

Baker, M. C., 1983, The behavioral response of female Nuttall's White-crowned Sparrows to song of the natal dialect and alien dialect, *Behav. Ecol. Sociobiol.* **12**:309–315.

Baker, M. C., and Mewaldt, L. R., 1978, Song dialects as barriers to dispersal in White-crowned Sparrows, Zonotrichia leucophrys nuttalli, Evolution 32:712–722.

Baker, M. C., and Mewaldt, L. R., 1979, The use of space by White-crowned Sparrows: Juvenile and adult ranging patterns and home range versus body size comparisons in an avian granivore community, Behav. Ecol. Sociobiol. 6: 45–52.

Baker, M. C, and Mewaldt, L. R., 1981, Response to "Song dialects as barriers to dispersal: A re-evaluation," Evolution 35:189–190.

Baker, M. C., Mewaldt, L. R., and Stewart, R. M., 1981a, Demography of White-crowned Sparrows (Zonotrichia leucophrys nuttalli), Ecology 62:636–644.

Baker, M. C., Spitler-Nabors, K.J., and Bradley, D. C., 1981b, Early experience determines song dialect responsiveness of female sparrows, Science 214: 819–820.

Baker, M. C., Thompson, D. B., and Sherman, G. L., 1981c, Neighbor/stranger song discrimination in White-crowned Sparrows, Condor 83:265 –267.

Baker, M. C., Thompson, D. B., Sherman, G. L. and Cunningham, M. A., 1981d, The role of male-versus-male interactions in maintaining population dialect structure, Behav. Ecol. Sociobiol. 8:65–69.

Baker, M. C., Sherman, G. L., Theimer, T. C., and Bradley, D. C., 1982a, Population biology of White-crowned Sparrows: Residence time and local movements of juveniles, Behav. Ecol. Sociobiol. 11:133–137.

Baker, M. C., Spitler-Nabors, K. J., and Bradley, D. C., 1982b, The response of female mountain White-crowned Sparrows to songs from their natal dialect and an alien dialect, Behav. Ecol. Sociobiol. 10:175–179.

Baker, M. C., Thompson, D. B., Sherman, G. L., Cunningham, M. A., and Tomback, D. F., 1982c, Allozyme frequencies in a linear series of song dialect populations, Evolution 36:1020–1029.

Baker, M. C., Baker, A. E. M., Cunningham, M. A., Thompson, D. B. and Tomback, D. F., 1984, Reply to "Allozymes and song dialects: A reassessment," Evolution 38:449–451.

Banks, R. C., 1964, Geographic variation in the White-crowned Sparrow, Zonotrichia leucophrys, Univ. Calif. Publ. Zool. 70:1–123.

Baptista, L. F., 1974, The effects of songs of wintering White-crowned Sparrows on song development in sedentary populations of the species, Z. Tierpsychol. 34:147–171.

Baptista, L. F., 1975, Song dialects and demes in sedentary populations of the White-crowned Sparrow (Zonotrichia leucophrys nuttalli), Univ. Calif. Publ. Zool, 105:1–52.

Baptista, L. F., 1977, Geographic variation in song and dialects of the Puget Sound White-crowned Sparrow, Condor 79:356–370.

Baptista, L. F., and King, J. R., 1980, Geographical variation in song and song dialects of montane White-crowned Sparrows, Condor 82:267–284.

Baptista, L. F., and Morton, M. L., 1982, Song dialects and mate selection in montane White-crowned Sparrows, Auk 99:537–547.

Baptista, L. F., and Petrinovich, L., 1984, Social interaction, sensitive phases, and the song template hypothesis in the White-crowned Sparrow, Anim. Behav. 32:172–181.

Bart, J., 1977, Impact of human visitation on avian nesting success, Living Bird 16:187–192.

Bart, J., and Robson, D. S., 1982, Estimating survivorship when the subjects are visited periodically, Ecology 63:1078–1090.

Bateson, P. P. G., 1978, Sexual imprinting and optimal outbreeding, Nature (London) 273:659–660.

Bateson, P., 1979, How do sensitive periods arise and what are they for? Anim. Behav. 27:470–486.

Brooks, R. J., and Falls, J. B., 1975, Individual recognition by song in White-throated Sparrows. I. Discrimination of songs of neighbors and strangers, Can J. Zool. **53**:879–888.

Cunningham, M. A., and Baker, M. C., 1983, Vocal learning in White-crowned Sparrows: sensitive phase and song dialect, Behav. Ecol. Sociobiol. **13**:259–269.

Emlen, S. T., Rising, J. D., and Thompson, W. L., 1975, A behavioral and morphological study of sympatry in the Indigo and Lazuli Buntings of the Great Plains, Wilson Bull. **87**:145–179.

Falls, J. B., 1978, Bird song and territorial behavior, in: Aggression, Dominance, and Individual Spacing in the Study of Communication and Affect, Volume 4 (L. Krames, P. Pliner, and T. Alloway, eds.), Plenum Press, New York, pp. 61–89.

Ficken, M. S., and Ficken, R. W., 1968, Reproductive isolating mechanisms in the Blue-winged Warbler -Golden-winged Warbler complex, Evolution **22**:166–179.

Gottfried, B. M., and Thompson, C. F., 1978, Experimental analysis of nest predation in an old field habitat. Auk **95**:304–312.

Hafner, D. J., and Petersen, K. E. 1984, Song dialects and gene flow in the White-crowned Sparrow, Zonotrichia leucophrys, Evolution.

Handford, P., and Nottebohm, F., 1976, Allozymic and morphological variation in population samples of Rufous-collared Sparrow, Zonotrichia capensis, in relation to vocal dialects, Evolution **30**:802–817.

Harvey, P. H., Greenwood, P. J., and Perrins, C. M., 1979, Breeding area fidelity of Great Tits (Parus major), J. Amin. Ecol. **48**:305–313.

Kern, M. D., and King, J. R., 1972, Testosterone-induced singing in female White-crowned Sparrows, Condor **74**:204–208.

King, A. P., and West, M. J., 1983, Epigenesis of cowbird song: A joint endeavor of males and females, Nature, (London) **305**: 704–706.

Konishi, M. 1965, The role of auditory feedback in the control of vocalization in the White-crowned Sparrow, Z. Tierpsychol. **22**:770–783.

Kroodsma, D. E., 1982, Learning and the ontogeny of sound signals in birds, in: Acoustic Communication in Birds. II. Song Learning and Its Consequences (D. E. Kroodsma and E. H. Miller, eds.), Academic Press, New York, pp. 1–23.

Lack, D., 1965, Evolutionary ecology, J. Anim. Ecol. **34**:223–231.

Lande, R., 1980, Genetic variation and phenotypic evolution during allopatric speciation, Am. Natural. **116**:463–479.

Lanyon, W. E., 1957, The comparative biology of the meadowlarks (Sturnella) in Wisconsin, Publ. Nuttall Ornithol. Club no. 1.

Lanyon, W. E., 1978, Revision of the Myiarchus flycatchers of South America, Bull. Am. Mus. Nat. Hist. **161**:429–627.

Margoliash, D., 1983, Acoustic parameters underlying the responses of song-specific neurons in the White-crowned Sparrow, J. Neurosci. **3**:1039–1057.

Marler, P., 1970, A comparative approach to vocal learning: Song development in White-crowned Sparrows, J. Comp. Physiol. Psychol. **71:1–25.**

Marler, P., and Peters, S., 1981, Sparrows learn adult song and more from memory, Science **213**:780–782.

Marler, P., and Tamura, M., 1962, Song "dialects" in three populations of White-crowned Sparrows, Condor **64**:368–377.

Marler, P., and Tamura, M., 1964, Culturally transmitted patterns of vocal behavior in sparrows, Science **146**:1483–1486.

McGregor, P. K., and Krebs, J. R., 1982, Mating and song types in the Great Tit, Nature (London), **297**:60–61.

Milligan, M., and Verner, J., 1971, Inter-population song discrimination in the White-crowned Sparrow, *Condor* 73:208–213.

Mundinger, P. C., 1982, Microgeographic and macrogeographic variation in the acquired vocalizations of birds, in: *Acoustic Communication in Birds. II. Song Learning and Its Consequences* (D. E. Kroodsma and E. H. Miller, eds.), Academic Press, New York, pp. 147–208.

Murray, B. G., Jr., and Gill, F. B., 1976, Behavioral interactions of Blue-winged and Golden-winged Warblers, *Wilson Bull.* 88:231–254.

Neel, J. V., and Ward, R. H., 1972, The genetic structure of a tribal population, the Yanomama Indians. VI. Analysis by F-statistics (including a comparison with the Makiritare and Xavante), *Genetics* 72:639–666.

Nei, M., Maruyama, T., and Chakraborty, R., 1975, The bottleneck effect and genetic variability in populations, *Evolution* 29:1–10.

Nottebohm, F., 1969, The song of the Chingolo, *Zonotrichia capensis*, in Argentina: Description and evaluation of a system of dialects, *Condor* 71:299–315.

Nottebohm, F., and Selander, R. L. K., 1972, Vocal dialects and gene frequencies in the Chingolo sparrow (*Zonotrichia capensis*), *Condor* 74:137–143.

Patterson, T. L., and Petrinovich, L., 1979, Field studies of habituation: II. Effect of massed stimulus presentation, *J. Comp. Physiol. Pschol.* 93:351–359.

Payne, R. B., 1981, Population structure and social behavior: Models for testing the ecological significance of song dialects in birds. in: *Natural Selection and Social Behavior: Recent Research and New Theory* (R. D. Alexander and D. W. Tinkle, eds.) Chiron Press, New York, pp. 108–120.

Petrinovich, L., and Baptista, L. F., 1984, Song dialects, mate selection, and breeding success in White-crowned Sparrows, *Anim. Behav.* (in press).

Petrinovich, L., and Patterson, T. L., 1979, Field studies of habituation: I. Effect of re-productive condition, number of trials, and different delay intervals on responses of the White-crowned Sparrow, *J. Comp. Physiol. Psychol.* 93:337–350.

Petrinovich, L., and Patterson., T. L., 1981, The response of White-crowned Sparrows to songs of different dialects and subspecies, *Z. Tierpsychol.* 57:1–14.

Petrinovich, L., and Patterson, T. L., 1982, The White-crowned Sparrow: Stability, re-cruitment, and population structure in the Nuttall subspecies (1975–1980), *Auk* 99:1–14.

Petrinovich L., and Patterson, T. L., 1983, The White-crowned Sparrow: Reproductive success (1975–1980), *Auk* 100:811–825.

Petrinovich, L., Patterson, T., and Baptista, L. F., 1981, Song dialects as barriers to dispersal: A re-evaluation, *Evolution* 35:180–188.

Ricklefs, R. E., 1977, Fecundity, mortality, and avian demography, in: *Breeding Biology of Birds* (D. S. Farner, ed.), Natl. Acad. Sci., Washington, D.C., pp. 366–447.

Searcy, W. A., McArthur, P. D., Peters, S. S., and Marler, P., 1981, Response of male Song and Swamp Sparrows to neighbour, stranger, and self songs, *Behaviour* 77:152–163.

Sibley, C. G., and Sibley, F. C., 1964, Hybridization in the Red-eyed Towhees of Mexico: The populations of the southeastern plateau region, *Auk* 81:479–504.

Spitler-Nabors, K. J., and Baker, M. C., 1983, Reproductive behavior by a female songbird: Differential stimulation by natal and alien song dialects, *Condor* 85:491–494.

Tomback, D. F., and Baker, M. C., 1984, Assortative mating by White-crowned Sparrows at song dialect boundaries, *Anim. Behav.* 32:465–469.

Tomback, D. F., Thompson, D. B., and Baker, M. C., 1983, Dialect discrimination by White-crowned Sparrows: Reactions to near and distant dialects, *Auk* 100:452–460.

Trainer, J. M., 1983, Changes in song dialect distributions and microgeographic variation in song of White-crowned Sparrows (*Zonotrichia leucophrys nuttalli*), *Auk* **100**:568–582.

Willis, E. O., 1973, Survival rates for visited and unvisited nests of Bicolored Antbirds, *Auk* **90**:263–267.

Zink, R. M., and Barrowclough, G. F., 1984, Allozymes and song dialects: A reassessment, *Evolution* **38**:444–448.

CHAPTER 4

ON THE NATURE OF GENIC
VARIATION IN BIRDS

GEORGE F. BARROWCLOUGH,
NED K. JOHNSON, and ROBERT M. ZINK

1. INTRODUCTION

Although papers concerned with electrophoresis and systematics have
been appearing in the ornithological literature since the early 1950s
(see review in Sibley *et al.*, 1974), a period of rapid development of
biochemical techniques and methods of analysis began only in the
middle 1960s when empirical and theoretical population geneticists
devoted increasing attention to the field. These workers were concerned
not with questions of systematics, but rather with problems of the nature
of evolutionary processes (Lewontin, 1974). Ornithologists did not adopt
these new techniques and analyses as quickly as other evolutionary
biologists; the first quantitative studies of individual heterozygosity and
genetic differentiation in natural populations of birds did not appear
until the late 1970s (see, for example, Smith and Zimmerman, 1976;
Barrowclough and Corbin, 1978). Now, however, a considerable amount
of avian electrophoresis is being performed, not only in relation to

GEORGE F. BARROWCLOUGH • Department of Ornithology, American Museum of Natural History, New York, New York 10024.   NED K. JOHNSON • Museum of Vertebrate Zoology and Department of Zoology, University of California, Berkeley, California
94720   ROBERT M. ZINK • Department of Ornithology, American Museum of Natural
History, New York, New York 10024. *Present Address*: Museum of Zoology, Louisiana
State University, Baton Rouge, Louisiana 70803.

taxonomy and systematics (e.g., Zink, 1982; Gutierrez et al., 1983), but also in association with investigations of evolutionary mechanisms (e.g., Avise et al., 1980b; Barrowclough, 1980b), ecology (Baker, 1975), and behavior (Baker and Fox, 1978; Sherman, 1981). Thus, a quantitative assessment of the nature of genic variation in birds seems appropriate at this time. Such an analysis may be of use to workers trying to interpret patterns of genic variation in their own data, and in planning eletrophoretic studies of behavioral problems.

One advantage of studying electrophoretic variation comes from the simple relationship between such variation and an organism's genotype. This relationship allows one to analyze allozymic variation using the tools of population genetics—a large body of method and theory relating evolutionary processes to patterns of genetic variation. Thus, it is possible to investigate the nature of electrophoretic variation with a detail and a statistical power not easily achieved with most other suites of characters. In this paper, we investigate the patterns and processes associated with intrapopulational and interpopulational genic variation.

2. THE MUTATION-DRIFT THEORY OF GENIC VARIATION

The mutation-drift (i.e., neutral) hypothesis of genetic variation was developed by a number of workers over the last 20 years (King and Jukes, 1969; Kimura and Ohta, 1971; Nei, 1975, 1983; Kimura, 1983). For a nontechnical overview of the theory and the empirical evidence bearing on it, see Kimura (1979). The underlying premise is that the large quantity of genic variation uncovered by electrophoretic surveys is not being maintained by natural selection; that is, the variation is not adaptive. Instead, proponents of the neutral school argue that alternative alleles are selectively equivalent and are maintained in natural populations by a balance between new mutations and random genetic drift. Hence, selection is "neutral" with respect to the segregating alleles. Likewise, most of the differences found among populations and species, both differences in frequency and allelic fixations, are thought to have arisen predominantly due to stochastic processes, such as mutation and genetic drift, rather than as a result of selection, a deterministic process. One of the attractive aspects of this theory is that it permits one to make a number of quantitative predictions about patterns of genic variation, both within and among populations. It is a straightforward task to test the theory by checking for empirical consistency with predictions; falsification of selectionist hypotheses, such as overdominance, are not as simple because of the number of different alternatives and the lack of precise quantitative models. Hence, we favor the epistemological position that a null hypothesis of neutrality should be held unless such an initial supposition can be rejected in a given case. Here we investigate the fit of the results of avian electrophoretic surveys to the predictions of the mutation-drift theory following the procedures outlines by Chakraborty and his colleagues (Fuerst et al., 1977; Chakraborty et al., 1978, 1980).

3. INTRAPOPULATIONAL GENIC VARIATION

3.1. Methods

If the genic variation uncovered by electrophoresis is selectively neutral, or approximately so, then the patterns of distributions of alleles within breeding populations should be the product of stochastic processes. Because of the nature of such processes, it is not possible to predict the pattern at a single locus; it is the distribution across many alleles and loci that is predicted. Thus, the predictions of the various mutation-drift models for intrapopulational genic variation are based on large sample sizes across many loci. Consequently, it is essential to have reasonably large samples of both loci and individuals. In this study we attempted to examine the majority of avian studies in which on the order of 1000 or more total genes have been examined from single breeding populations. A further qualification was a restriction to those studies in which all individuals were examined for all loci. In some studies, if a locus proves invariant after the first ten or so individuals are scored, then further examination of the locus is stopped. Such a regimen is not suitable for our purposes because it introduces a bias against rare alleles. Of course, such a procedure may be perfectly reasonable for other purposes, for example, in the estimation of both genetic distance and the among-population component of genetic variance.

Many of the predictions of the mutation-drift theory depend on the values of two parameters, the effective size of a breeding population, N_e, and the mutation rate of a gene to alternative alleles, μ. Unfortunately, very little is known about the actual values of either of these numbers. Some work has been done on estimating the effective sizes of local populations, or demes, of birds (Barrowclough, 1980a; Barrowclough and Shields, 1984), but for the neutralist equations we need to know the effective size of the entire species' population, probably a very large number. The mutation rate is similarly unknown, but is probably a very small number. Fortunately these two unknowns enter many of the relevant equations in the form of a product, $M = 4N_e\mu$, which determines not only the distribution of allelic frequencies across loci, but also the overall heterozygosity. Thus, the approach we have taken here is to use the actual empirical value of overall heterozygosity for a study, $\hat{H}$, to estimate $\hat{M}$, in particular $\hat{M} = 4N_e\,\mu = \hat{H}/(1 - \hat{H})$. For each sample of individuals representing a single breeding population, $\hat{H}$ was estimated as

$$1/L \sum_{i=1}^{L} \left[1 - \sum_{j=1}^{k_i} \hat{x}_{ij}^2 \right]$$

where $\hat{x}_{ij}$ is the frequency of the jth allele at the ith locus, k_i is the number of alleles at the ith locus, and L is the total number of loci examined. This value of H is that expected if there are no significant departures from Hardy-Weinberg equilibrium (e.g., Nei, 1975). We used our estimate of the variable, M, to predict several detailed aspects of the distribution of allelic frequencies. This procedure is not circular because there are infinitely many distributions of allelic frequencies that will yield a single value of heterozygosity; the neutral theory we employ is consistent with only one of these. We tested whether the actual distribution is significantly different from that theoretical one.

Several variants of the mutation-drift models have been developed. Here we tested the fit of the avian data to the simplest of these, the *Infinite allele–Constant mutation rate* (IC) model (Chakraborty et al., 1980). This model is powerful in the sense of Popper (1968); it leads to specific prediction of a number of aspects of the distribution of genetic variation within populations. As indicated above, the only data required to generate the predictions are estimates of average heterozygosity; the data required to test the predictions are a table of allelic frequencies for all loci.

The predictions examined with our data involve: (1) the distribution of number of alleles versus allelic frequency, (2) the distribution of numbers of loci by individual-locus heterozygosity, (3) the numbers of common alleles per locus, and (4) the numbers of rare alleles per locus. The degree of concordance of the observed and the expected distributions was examined using the Kolmogorov–Smirnov test (Sokal and Rohlf, 1969).

The specific formulas and calculations used in generating the predictions are described in two recent papers (Fuerst et al., 1977; Chakraborty et al., 1980). The number of alleles, n, expected in the frequency interval between p and q is:

$$n(p,q) = M \sum_{i=Np+1}^{Nq} \frac{1}{i} \prod_{j=1}^{i} \frac{N + 1 - j}{N + M - j}$$

where N is the total number of genes sampled and M is defined as above. This formula was used to generate the overall distribution of numbers of alleles versus frequency as well as the numbers of rare and common alleles. The expected distribution of single locus heterozygosity was tabulated in Fuerst et al. (1977). One of us (G.F.B.) has written a computer program to perform these calculations.

3.2. Results

We found electrophoretic data meeting our qualifications for 24 species of birds (Table I). The numbers of loci and individuals examined from single populations ranged from 27 to 48 and from 14 to 51, respectively. The total number of genes examined (for diploid individuals, this is twice the product of the number of individuals and the number of loci) varied from 918 to 3978.

The most general tests of the mutation-drift hypothesis are comparisons of the overall observed and expected distributions of alleles by frequency. That is, how many alleles occur with a frequency between 0 and 0.05, how many between 0.05 and 0.10, and so forth. We had 24 such comparisons, one for each taxon, and in general the agreement of the actual distributions to those expected was quite good. In Fig. 1 are shown four of these observed and expected distributions. The general expectation of all the neutral models, when heterozygosity is low (<0.30 or so), is a U- or J-shaped distribution with few alleles segregating at intermediate frequencies. The Kolmogorov–Smirnov test was used to examine the fit of the distributions; only one of the 24 tests was significant at the 0.05 level, that for the Fox Sparrow (Passerella iliaca) (Table I). One significant value out of 24 might simply represent a chance event; however, an examination of the Passerella results suggests the discrepancy between the observed and expected distributions may have been due to an excess of rare alleles (alleles with frequencies less than or equal to 0.05). Similar trends, although not as pronounced, exist in several of the distributions from other species. Hence we examined this aspect in more detail.

The formula used to compute the expected distribution of alleles by frequency (Equation 5 of Chakraborty et al., 1980) also can be used to compute the expected number of alleles in any specific frequency range. We have computed the numbers of expected common and rare alleles per locus for the 24 studies. In Figs. 2 and 3 we compare these results with the observed values. Unfortunately there is no convenient way to test these results for significance. Chakraborty et al. (1980) pointed out that chi-square tests are not appropriate due to the nature of the

TABLE I

Sample Sizes, Heterozygosities, and Tests of Neutral Predictions for Selected Avian Electrophoresis Studies.[a]

Species	Individuals	Loci	$\hat{H}$	D_{max}(alleles)	$D_{max}(h)$	Reference
Oceanodroma leucorhoa	39	48	0.050	0.080	0.047	R. Baker (unpublished)
Cyrtonyx montezumae	23	27	0.033	0.092	0.075	Gutierrez et al. (1983)
Callipepla squamata	20	27	0.054	0.045	0.048	Gutierrez et al. (1983)
Lophortyx gambelii	19	27	0.032	0.061	0.078	Gutierrez et al. (1983)
Larus californicus	30	35	0.037	0.035	0.050	Zink and Winkler (1983)
Sphyrapicus nuchalis	15	37	0.039	0.033	0.054	Johnson and Zink (1983)
Empidonax difficilis	28	45	0.053	0.100	0.028	N. K. Johnson (unpublished)
Empidonax euleri	18	38	0.071	0.103	0.154	N. K. Johnson (unpublished)
Eremophila alpestris	17	35	0.102	0.086	0.067	N. K. Johnson (unpublished)
Parus atricapillus	20	35	0.044	0.122	0.057	M. J. Braun (unpublished)
Parus carolinensis	20	35	0.043	0.160	0.057	M. J. Braun (unpublished)
Parus gambeli	15	33	0.020	0.167	0.114	M. J. Braun (unpublished)

Vireo chivii	14	38	0.067	0.138	0.092	N. K. Johnson (unpublished)
Vireo solitarius	20	42	0.045	0.042	0.029	N. K. Johnson (unpublished)
Dendroica coronata	27	32	0.034	0.056	0.078	Barrowclough (1980b)
Pipilo erythrophthalmus	17	39	0.146	0.030	0.154	M. J. Braun (unpublished)
Piplo ocai	15	39	0.129	0.101	0.154	M. J. Braun (unpublished)
Amphispiza belli	20	43	0.047	0.138	0.066	N. K. Johnson (unpublished)
Passerella iliaca	51	39	0.049	0.203[b]	0.136	Zink (1982, unpublished)
Zonotrichia capensis	14	41	0.054	0.099	0.048	Zink (1982, unpublished)
Junco hyemalis	25	37	0.061	0.166	0.126	Barrowclough (unpublished)
Junco phaeonotus	24	36	0.056	0.096	0.059	Barrowclough (unpublished)
Geospiza fortis	19	27	0.070	0.167	0.199	Yang and Patton (1981, unpublished)
Geospiza conirostris	17	27	0.067	0.096	0.050	Yang and Patton (1981, unpublished)

[a] D_{max}(alleles) is the Kolmogorov-Smirnov statistic for the comparison of the distributions of alleles by frequency, empirical versus predicted; D_{max}(h) is the statistic for the comparison of the distributions of single locus heterozygosity.
[b] $P < 0.05$.

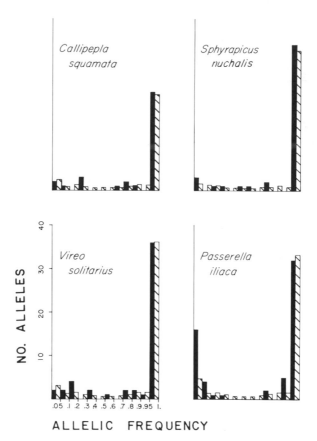

FIGURE 1. Observed (solid) and expected (hatched) distributions of numbers of alleles by frequency for four species of birds. Observed distributions based on starch-gel electrophoresis of tissue extracts from individuals taken from single breeding populations. Expected distributions calculated using the Infinite allele–Constant mutation rate model of the mutation-drift theory.

sampling distribution; in addition, the distribution of the R^2 test they suggested has not been tabulated for smaller sample sizes such as ours. However, qualitative analysis of Figs. 2 and 3 suggests a pattern of excess rare alleles compared with the pattern of common alleles.

Heterozgyosity can be estimated separately for each locus in a population. That is, monomorphic loci have a heterozygosity of 0.0; polymorphic loci have a heterozygosity of $1 - \Sigma x_i^2$, where x_i is the frequency of the ith allele segregating at the locus. The neutral model leads to predictions about the distribution of this interlocus heterozygosity,

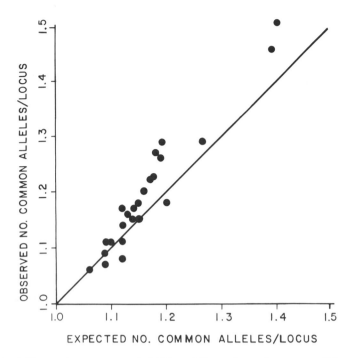

FIGURE 2. Observed versus expected (IC model) numbers of common alleles per locus for 24 studies of avian genic variation. Common alleles are those occurring with a frequency greater than 0.05.

given the overall average heterozygosity. The general expectation is that most loci will show no heterozygosity, most that are heterozygous will be only moderately so, and a small fraction will have values ranging up to 0.5 to 0.6. We have compared the data from all 24 avian studies with the expected distributions. An example is shown in Fig. 4. Kolmogorov–Smirnov tests indicate no significant departures of the observed distributions from the expected ones (Table I).

3.3. Remarks

Our intrapopulational results are in good agreement with the predictions of the *Infinite allele–Constant mutation rate* (IC) model of neutral theory. The only discrepancy we find is an apparent excess number of rare alleles in several avian populations (Table I) over the number predicted by the neutral model. Chakraborty *et al.* (1980) noted a similar phenomenon in their examination of protein variation in *Dro-*

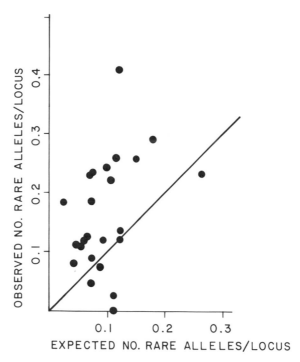

FIGURE 3. Observed versus expected (IC model) numbers of rare alleles per locus for 24 studies of avian genic variation. Rare alleles are those occurring with a frequency less than or equal to 0.05.

sophila. These results should be quantitatively checked when larger sample sizes (> 100 individuals) become available from single breeding populations of birds. Although the reason for the existence of these excess alleles is unclear, several points can be made. For example, although many different models for the maintenance of allelic variation by natural selection are possible, an excess of low frequency alleles is not characteristic of such phenomena as overdominance and genic selection (Li, 1978). Instead of a U-shaped distribution, overdominance yields a prediction of a peak of alleles with a mean frequency near 0.5 (Nei, 1983); this is clearly not consistent with the avian data. Chakraborty *et al.* (1980) noted that reduced population numbers in the past (bottlenecks), would produce an excess of rare alleles, as would the slightly deleterious allele hypothesis of Ohta (1974). Ohta claims that the selective values of mutations form a continuum from lethal to slightly deleterious to neutral to advantageous. In her view most electrophoretic variation is composed of the subset of mutations that are very slightly

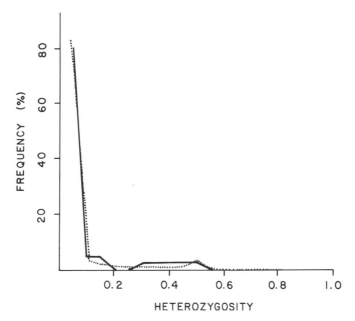

FIGURE 4. Observed (solid) and expected (broken) distribution of single locus hetero-zygosity for an electrophoretic study of *Empidonax difficilis*. Expectation calculated using the IC model. Most loci (80%) have heterozygosities between 0.0 and 0.05, whereas there are only single loci (2.2%) in each of the classes from 0.25 → 0.30 to 0.45 → 0.50.

deleterious and that exist in mutation–selection balance in large populations. However, as pointed out by Chakraborty *et al.* (1980), Ohta's theory results in the prediction of an excess of rare alleles, over the IC predictions, in all species and a positive correlation with population size and heterozygosity. The bottleneck hypothesis yields a prediction of an excess of rare alleles only in those populations that have gone through a bottleneck, not in all populations. We have no way of knowing which, if any, of these avian populations have gone through bottlenecks. Nevertheless, the excess of rare alleles is not present in all populations; moveover, the correlation of these excess rare alleles on heterozygosity is not significantly different from zero and in fact, is negative ($r = -0.104$, $df = 22$). Thus our results are consistent with bottlenecking, but apparently not with the slightly deleterious allele theory.

Recently, there have been investigations of the extent to which electrophoretic surveys reveal the full level of genic variation in populations. For example, Aquadro and Avise (1982) found some additional alleles in passerine birds by resurveying selected taxa using varying electrophoretic conditions. In some cases, then, it is possible that there are more alleles segregating than we have identified. For such cases this would increase both the level of heterozygosity and the numbers of alleles in the lower frequency classes. Although we suspect this would not alter the fit to the neutral model, the question ought to be investigated when sufficient data become available.

4. INTERPOPULATIONAL GENIC VARIATION

4.1. Methods

Consistency with the mutation-drift hypothesis within populations does not necessarily mean that such consistency will be observed between populations. Hence, we examined patterns of genic differentiation between populations of birds in relation to the neutralist expectations. As with variation within populations, if the processes resulting in genic differentiation among populations, either within or between species, are primarily stochastic, then the mutation-drift theory will yield accurate predictions concerning the distributions across loci of the patterns of differences. In particular, if there is no selection acting across populations or species to maintain alleles at some global optimum frequency or alternatively acting to promote different alleles in different environments, then genetic differences should evolve as a function of period of isolation, population sizes, and mutation rates. Because the accumulation of genetic distance will vary randomly from locus to locus, the variance of genetic distance across loci will increase with time. However, as time increases further and alternate alleles start to become fixed in different populations, genetic distance at increasing numbers of loci will reach one, and the variance across loci will start decreasing. Thus, the mutation-drift theory leads to the prediction of a first increasing, then decreasing variance of genetic distance across loci with time of isolation. This pattern should hold regardless of taxonomic rank; the only important variables are time of isolation and level of heterozygosity. In reality, of course, the time of isolation is never known. However, average genetic distance can be used in place of time, because if the mutation-drift model does hold, genetic distance will be a monotonically increasing function of time. Therefore, our method was as follows: for each pairwise comparison of population samples, the genetic distance was calculated for each locus. The minimum genetic distance was used (Chakraborty *et al.*, 1978); it varies from zero to one. For example, if two populations were compared for

30 loci, there would be 30 single locus genetic distances. A mean and a variance were calculated from these values. In this way every pair of populations in a study can be represented by a point on a plot of mean versus variance of genetic distance. We compared these points with the theoretical curve calculated by Chakraborty et al. (1978, pp. 377–378). This distinctive, inverted U-shaped curve represents the expectation if the mutation-drift hypothesis holds for interpopulational or transpecific genic evolution. Empirical results are compared by inspection owing to the "semiquantitative" nature of the test (Chakraborty et al., 1978).

In the examination of transpecific genic variation, the number of loci is a critical factor. However, in contrast to the within-population analysis, the number of individuals is not as critical. Where possible, we attempted to employ only samples with five or more individuals; Nei (1978) and Gorman and Renzi (1979) indicated that these sample sizes are sufficient to estimate genetic distances and heterozygosities if heterozygosities are low (<15%), genetic distances are high (>0.10), and the number of loci studied is fairly large. These conditions were generally met in the avian studies employed here.

4.2. Results

Examination of interpopulation genic variation, using the methods derived from the mutation-drift hypothesis, requires a fairly wide range of genetic distances. However, protein differentiation between avian populations tends to be conservative (Avise, 1983). Most electrophoretic studies of birds have resulted in estimates of genetic distances below 0.20. Therefore, testing neutralist predictions with avian data is hampered. For example, the greatest potential for departure from the pattern predicted by the neutralist model, for the plot of variance versus mean genetic distance, occurs at intermediate values of mean distance. Thus, many studies of avian protein variation were not especially useful for this analysis, including all studies of intraspecific genic variation.

In order to cover as broad a spectrum of genetic distances as possible, we chose the following taxa for study: Procellariiformes (Barrowclough et al., 1981), Galliformes (Gutierrez et al., 1983), Furnariidae (Braun and Parker, 1985), and Geospizinae (Yang and Patton, 1981). This list is not exhaustive, but does cover a wide range of passerine and nonpasserine birds, and includes most of the larger values of genetic distance (distances in the range 0.3 to 0.7) known for birds.

The pattern of among-locus variance of genetic distance versus mean genetic distance for pairs of species (Fig. 5) is in good agreement with the pattern expected under the IC model of the mutation-drift theory.

4.3. Remarks

The sampling distribution of the relationship between the mean and the variance of genetic distance is unknown. Thus, there is no quantitative test of the fit. Nevertheless, Fig. 5 indicates a very good qualitative fit of avian electrophoretic data to predictions derived from the mutation-drift theory. This fit is inconsistent with such paradigms as stabilizing selection maintaining alleles at a number of loci. Such selection would result in a reduced variance from that expected under the neutral hypothesis because of a conservative pattern of allelic frequencies across taxa at the selected loci. Some models of selection involving an advantage for different alleles in new environments might mimic the inverted U-shaped pattern. However, whether the curve would be so close to the neutral one is unclear. Further, we find it telling that the precise distributions of genic variation both within breeding populations and among species can be explained by a single variable, $M = 4N_e\mu$.

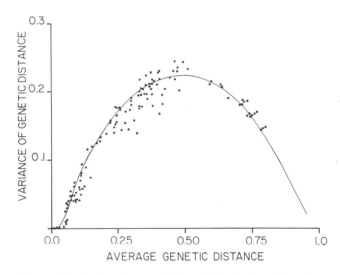

FIGURE 5. Observed (points) and expected (curve) distribution of interlocus variance of genetic distance as a function of mean genetic distance between species of birds. Expectation computed using the IC model with the assumption that overall heterozygosity averages 0.05.

5. DISCUSSION

5.1. Interpretation

Our results suggest that patterns of both intrapopulational and interspecific genic variation are consistent with the mutation-drift theory in a wide assortment of avian species. We suspect that this might be true of birds in general. Unfortunately, the taxa that met the requirements of our analysis are not a random sample of all birds. They are biased geographically toward temperate zone species and taxonomically toward specific taxa, especially oscines. Nevertheless, these results from diverse lineages are essentially independent of each other, and represent most of the largest avian electrophoretic surveys to date. They suggest that consistency with neutrality is a widespread phenomenon.

The sample sizes we have been able to use, on the order of a few thousand genes per study, are small to modest by population genetics standards. Discrimination between the IC model and other neutral models, including ones in which mutation rates vary among loci, would require much larger numbers of individuals and loci. Such samples are not available at the present time. In addition, the statistical tests we have used are nonparametric and not particularly powerful. However, we believe the observed patterns are sufficiently different from the general expectations of a number of adaptive models (e.g., overdominance–balancing selection) to be suggestive.

To avoid possible misinterpretation, some clarifications may be necessary. First, the fit to the mutation-drift model does not imply that the alleles uncovered by electrophoresis have no function. Each enzyme has a reaction that it catalyzes. In fact, the alleles are detected by observing the position on a gel at which they cause a substrate to be chemically altered. Rather, the word "neutral" means that the various alleles found segregating at a locus are equally good at performing their function. More precisely, "neutral" implies that the difference between selection coefficients for alternative alleles is not greater than the inverse of the effective population size. For such cases, polymorphisms within populations originate and are maintained by the interaction of stochastic factors such as mutation and drift. Second, the agreement of the avian data to the predictions of the mutation-drift model does not imply that all mutations are "neutral" in this same sense. Many mutations are deleterious, but they are quickly eliminated from a population by natural selection. Again, the analysis here only refers to that subset of all possible alleles that our surveys find—the ones that are adequately functional physiologically.

A quantitative characterization of avian genetic variation such as this can serve a number of useful purposes. For instance, the selectionist–neutralist arguments of the early and mid-1970s were based on consistency with verbal or qualitative models (see Lewontin 1974) or with the use of such techniques as correlation (see Schnell and Selander, 1981). We think that with the development of relatively precise and successful quantitative models, however, the burden of proof has shifted, and electrophoretic variation now ought to be considered neutral unless a null hypothesis of selective neutrality can be rejected. The view that this variation may be a panacea useful in tying together selection, behavior, ecology, and genetics through key marker alleles and correlations is not very plausible on probabilistic and physiological grounds (Schnell and Selander, 1981). Rather, if much of the variation observed is neutral or near neutral, then the greatest utility of studies of electrophoretic variation within species will be to allow one to investigate demographic processes that leave their characteristic imprints on the patterns of variation among populations. Gene flow, periods of isolation, and population bottlenecks are examples of such processes (Yang and Patton, 1981).

We do not doubt that some allozymic polymorphisms are maintained by natural selection. But we do think that locus-by-locus attempts to correlate enzyme polymorphisms with environmental parameters can be misleading. It would be surprising if level of polymorphism, or specific single-locus genotypes, could not be correlated with something. However, genetic variation should be screened initially for consistency with neutralist expectations before adaptive explanations are constructed (Gould and Lewontin, 1979).

5.2. Application to Behavioral Problems

Some researchers use genic markers to investigate such genetic patterns as relationships among individuals and problems of paternity exclusion (Sherman, 1981). However, the general patterns of Figs. 1 and 4 suggest that in studies of many avian populations, involving on the order of 40 loci, there will be on average only about three to five alleles segregating with frequencies in the range of 0.20 to 0.80. It is intermediate-frequency alleles that are of greatest use in determining genealogical relationship. For example, consider the simple situation in which there is a single locus with two alleles segregating, each at a frequency of 0.50. Suppose the genotypes of a female, that of an offspring, and that of two males are known and there is an interest in knowing which of the two males is the parent of the offspiring. Under these conditions (genotypes of all individuals known and alleles seg-regating at the optimal frequencies), the probability of determining pa-ternity is 0.188. With two such loci the probability is 0.340; with three it rises to 0.464, still less than 0.50. Fifteen loci, each with two alleles segregating with frequencies of approximately 0.50, would be required to have a probability of 0.95 of determining paternity. Based on our studies of avian within-population genic variation, it would be nec-essary to screen on the order of 100 loci, currently an almost impossible task, in order to find 15 highly variable ones. Consequently, although we agree with Sherman (1981) that electrophoresis might be a useful tool for the investigation of behavioral problems, we suggest that the following caveat is in order. Many loci will have to be surveyed in most birds to find a sample of sufficiently polymorphic ones for the exam-ination of behavioral questions and even then many such attempts will not be entirely successful (e.g., Schwartz and Armitage, 1983).

Rare alleles also may be helpful for investigations of paternity, relatedness, and similar kinds of behavioral studies. However, because of its rarity, a single rare allele will only be of use in one or two instances; it will take a large number of rare alleles to provide more than a few anecdotal results.

5.3. Application to Systematic Problems

Our results indicate general agreement of patterns of genetic dif-ferentiation among avian species with the predictions of the mutation-drift hypothesis. Clearly, additional data are needed owing to the rel-atively small number of studies currently available of taxa differentiated to a substantial degree. However, some useful implications emerge from our preliminary investigations. Several studies have attempted to cor-relate genic differentiation with time and date of evolutionary diver-gence (e.g., Gutierrez et al., 1983). This procedure, based on the as-sumed existence of a molecular clock (Wilson et al., 1977; Thorpe, 1982), is contingent on a stochastically regular accumulation of allelic frequency differences and fixations, as would obtain if genic variation were neutral and population sizes were constant.

A second area in which transpecific neutrality is relevant is in the inference of phylogeny, an area addressed empirically in several recent avian studies (e.g., Barrowclough and Corbin, 1978; Avise et al., 1980a; Yang and Patton, 1981). As Felsenstein (1978) has shown, all methods of phylogenetic inference implicitly depend on evolutionary rates being not too divergent among lineages and some specific models, such as his maximum likelihood method (Felsenstein, 1981), explicitly involve the assumption of stochastic differentiation of isolated populations. Thus, both phylogenetic inference and estimating dates of isolation, based on electrophoretic studies of protein divergence, are rendered more plausible if transpecific allozymic evolution is the result of a neutral, mutation-drift process.

ACKNOWLEDGMENTS. Jill A. Marten performed the electrophoretic analyses of several species. John C. Avise, John E. Cadle, Kendall W. Corbin, Joanne C. Daly, Allan Larson, Jill Marten, James L. Patton, Margaret F. Smith, and Francois Vuilleumier provided useful comments on the manuscript. James L. Patton supplied some unpublished data on the Geospizinae. Michael J. Braun provided unpublished data from his studies of *Pipilo* and *Parus*, Richard Baker from his study of *Oceanodroma*, and Michael J. Braun and Theodore A. Parker, III from their study of furnariids. We are grateful to all these persons.

This research was supported in part by National Science Foundation grant DEB79-20694 to N. K. Johnson. R.M.Z.'s collection of *P. iliaca* was supported in part by a grant from the Frank M. Chapman Memorial Fund of the American Museum of Natural History.

REFERENCES

Aquadro, C. F., and Avise, J. C., 1982, Evolutionary genetics of birds. VI. A reexamination of protein divergence using varied electrophoretic conditions, *Evolution* **36**:1003–1019.

Avise, J. C., 1983, Commentary, in: *Perspectives in Ornithology* (A. H. Brush and G. A. Clark, Jr., eds.), Cambridge University Press, New York, pp. 262–270.

Avise, J. C., Patton, J. C., and Aquadro, C. F., 1980a, Evolutionary genetics of birds I. Relationships among North American thrushes and allies, *Auk* **97**:135–147.

Avise, J. C., Patton, J. C., and Aquadro, C. F., 1980b, Evolutionary genetics of birds. Comparative molecular evolution in New World warblers and rodents, *J. Hered.* **71**:303–310.

Baker, M. C., 1975, Song dialects and genetic differences in White-crowned Sparrows (*Zonotrichia leucophrys*), *Evolution* **29**:226–241.

Baker, M. C., and Fox, S. F., 1978, Dominance, survival, and enzyme polymorphism in Dark-eyed Juncos, *Junco hyemalis*, *Evolution* **32**:697–711.

Barrowclough, G. F., 1980a, Gene flow, effective population sizes, and genetic variance components in birds, *Evolution* **34**:789–798.

Barrowclough, G. F., 1980b, Genetic and phenotypic differentiation in a wood warbler (genus *Dendroica*) hybrid zone, *Auk* **97**:655–668.

Barrowclough, G. F., and Corbin, K. W., 1978, Genetic variation and differentiation in the Parulidae, *Auk* **95**:691–702.

Barrowclough, G. F., and Shields, G. F., 1984, Karyotypic evolution and long-term effective population sizes in birds, *Auk* **101**:99–102.

Barrowclough, G. F., Corbin, K. W., and Zink, R. M., 1981, Genetic differentiation in the Procellariiformes, *Comp. Biochem. Physiol.* **69B:** 629–632.

Braun, M. J., and Parker, T. A., III, 1985, Molecular, morphological, and vocal evidence on the taxonomic relationships of "*Synallaxis*" *gularis* [Furnariidae] and other synallaxines, in: *Neotropical Ornithology* (M. S. Foster and P. A. Buckley, eds.), Amer. Ornithol. Union, Allen Press, Lawrence, Kansas (in press).

Chakraborty, R., Fuerst, P. A., and Nei, M., 1978, Statistical studies on protein polymorphism in natural populations. II. Gene differentiation between populations, *Genetics* **88:**367–390.

Chakraborty, R., Fuerst, P. A., and Nei, M., 1980, Statistical studies on protein polymorphism in natural populations. III. Distribution of allele frequencies and the number of alleles per locus, *Genetics* **94:**1039–1063.

Felsenstein, J., 1978, Cases in which parsimony or compatibility methods will be positively misleading, *Syst. Zool.* **27:**401–410.

Felsenstein, J., 1981, Evolutionary trees from gene frequencies and quantitative characters: Finding maximum likelihood estimates, *Evolution* **35:**1229–1242.

Fuerst, P. A., Chakraborty, R., and Nei, M., 1977, Statistical studies on protein polymorphism in natural populations. I. Distribution of single locus heterozygosity, *Genetics* **86:**455–483.

Gorman, G. C., and Renzi, J., 1979, Genetic distance and heterozygosity estimates in electrophoretic studies: Effects of sample size, *Copeia* **1979:**242–249.

Gould, S. J., and Lewontin, R. C., 1979, The spandrels of San Marco and the Panglossian paradigm: A critique of the adaptationist programme, *Proc. R. Soc. Lond. B* **205:**581–598.

Gutierrez, R. J., Zink, R. M., and Yang, S. Y., 1983, Genic variation, systematic, and biogeographic relationships of some galliform birds, *Auk* **100:**33–47.

Johnson, N. K., and Zink, R. M., 1983, Speciation in sapsuckers (*Sphyrapicus*): I. Genetic differentiation, *Auk* **100:**871–884.

Kimura, M., 1979, The neutral theory of molecular evolution, *Sci. Am.* **241:**98–126.

Kimura, M., 1983, The neutral theory of molecular evolution, in: *Evolution of Genes and Proteins* (M. Nei and R. K. Koehn, eds.), Sinauer Assoc., Sunderland, Massachusetts, pp. 208–233.

Kimura, M., and Ohta, T., 1971, *Theoretical Aspects of Population Genetics*, Princeton Univ. Press, Princeton, New Jersey.

King, J. L., and Jukes, T. H., 1969, Non-Darwinian evolution, *Science* **164:**788–798.

Lewontin, R. C., 1974, *The Genetic Basis of Evolutionary Change*, Columbia University Press, New York.

Li, W. H., 1978, Maintenance of genetic variability under the joint effect of mutation, selection and random drift, *Genetics* **90:**349–382.

Nei, M., 1975, *Molecular Population Genetics and Evolution*, American Elsevier, New York.

Nei, M., 1978, Estimation of average heterozygosity and genetic distance from a small number of individuals, *Genetics* **89:**583–590.

Nei, M., 1983, Genetic polymorphism and the role of mutation in evolution, in: *Evolution of Genes and Proteins*, (M. Nei, and R. K. Koehn, eds.), Sinauer Assoc., Sunderland, Massachusetts, pp. 165–190.

Ohta, T., 1974, Mutation pressure as the main cause of molecular evolution and polymorphism, *Nature* **252:**351–354.

Popper, K. R., 1968, *The Logic of Scientific Discovery*, Harper and Row, New York.

Schnell, G. D., and Selander, R. K., 1981, Environmental and morphological correlates of genetic variation in mammals, in: *Mammalian Population Genetics* (M. H. Smith, and J. Joule, eds.), University Georgia Press, Athens, Georgia, pp. 60–99.

Schwartz, O. A., and Armitage, K. B., 1983, Problems in the use of genetic similarity to show relatedness, *Evolution* **37:**417–420.

Sherman, P. W., 1981, Electrophoresis and avian genealogical analyses, *Auk* **98:**419–422.

Sibley, C. G., Corbin, K. W., Ahlquist, J. E., and Ferguson, A., 1974, Birds, in: *Biochemical and Immunological Taxonomy of Animals* (C. A. Wright, ed.), Acad. Press, London, pp. 89–176.

Smith, J. K., and Zimmerman, E. G., 1976, Biochemical genetics and evolution of North American blackbirds, family Icteridae, *Comp. Biochem. Physiol.* **53B:**319–324.

Sokal, R. R., and Rohlf, F. J., 1969, *Biometry*, W. H. Freeman, San Francisco, California.

Thorpe, J. P., 1982, The molecular clock hypothesis: Biochemical evolution, genetic differentiation and systematics, *Annu. Rev. Ecol. Syst.* **13:**139–168.

Wilson, A. C., Carlson, S. S., and White, T. J., 1977, Biochemical evolution, *Annu. Rev. Biochem.* **46:**573–639.

Yang, S. Y., and Patton, J. L., 1981, Genic variability and differentiation in the Galapagos finches, *Auk* **98:**230–242.

Zink, R. M., 1982, Patterns of genic and morphologic variation among sparrows in the genera *Zonotrichia, Melospiza, Junco,* and *Passerella, Auk* **99:**632–649.

Zink, R. M., and Winkler, D. W., 1983, Genetic and morphological similarity of two California Gull populations with different life history traits, *Biochem. Syst. Ecol.* **11:**397–403.

CHAPTER 5

ECOMORPHOLOGY

BERND LEISLER and HANS WINKLER

1. INTRODUCTION

Students of ecomorphology attempt to understand the interrelationships between morphological variation among individuals, populations, species and higher taxa, and communities and the corresponding variation in their ecology. Few biologists would deny there are differences between organisms that are related to differences in their ecology. Going beyond a naive inspection of the living world and pursuing a more scientific approach to the subject makes it more controversial. Discussion among morphologists has mainly centered on problems such as the adaptive significance of certain morphological features and by what methods one can detect adaptations. Some modern papers (Peters *et al.*, 1971; Bock, 1977) deal with these problems in detail and provide ample discussion of the difficulties involved in the study of adaptation. Another group of studies put major emphasis on the morphological variation within communities (e.g., Karr and James, 1975; Baker, 1979). An especially interesting aspect of ecomorphology is the examination of adaptive radiation of genera within families, since genera should be both systematic as well as ecological units (Inger, 1958; Illies, 1970; Leisler, 1980).

BERND LEISLER • Max-Planck Institute for Behavioral Physiology, Radolfzell Ornithological Station, Am Obstberg, D-7760 Radolfzell-Möggingen, Federal Republic of Germany. HANS WINKLER • Institute for Limnology of the Austrian Academy of Sciences, Gaisberg 116, A-5310 Mondsee, Austria. Order of authors determined at random.

Following the ideas of Bock and Wahlert (1965), we can differentiate three approaches to the study of adaptation: first, asking the "what" questions results in descriptions of form; second, "how" questions lead to the analysis of the various properties arising from the form; and third, answering "why" questions yields information about biological role. Ecomorphology has to go through all these stages, although the main interest certainly lies in solving the questions of biological role. Characteristically, in such an enterprise, much information about the natural history of the organisms of interest has to be gathered and included in the analysis. Consequently, ecomorphologists try to address more general questions. One question is how variation of morphological features or whole organisms is subjected to environmental constraints. This forces one to deal with such controversial subjects as "Lebensform" and convergence. Another problem, symmetrical to the preceding one, is the role of internal constraints. As animals in their interaction with the environment cannot allocate unrestricted energy, they are naturally limited in the development of morphological adaptations. Besides this trivial constraint there are many selective forces that under a given morphological construction lead to conflicting demands that can only be met by compromise. Our approach is a morphometric one. We measure various external and internal (skeletal) characters of known function. In this respect, therefore, we rely on studies of morphology and function already available. Examples for such studies would be Rüggeberg (1960) on the functional morphology of the foot or the theoretical analysis of Winkler and Bock (1976) dealing with aspects of hind limb proportions. Relevant studies of the flight apparatus are those of Brown (1963), Nachtigall and Kempf (1971), Kokshaysky (1973), Greenwalt (1975), and Hummel (1980). For interpretations concerning the feeding apparatus, the analyses of bill construction by Beecher (1962), Bock (1964, 1966), and Lederer (1975, 1980) and the studies of the function of rictal bristles by Lederer (1972), Dyer (1976), and Conover and Miller (1980) were consulted.

Inherent to the analysis of a number of morphological features is the problem that the same measurement may define very different aspects of a morphological structure in various taxonomic groups. Thus, in the course of the development of ecomorphological hypotheses, the analysis should be restricted to a closely knit taxonomic unit until further evidence justifies broader application (Lederer, 1984).

The reason for using more than one character in our approach is merely to include more information about morphological variation. As in the case of single characters, variation is what has to be explained by the pertinent theories. We form no a priori hypotheses about the meaning of covariation as the concept of "morphological integration" (Olson and Miller, 1958) does. We also do not attempt to go beyond our data set: the variation we describe is that observed among the species measured and the characters selected. Therefore, problems concerning sampling theory or statistical inference do not affect analysis. Thus it also seems justifiable to select sets of characters for particular analyses since it is legitimate to study a single character thought to be interesting. The most promising method of selection to us was to divide the characters into functional complexes (Oxnard, 1973, 1975).

We will show that description of various aspects of morphology leads to questions that can be successfully answered and extended to further problems. Consequently, a methodology can be outlined that should bring about substantial progress in ecomorphology.

2. METHODS

For the analyses of the data, we have used multivariate methods throughout. Although these methods imply the use of many measurements simultaneously, the degree of analytical sophistication is still low, for we consider only linear compounds of the data. In fact, most data have been transformed into logarithms to allow for allometric relationships; thus strictly speaking, the analyses are not linear with respect to the original measurements. Further restriction stems from our attempt to employ only commonly available and popular data analytic procedures. Hence, we claim neither that our data-analytic methods are in any sense the only possible ones nor that they are even the optimal ones. Still, by defining the procedures, the analyses can be repeated by others and interpretation is explicitly constrained to a well-defined subset of possible interpretations. Besides these more general considerations, each of the various methods has its own set of constraints that will be discussed.

The primary step in virtually all analyses was to reduce character space through the application of principal component analysis (PCA). Here, factorization of the correlation matrix was carried out rather than with the covariance matrix. With the possible shortcomings of such an approach (Morrison, 1976, p. 268), this is the simplest way to overcome the problem of character scaling. Because the characters used were of widely varying magnitude, standardization seemed to be unavoidable. Computation of the correlation matrix is equivalent to the analysis of the covariance matrix of standardized data (unit variance, zero mean). It should be kept in mind that this method leads to components that point geometrically in the direction of the main variation. They are also orthogonal, which means that they are not correlated among themselves and therefore may not reflect the character of overall variation truthfully.

Linear correlation between linear compounds of data sets (typically one for ecology and one for morphology) were analyzed with canonical correlation analysis (CCA). Sensible results from such an analysis can only be expected when the number of items significantly outnumbers the number of characters. Multiple regression analysis (MRA), the attempt to find interrelationships between one variable and a set of some other variables, was employed only as part of a method suggested by Albrecht (1979) to aid in the interpretation of morphological variation as represented, for instance, by principal components with respect to some environmental parameter (compare Johnston and Selander, 1971; Johnston, 1972). The method essentially consists of regressing the PCA scores of several components simultaneously with a variable of interest (e.g., vegetation height) and examining the angular relationships between this variable and the major axes of morphometric variation. Graphic presentation of the corresponding results is straightforward and renders useful support for ecomorphological interpretation.

The data were not only standardized to make characters comparable, but also adjusted to allow comparisons between species in a way thought to be of interest. This was achieved by dividing all the characters (with the exception of character 53) by the cube root of the lean weight (Amadon, 1943). To minimize errors due to seasonal fluctuations in body weight, data were taken for males mainly in the breeding season. The basic entity of all the analyses to follow, therefore, is shape with respect to body weight (see Mosimann, 1970, for a discussion of this particular meaning of "size" and "shape"). Another measure of "size" frequently used is a compound of various skeletal characters as a compound measurement of the body core skeleton: (sternum length + pelvis length + coracoid length) × (sternum width + pelvis width) × height of crista sterni (Hoerschelmann, 1966). This measure was not used to avoid possible problems inherent to ratios in general (Atchley et al., 1976), especially because characters that are part of this compound size variable were also part of the data sets to be analyzed. However, weight and compound measure correlate well and can be used interchangably in practice (Fig. 1).

To summarize, the steps prior to analysis were:

1. Character selection based on present knowledge of functions and the intention of the analysis.

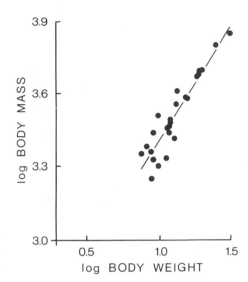

FIGURE 1. The correlation between lean body weight and a measure of body mass derived from measurements of the body core skeleton (n = 25, r = 0.93, $P \le$ 0.01%).

2. Computation within species means for all characters.
3. Calculation of species "shape" vectors by dividing within species means by the cube root of mean lean weight of respective species.
4. Transformation to logarithms.
5. Standardization across characters by confining analysis to correlation matrices.

3. RESULTS AND DISCUSSION

3.1. Analysis of the Morphological *Lebensform*

3.1.1. Ecological Group

The most popular concept concerning ecological groups, since its revival by Root (1967), is the guild. It is defined as a group of species that utilizes the same class of resources in a similar manner. Taxonomic relationships among the members of a guild are of no concern. Definition of a particular guild is somewhat arbitrary and bears some analogies with the genus in taxonomy. Intraguild comparisons are important since they reveal general rules for solving a given ecological problem (Winkler, 1971). If taxonomic diversity is not too great, it is likely that a particular ecological demand is met with the same morphological answer, rather than with radically different ones. This is especially true for a morphometric approach such as ours in which measurements refer to homologous structures. To see this point, one has only to envisage the problems arising when trying to study the nectar-feeding guild, which would include bats and birds as well as large insects.

The guild we chose comprises eight turdid chats, living in grass-lands of East Africa. Higher vegetation is absent in these areas, and comparisons have been made in areas where the vegetation has been burned. Food resources are scarce and unstable. In this combination, the guild exists only in winter when the resident species are united with Palaearctic migrants. All the members of the guild eat arthropods and may include small vertebrates in their diet (Leisler et al., 1983). For the analysis 18 external characters were included (Table I).

TABLE I

Correlations[a] of the Components with the Original Characters in a Principal Component Analysis of 18 External Characters of 8 Species of Turdid Chats

	Component		
	1	2	3
Wing length	0.46	−0.79	−0.29
Tail length	0.02	−0.32	−0.84
Tarsus	−0.16	0.57	−0.47
Bill length	0.57	−0.73	0.31
Bill depth	−0.29	0.58	0.70
Bill width	−0.82	−0.05	0.10
Hind toe	−0.94	−0.03	0.28
Inner toe	−0.96	−0.21	0.16
Middle toe	−0.82	−0.48	0.26
Outer toe	−0.81	−0.50	0.08
Hind claw	−0.82	0.29	−0.15
Inner claw	−0.69	0.27	−0.37
Middle claw	−0.84	−0.30	−0.22
Outer claw	−0.85	−0.07	−0.42
Foot span	−0.91	−0.16	0.38
Foot span with claws	−0.91	−0.38	0.10
Rictal bristle length	−0.59	0.05	−0.66
Kipp's distance	0.46	−0.81	0.04
Percentage of total variance	51.25	19.54	15.36
Cumulative percentage of variance	51.25	70.79	86.15

[a]Values in boldface type signify characters used for interpretation.

Principal component analysis extracted three components of interest. Ordination of the species in the space defined by the first and the second component is shown in Fig. 2. Correlations (loadings) between original characters and the respective components (Table I) allow the following interpretation. The first component mainly represents foot size (i.e., progressing along the first axis in positive direction is associated with a decrease in foot size), slenderness of bill, and length of rictal bristles. Along this axis the whinchat (*Saxicola rubetra*) is clearly separated from all the other chats. This species often clings to herbaceous vegetation and dry stalks, whereas the other species dwell more on the ground.

The second component extracted by PCA is characterized by shorter, more rounded wings, longer tarsi, and also a stronger bill. High scores in this component are attained by species that indulge in more cursorial habits; that is, species with pronounced pedal locomotion. Along this axis (Fig. 2), the species that use the dash and jab as a predominant foraging strategy are separated from those that follow a perch and pounce strategy (Cornwallis, 1975). Long tarsi are an obvious advantage to cursorial birds since they allow them to cover ground more quickly and increase their field of view (Engels, 1940; Dilger, 1956; Grant, 1966; Newton, 1967; Gaston, 1974).

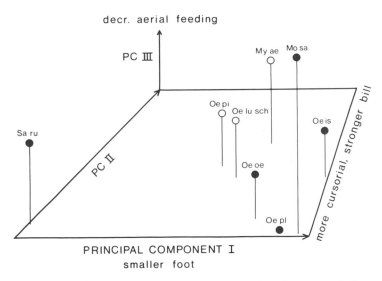

FIGURE 2. Plot of the first three component axes in a PCA of 18 external characters of a guild of 8 turdid chats wintering in East Africa. See also Table I. Full circles denote migrant species. Sa ru, *Saxicola rubetra*; My ae, *Myrmecocichla aethiops*; Oe is, *Oenanthe isabellina*; Oe oe, *O. oenanthe*; Oe lu sch, *O. lugens schalowi*; Oe pl, *O. pleschanka*; Oe pi, *O. pileata*; Mo sa, *Monticola saxatilis*.

So far the analyses have only disclosed the fine structure within the guild without pointing to further possible routes of interpretation and research. The results are well within expectation and PCA merely yields an instructive graphic interpretation. However, in addition to the strong connection of characters intimately linked with locomotion, as delineated by the second component, there is a remarkable association with bill shape that was not anticipated. A posteriori interpretation would be that more ground-feeding species breeding in sheltered ground sites (*Oenanthe pileata, O. lugubris schalowi, O. isabellina, Myrmecocichla aethiops*) have a stronger bill as a possible adaptation for frequent pecking and digging into the ground.

The third component represented only a small portion of the total variation (Table I) and was associated with decreasing aerial feeding. Characters such as tail length (aerial maneuverability), and length of rictal bristles together with bill height (flycatching) determine the morphological variation.

3.1.2. Taxonomic Group

Comparisons within taxonomic units, as long as they also reflect evolutionary units, lead to a better understanding of how ecological factors shape the morphological building material. Because of their historical similarities, possible internal constraints may be found in this way if they are present.

Old World warblers (Sylviidae) form an ideal group for such comparisons. This family of insectivorous birds consists of some species-rich genera that inhabit different habitats generated in the process of terrestrialization of aquatic ecosystems, as well as shrubs, forests, and sclerophyllous plant communities. Most species are common within their ranges and show pronounced differences in migratory habits.

In this analysis we concentrate on two points. First, what specializing adaptations (Wright, 1941; Huxley, 1943) were developed by the species in response to the different environmental conditions? From this we may proceed to discuss basic adaptations of general importance for whole genera. Second, the problem of radiation is addressed. In this connection, diversification within the genera with respect to functionally defined character sets (functional complexes) is examined.

The criterion for membership of a character in a functional complex is derived from gross functional considerations. Only after the morphometric space within groups has been explored can we procede to the analysis of the entire character set, thus permitting the evaluation of cross connections between the previously defined subgroups and a reappraisal of them.

3.1.2.a. *Bill (Feeding Apparatus).* Seven characters belong to this character set. Besides bill measurements, the length of the rictal bristles was included since they are accessory food-capturing structures that are involved in directing food into the gape as well as in the protection of the eyes.

Through PCA (Fig. 3), three meaningful character axes have been extracted. The first component shows a positive correlation with slender bills and shorter rictal bristles. The second component treats curvature of bills, and the third, shorter bills with shorter rictal bristles.

We find no clear group of species, *Acrocephalus* warblers are widely spread over the plane defined by the first and the second principal components. The more or less short-billed *Sylvia* warblers, with their predominantly gleaning habits, are somewhat separated along the third

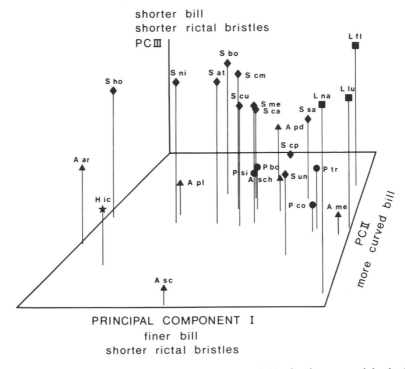

FIGURE 3. Plot of the first three component axes in a PCA of 7 characters of the feeding apparatus in 25 sylviid species. L lu, *Locustella luscinioides*; L fl, *L. fluviatilis*; L na, *L. naevia*; A me, *Acrocephalus melanopogon*; A pd, *A. paludicola*; A sch, *A. schoenobaenus*; A sc, *A. scirpaceus*; A pl, *A. palustris*; A ar, *A. arundinaceus*; H ic, *Hippolais icterina*; S ni, *Sylvia nisoria*; S ho, *S. hortensis*; S bo, *S. borin*; S at, *S. atricapilla*; S cm, *S. communis*; S cu, *S. curruca*; S me, *S. melanocephala*; S ca, *S. cantillans*; S cp, *S. conspicillata*; S un, *S. undata*; S sa, *S. sarda*; P tr, *Phylloscopus trochilus*; P co, *P. collybita*; P bo, *P. bonelli*; P si, *P. sibilatrix*.

component. The slender bills of the *Locustella* warblers is also reflected in the analysis. The reed warbler (*Acrocephalus scirpaceus*) is a frequent flycatcher, especially at dusk, and its bill is obviously adapted for this feeding method. Separation of this species from all the others is distinct (Fig. 3). However, close inspection of the numerical results of the PCA shows that variation in this character set is quite uniform, and no obvious patterns or predominant modes of variation seem to be present. Geometrically, this also can be seen (Fig. 3) in the loose scatter of points covering almost the same space along all three axes (a strong correlation structure would result in an increase of compression with the order of the component). Morphologically this means that there are no significant internal constraints; bill shape is furnished with many degrees of freedom of variation. Consequently, one would expect high variability in bill shape. Further support to this interpretation comes from the work of Lederer (1975), who offered evidence for the existence of a looser relationship between bill morphology and diet in insectivorous birds than in seedeaters.

3.1.2.b. *Hind Limb.* The 15 characters included in the set all belong to the leg or the pelvic girdle and are functionally related to pedal locomotion.

Two main components of morphometric variation as detected by PCA can be discussed in detail. The first component is correlated with foot size, especially with the length of the hind toe including the claw, and width of pelvis. Species scoring high on this component (Fig. 4) have small feet and a broad pelvis. This combination of characters can be interpreted as an adaptation to perching. Arboreal perchers and upper-story warblers such as the garden warbler (*Sylvia borin*), the blackcap (*S. atricapilla*), the Mediterranean Orphean warbler (*S. hortensis*), the wood warbler (*Phylloscopus sibilatrix*), and Bonelli's warbler (*P. bonelli*) fit into this category. Low scores characterize species having excellent clinging abilities. They have, as would be expected (Winkler and Bock, 1976), long legs and (hind) toes, as well as slim pelves. *Acrocephalus* warblers, with their adaptations for negotiating dense, vertical structures (as found in reed belts) therefore score low on the first axis. The reed warbler (*A. scirpaceus*) and the moustached warbler (*A. melanopogon*) are typical representatives of this group. Between them, as the one extreme, and the typical perchers, as the other, are birds that live in high to low bushes, scrub, and on forest edges, as well as in dense (herbaceous) vegetation with many horizontal components.

The second principal component is mainly correlated with the length of the inner toe, the length of the middle claw, and tarsus length. Short inner toes and long middle claws and tarsi are features associated

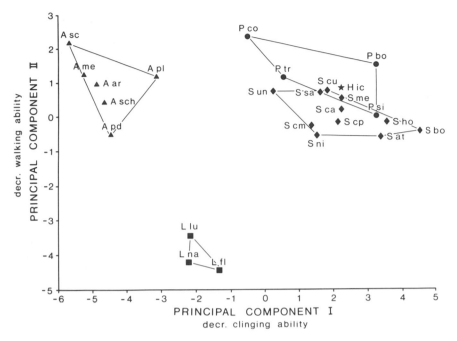

FIGURE 4. Plot of the first two compenent axes in a PCA of 15 hindlimb characters of 25 sylviid species. Polygons enclose genera. Species abbreviations as in Fig. 3.

with impaired walking abilities, so species with high walking abilities score low on this axis (Fig. 4). They are also characterized by long pelves. The group adapted best for running and walking in our species assemblage is the *Locustella* warblers (Leisler, 1977a). The Savi's warbler (*L. luscinioides*), a hopping species that frequently uses reed stems as song perches, has the weakest walking foot within the genus.

Overall evaluation of the PCA results shows a picture of morphological variation distinctly different from the one found in the analysis of the feeding apparatus. Strong correlations between characters suggest the existence of pronounced functional (e.g., mechanical) and internal constraints. Genera group themselves into disjunct groups (Fig. 4). Within them there seem to be few possibilities for differentiation. An historical "load" seems to prevail that expresses itself in genus-specific locomotion. For example, even though Savi's warblers (*L. luscinioides*) occupy the same habitats as the *Acrocephalus* warblers, they show little tendency to develop corresponding hind limbs. The same is true in an opposite direction for the aquatic warbler (*A. paludicola*). This species lives in habitats typical for *Locustella* warblers, but almost no concurrent change in morphology has taken place (Leisler, 1975, 1977b).

3.1.2.c. Forelimb. Eighteen characters involving the upper extremities served as a data base for the PCA as depicted in Fig. 5. Again, two components are sufficient to discuss the corresponding morphometric variation in our set of species.

Component one is positively correlated with less pointed, more rounded wings. In terms of the characters actually used, it is correlated negatively with forearm and manus length and positively with wing-loading and graduation of the tail. This axis separates the icterine warbler (*Hippolais icterina*) together with the *Phylloscopus* warblers from the other three genera.

The second major component is related to maneuverability. Birds highly gifted in this respect possess low aspect ratios, long tails, rounded wings, many and longer midwing slots, and low wing loadings. Due to the negative correlation between maneuverability and the second component, the high performance fliers *Phylloscopus* and *Hippolais* score low on this component, as well as on the first one. This indicates that

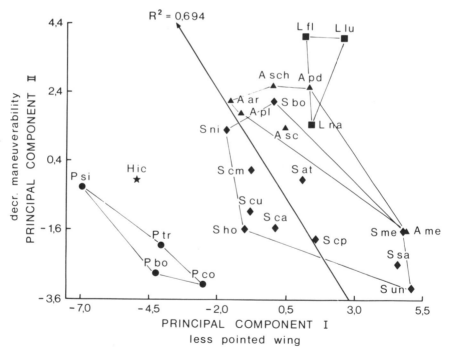

FIGURE 5. Plot of the first two component axes in a PCA of 18 forelimb characters of 25 sylviid species; also shown is the relationship between PCA and migratory distance, with its multiple correlation coefficient. Species abbreviations as in Fig. 3.

their wings combine adaptations for high speed and maneuverability. The combination of low speed with high maneuverability, found on the lower right of Fig. 5, characterizes flight structure adapted for thickets. Typical representatives would be the Dartford warbler (*Sylvia undata*), Marmora's warbler (*S. sarda*), the Sardinian warbler (*S. melanocephala*), and also the moustached warbler (*Acrocephalus melanopogon*). The highest scores in component two are attained by species that are poor fliers and are even reluctant to use their wings except in migration. Species such as the river warbler (*Locustella fluviatilis*) and Savi's warbler conform to this pattern.

3.1.2.d. Total Morphology. To reveal some basic morphological differences within and between genera, a PCA of 48 characters that included all the characters found to be of importance in the previous analyses was carried out. The three components of interest obtained correspond to patterns of morphological variation that uncover functional relationships not anticipated when the division into functional groups was done.

On the negative side of the first axis (Fig. 6), species are found that would be designated as flight specialists, as opposed to those well equipped for employing the hindlimbs. This corresponds to variation in foot size, wing dimensions, slotting of the wings, wing loading, and tail graduation.

The second component mainly involves wing characters. In functional terms, it is positively correlated with maneuverability and morphologically with rounded wings and larger skulls.

Birds with limited vertical clinging abilities and a more gleaning mode of foraging reach high scores on the third axis. The length of the distal leg segments, the bill, and the rictal bristles correlate negatively with the corresponding component and thus show the most variation in this direction.

In the space spanned by the first two components, the genera are already well separated. *Phylloscopus*, *Hippolais*, and *Sylvia* are flight specialists, and *Acrocephalus* and *Locustella*, are hindlimb specialists. The third component separates *Sylvia* and *Locustella*, with their limited clinging abilities and their gleaning habits, from *Phylloscopus* and *Acrocephalus*. Of some interest is the inverse relationship in the variation of the upper and lower extremities. Similar results have been obtained in European thrushes (Stork, 1968) other turdids (Gibson et al., 1976), Parulid warblers (Eaton et al., 1963), and ducks (Raikow, 1973).

3.1.2.e. Radiation. The previous analyses indicate that groups defined taxonomically and ecologically exhibit different degrees of differentiation in single, functionally defined character sets. Radiation can

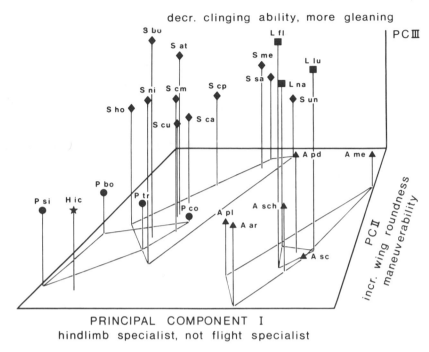

FIGURE 6. Plot of the first three component axes in a PCA of 48 morphological characters of 25 sylviid species. Species abbreviations as in Fig. 3.

involve divergence in both size and shape. Our analytical approach to problems of this sort was to compare the contribution of single characters with the total variance within a genus. Again, characters were combined into functional groups and standardized with respect to weight. To answer the main question, namely whether certain genera diverged differently with respect to some parameters, standard deviations of the standardized data were summed within functional sets and compared with the total of all characters. Note that standardization only involved bird size and not character states. A total of 41 characters was used. They were divided among those of the bill (6 characters), the forelimb (14), the hindlimb (15), and the skull (6). Analysis was carried out only for the species-rich genera *Acrocephalus*, *Phylloscopus*, and *Sylvia*.

As can be seen in Table II, *Sylvia* and *Acrocephalus* diverged mostly with respect to bill, *Sylvia* and *Phylloscopus* with respect to the forelimb, and *Phylloscopus* in the hindlimb. Skull dimensions vary little in all three genera.

Dividing the *Sylvia* warblers into those restricted to the Mediter-

TABLE II
Percent Contribution of Variation in Character Sets to Total Variation in
Some Sylviid Genera

	Bill (6 characters)	Forelimb (14 characters)	Hindlimb (15 characters)	Skull (6 characters)
Acrocephalus (6 species)	24.6	40.2	24.7	10.5
Phylloscopus (4 species)	13.3	43.1	35.3	8.3
Sylvia (11 species)	26.6	44.6	18.0	10.8
Mediterranean *Sylvia* (6 species)	38.5	33.2	19.4	8.9
Northern *Sylvia* (5 species)	14.9	40.7	30.1	14.3

ranean region and those that occur mainly in the North, further lines of radiation within this genus can be shown (Table II). The Mediterranean warblers have diverged more with respect to bill construction than the northern ones. Conversely, northern warblers occupy more morphological space defined by characters of the hindlimb than Mediterranean warblers. Simple analyses of this sort may hint at the morphological features involved in the ecological segregation of the respective genera.

3.2. The Relation between Morphology and Ecology

After having explored the morphometric space and discussed its variation in relation to known functions and biological roles, we now attempt to relate morphological and ecological variation in a more quantitative and precise manner. Presently, only few quantitative ecological data exist. Still, those available will be used in a way outlined by Albrecht (1979) and described in more detail in the methods section.

3.2.1. Overall Morphology and Distance of Migration

For birds of the temperate zones, migration is a most demanding factor, affecting their entire existence. Sylviids use wintering grounds that often are in areas south of the African Sahel zone. The distances the populations of our study have to migrate are variable, and were estimated using information provided by Berthold (1973, 1979) and Curry-Lindahl (1981). By reducing the morphological space depicted by three dimensions in Fig. 6 to two dimensions (Fig. 7), we see a strong overall correlation between morphological and the unidimensional eco- logical variation. Ecological variation is only related to the second prin- cipal component of morphometric variation. In this direction both types of variation run parallel. This indicates that migration severely affects morphology in the direction defined by the second principal compo- nent. An important incompatibility clearly emerges. There are no spe- cies that have rounded wings and also travel far in migration. Habitats that require maneuverable flight are accessible only to species that are not required to travel far. Such species are the Dartford warbler, Mar- mora's warbler, the spectacled warbler (*Sylvia conspicillata*), and the moustached warbler. If we assume that the center of radiation for *Sylvia*

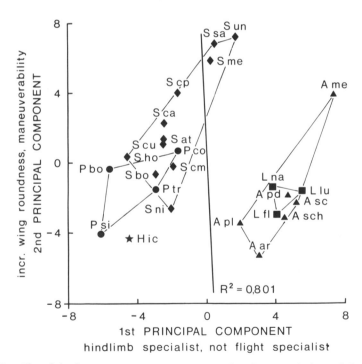

FIGURE 7. Plot of the first two component axes in a PCA of 48 morphological characters of 25 sylviid species (see also Fig. 6); also shown is the relationship between PCA and migratory distance. Species abbreviations as in Fig. 3.

lies in the Mediterranean region and that for *Phylloscopus* in the North (Gaston, 1974), the basically different migration distances may have determined the later pattern of evolution of these genera by any of a number of pre-existing adaptations for other ecological factors.

3.2.2. Forelimb and Distance of Migration

As the preceding analysis showed, migration mainly affects the forelimb. Exploring the space defined by the 18 characters related to flight (Fig. 5) shows that simple identification of a major axis of morphological variation with one external factor is not possible. Length of migratory distance affects both pointedness and maneuverability adversely, thus suggesting complex interactions with morphology and possibly other ecological factors.

3.2.3. Ecological Interpretation of the Overall Morphology of Genera

3.2.3.a. Sylvia. Principal component analysis of the 48 characters thought to characterize the total morphological variation demonstrates two major axes (Fig. 8). On the right side of the first axis species are

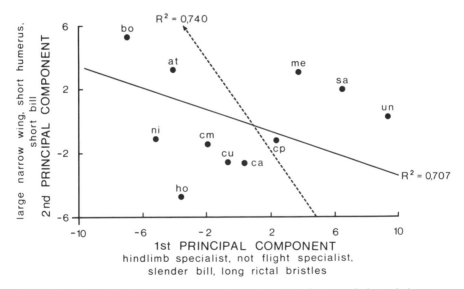

FIGURE 8. Plot of the first two component axes in a PCA of 48 morphological characters of the genus *Sylvia* (11 species). The heavy line shows the relationship between the components and migratory distance; the dashed line shows the relationship between the components and vegetation height. Species designations follow Fig. 3.

found that predominantly use pedal locomotion and have slender bills and long rictal bristles. The second component is positively correlated with narrowing of the wing and negatively with humerus and bill length.

The factor migratory distance mainly affects the first component (Fig. 8). The farther the birds fly on migration, the more developed the wing skeleton, the more wing slots are reduced, and the smaller are feet. The second component, although related to features of the wing, is seemingly not significantly related to migration.

Vegetation characteristics of *Sylvia* habitats have been measured by Cody and Walter (1976), Cody (1978), and Zbinden and Blondel (1981). Two main measures, namely vegetation height and vegetation density, are available for all species.

Vegetation height is correlated with the morphology of *Sylvia*, and both components of morphological variation show a relationship with this factor. Species, therefore, which live in higher vegetation have more developed wings and reduced feet and their wings are of high aspect ratio. Vertical movements appear to be important in these species, and they rely on flight for such movement. Similar conclusions were reached by Dilger (1956) in his study of certain thrush species.

Vegetation density does not show a significant relationship with overall morphology. The morphology of *Sylvia* warblers can be understood as being affected mostly by migration and vegetation height. Variation associated with these factors is not orthogonal. That means that the need to migrate constrains morphological variation in a way reasonably compatible with variation owing to vegetation height. The possible directions of morphological variation are set by the basic construction of this genus.

3.2.3.b. Acrocephalus, Marsh Warblers. The PCA of the overall morphology within this genus yields two major components. The first is negatively correlated with characters that correspond to pointed, less slotted wings and long, square tails. Species on the positive side of the first axis (Fig. 9) also have a reduced wing skeleton, large feet, thin bill, and short rictal bristles. The second component is negatively correlated with features pertaining to clinging abilities on thin, vertical structures such as reed stems.

Migratory distance runs practically parallel with the first component (Fig. 9): species that have more pointed and longer wings are also those covering longer distances in migration. There is no relation between migration and the character combination defined by the second principal component.

Vegetation height (data from Leisler, 1981) is negatively correlated with the second component. Species that score low on this axis are

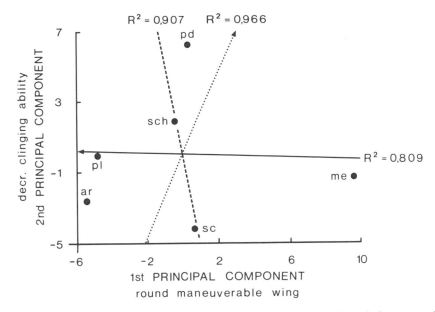

FIGURE 9. Plot of the first two components in a PCA of 48 morphological characters of the genus *Acrocephalus*. The heavy line shows the relationship between the components and migratory distance; the dashed line shows the relationship between PCA components and vegetation height, and the dotted line shows the relationship between PCA components and vegetation density. Species designations follow Fig. 3.

those living in higher vegetation. The relation between migration and vegetation height is not important.

Vegetation density is well correlated with morphology. Mainly, the character set correlated with component two is associated with this environmental parameter. Species that live in dense vegetation are less able to negotiate vertical structures. But, there is also an association with wing characters, reflecting the fact that denser vegetation promotes the development of more rounded, maneuverable wings.

Altogether, the strong interaction between vegetation height and migratory distance found in the *Sylvia* warblers is not present in *Acrocephalus*; the orthogonal axes of morphological variation largely correspond with the external forces of migration on one side and vegetation structure on the other. The main reason for the difference between these two genera lies in the different mode of locomotion in the habitat. *Acrocephalus* warblers mainly move within the vegetation, which accounts for the pronounced correlation between morphology and vegetation density; they cover vertical distances by climbing. They therefore gain some freedom for adapting the wings to other needs.

So far we have dealt only with single factors. For breeding habitats of *Acrocephalus*, enough data were available to characterize them in multivariate space. The method used was multiple discriminant analysis (DA), which finds those linear combinations of characters that best separate the various species. In the case of the *Acrocephalus* warblers (Leisler, 1981), one compound character resulting from the major component of the DA was sufficient to characterize the habitats of the warblers satisfactorially. This discriminant axis is positively correlated with ground cover and negatively with vegetation density in upper layers, vegetation height, number of emergent vegetation elements, and water depth. Again, a correlation of DA scores with the first principal component of morphological variation (Fig. 10) was carried out. No clear correlation seems to exist. Correlation with the second component is well marked (Fig. 11). This supports our earlier conclusion that the second component is largely relevant for coping with the structure of the habitat. The wing is of little importance for the habitat differentiation of this genus.

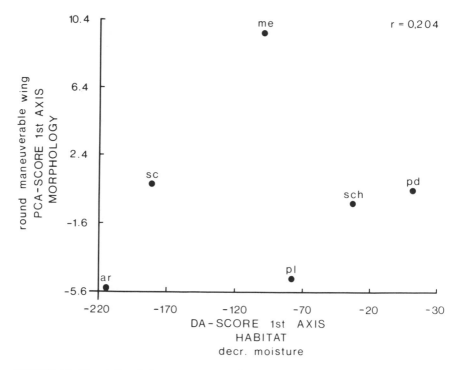

FIGURE 10. Correlation between scores on discriminant axis 1 of 12 breeding habitat characters and the scores on the first principal component of 48 morphological characters of *Acrocephalus* warblers.

3.3. The Relationship between Morphology and Behavior

A very important contribution to the understanding of morphological structures would result from having information about the actions immediately related to them. Unfortunately, there are little data available pertaining to the field of ecomorphology. Ecoethologists hardly seem to bother about morphology when building models or collecting field data. We can, however, demonstrate some of the possibilities arising from the use of classic multivariate methods. The data sets to be discussed describe six morphological characters related to feeding behavior and five behavioral characters. They were obtained from 17 species of neotropical birds belonging to the flycatcher guild by Sherry (unpublished data) in Nicaragua (Table III).

The method of choice in such instances is canonical correlation analysis (CCA). The mathematics of this method brings out a general problem in the analysis of multivariate responses. If the number of characters is close to or even greater than the number of species in-

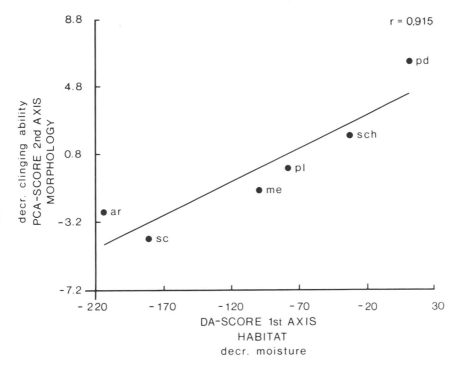

FIGURE 11. Correlation between scores on discriminant axis 1 of 12 breeding habitat characters and the scores on the second principal component of 48 morphological characters of *Acrocephalus* warblers.

vestigated, analysis is either not possible (mathematically singular matrices arise) or leads to tautological supercorrelations. Strong correlations within data sets have similar consequences. To circumvent these problems, the behavioral data were subjected to a PCA. It results in two (uncorrelated) components that describe feeding behavior. The first component signifies decreased hawking, increased snatching, and lower attack distances and the second component decreased pursuing in feeding behavior. Both components together account for 91% of the total variation. When the corresponding species scores are brought into a CCA as the character set characterizing behavior and the six morphological characters as the second set, analysis yields two canonical correlations of interest. The first (Fig. 12) shows that there is a relation between decreasing body mass, decreasing aspect ratio (i.e., increasingly rounded wings), and increasing length of the rictal bristles, with an increase in snatching and pursuing as feeding tactics. Birds at the positive side of the scatter (upper right, Fig. 12) are the genera *Nesotriccus*, *Terenotriccus*, and *Myiobius*, which pursue their prey on the wing in acrobatic maneuvers (Sherry, unpublished data). The hawking

TABLE III

Canonical Correlation Analysis of Behavior and Morphology of 17 Species of Neotropical Flycatching Birds

	Correlation with	
	Canonical variate I	Canonical variate II
Morphology		
Body mass	−0.68	
Aspect ratio	−0.71	
Tail area		−0.72
Lateral bill convexity		0.87
Bill length		
Rictal bristle length	0.55	
Behavior[a]		
PCA-component 1 (decreasing hawking, increasing snatching)	0.72	0.70
PCA-component 2 (decreasing pursuing)	−0.70	0.72
Canonical correlation	0.91	0.71

[a]Behavioral data (Sherry, unpublished) are PCA scores (see text for further detail). Only those correlations pertaining to interpretation are shown.

Colonia and Contopus species are at the negative end of the scatter. The second correlation (Fig. 13) associates increasing snatching (and decreasing pursuing) with decreasing tail area and increasing convexity of the bill. The highest scores on both axes are attained by Todirostrum and Platyrinchus species.

Thus, description of the complex interactions between behavior and morphology seems possible and promising. Heuristically, it also

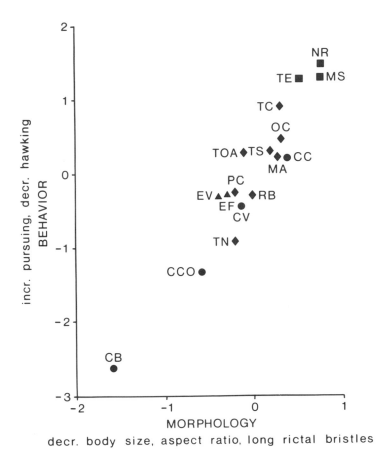

FIGURE 12. First canonical correlation between six morphological characters and two principal components derived from five behavioral characters in a tropical flycatcher guild. Data from Sherry (unpublished). *Abbreviations:* MA, *Myiornis atricapillus;* OC, *Oncostoma cinereigulare;* TC, *Todirostrum cinereum;* TN, *Todirostrum nigriceps;* TS, *Todirostrum sylvia;* RB, *Rhynchocyclus brevirostris;* TOA, *Tolmomyias assimilis;* PC, *Platyrinchus coronatus;* TE, *Terenotriccus erythrurus;* MS, *Myiobius sulphureipygius;* CC, *Contopus cinereus;* CV, *Contopus virens;* CB, *Contopus borealis;* EV, *Empidonax virescens;* EF, *Empidonax flaviventris;* NR, *Nesotriccus ridgwayi;* CCO, *Colonia colonus.*

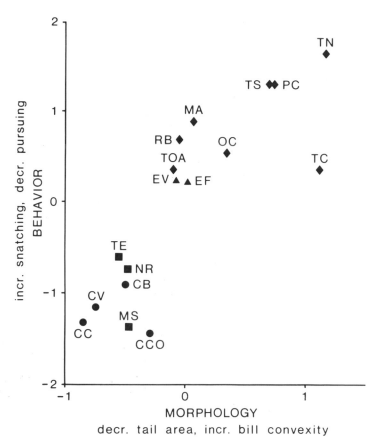

FIGURE 13. Second canonical correlation between six morphological characters and two principal components derived from five behavioral characters in a tropical flycatcher guild. Data from Sherry (unpublished). See Fig. 12 for defined abbreviations.

stresses the need to collect sufficient data from many species, rather than to indulge in interpretations of tautologies cropping up from expanding and refining morphological analysis of many characters in only a few species.

4. CONCLUSION

In our analyses we did not start from scratch, but tried to incorporate already established ideas and data. The arsenal of modern multivariate techniques of data analysis were used only to the level of sophistication necessary for useful results. Information was extracted by focusing attention either on ecologically defined groups (guilds) or on taxonomically defined ones. In both cases, groups were rather closely knit. In terms of characters, either large sets comprising many quite different aspects of gross morphology were selected or relatively few characters were used according to previous knowledge of possible functional significance. Similar refinement was also accomplished by splitting the taxonomic group into smaller units.

The results show that there is indeed a wide field of research opportunity available, which can lead to a deeper understanding of morphological structures and concrete testing of extant morphological concepts. Although analyses of the sort we presented may look attractive in themselves, they are only a second step towards the understanding of ecomorphology as envisaged by us. They definitely are not substitutes for detailed anatomical and morphological studies. They should, however, accompany them and should help morphologists perceive the specific problems of ecomorphology and perhaps function as a permanent challenge to those researchers who strive to understand adaptation. Besides these merely morphological aspects, it also should be stressed that much ecological work has yet to be done.

As has been mentioned before, the level of analysis to date is still preliminary. Future work may look quite different. What is needed next is not more sophistication in the machinery of multivariate analysis, but replication in corroboration of the correlations found. Work should be guided by the insights from previous analyses as well as by their failures. It is very important to design experiments testing functional interpretations. Such experiments, for example, have been conducted by Leisler (1977a) to study aspects of *Locustella* locomotion under controlled conditions.

Further analysis of this nature should help to define those characters and character complexes that have ecological relevance. This would help also to define those features of the environment that are pertinent to the animal. What we conceive is not an ecomorphology telling us again and again that there are wondrous correlations between morphology and conventionally defined environmental factors such as vegetation height, proportion of syrphids in the diet, or even migratory distance, but one providing abstract descriptions of physical features that can be handled and manipulated by certain morphological structures. From there, all possible environments that are accessible to the animal can be deduced. If, for instance, a particular combination of characters has been found to be relevant for locomotion in dense, vertically organized structures with a certain range of diameters, this would be of interest not only to morphologists, but also to ecologists wishing to understand why a species seems not to care whether the habitat consists of reed or rye. Clearly, concepts dealing with convergence or even assemblages of species in a particular case would benefit from such an approach.

Finally, careful analyses of both morphology and ethology of species and taxonomical or ecological groups of species should converge on a modern autecological niche concept. First attempts by morphologists (Bock and Wahlert, 1965) were overrun by the rapid development of new ecological models, focusing especially on ecosystems or communities. A return to the study of the complex interactions between organisms and environment from the viewpoint of the animal should, after its gentle depilation (Hurlbert, 1981), facilitate another, more successful attempt.

APPENDIX I
Characters Used in the Study

1. Weight
2. Wing length
3. Tail length
4. Tarsus
5. Bill length/skull
6. Bill depth
7. Bill width/base
8. Bill width/flanges
9. Hind toe
10. Inner toe
11. Middle toe
12. Outer toe
13. Hind claw
14. Inner claw
15. Middle claw
16. Outer claw
17. Foot span
18. Foot span with claws
19. Rictal bristle length
20. Wing span
21. Wing area
22. Wing chord
23. Alula
24. First primary (X)
25. First secondary (1)
26. Last tertial (3)
27. Tail, longest rectrix
28. Wing tip to primary X

APPENDIX I (Continued)

29. Wing tip to secondary 1
30. Graduation of tail
31. Notch on inner web of second primary
32. Number of notched primaries
33. Longest primary
34. Skull length
35. Length of brain case
36. Skull width
37. Width between quadrates
38. Interorbital width
39. Skull depth
40. Coracoid
41. Keel depth
42. Synsacrum/pelvis length
43. Synsacrum/pelvis width
44. Sternum length
45. Femur
46. Tibiotarsus
47. Tarsometatarsus
48. Humerus
49. Ulna
50. Carpometacarpus
51. Phalanx
52. Largest divergence of bill from a straight line
53. Bill angle (radians)

APPENDIX II
Characters Used in the Analysis of Single Functional Complexes

Forelimb (18 characters)

20. Wing span
21. Wing area
23. Alula
24. First primary (X)
25. First secondary (1)
26. Last tertial (3)
27. Tail, longest rectrix
28. Wing tip to primary X
29. Wing tip to secondary 1
30. Graduation of tail
31. Notch on inner web of second primary
32. Number of notched primaries
33. Longest primary
44. Sternum length
48. Humerus

(Continued)

APPENDIX II (Continued)

Hindlimb (15 characters)

  9. Hind toe
10. Inner toe
11. Middle toe
12. Outer toe
13. Hind claw
14. Inner claw
15. Middle claw
16. Outer claw
17. Foot span
18. Foot span with claws
42. Synsacrum/pelvis length
43. Synsacrum/pelvis width
45. Femur
46. Tibiotarsus
47. Tarsometatarsus

Feeding apparatus (7 characters)

  5. Bill length/skull
  6. Bill depth
  7. Bill width/base
  8. Bill width/flanges
19. Rictal bristle length
52. Largest divergence of bill from a straight line
53. Bill angle (radians)

Total morphological lifeform (48 characters)

  2. Wing length
  3. Tail length
  4. Tarsus
  5. Bill length/skull
  6. Bill depth
  7. Bill width/base
  8. Bill width/flanges
  9. Hind toe
10. Inner toe
11. Middle toe
12. Outer toe
13. Hind claw
14. Inner claw
15. Middle claw
16. Outer claw
17. Foot span
18. Foot span with claws
19. Rictal bristle length
20. Wing span
21. Wing area

APPENDIX II (Continued)

22. Wing chord
23. Alula
24. First primary (X)
26. Last tertial (3)
28. Wing tip to primary X
29. Wing tip to secondary 1
30. Graduation of tail
31. Notch on inner web of second primary
32. Number of notched primaries
34. Skull length
35. Length of brain case
36. Skull width
37. Width between quadrates
38. Interorbital width
39. Skull depth
40. Coracoid
41. Keel depth
42. Synsacrum/pelvis length
43. Synsacrum/pelvis width
44. Sternum length
45. Femur
46. Tibiotarsus
48. Humerus
49. Ulna
50. Carpometacarpus
51. Phalanx
52. Largest divergence of bill from a straight line
53. Bill angle (radians)

REFERENCES

Albrecht, G. H., 1979, The study of biological versus statistical variation in multivariate morphometrics: The descriptive use of mutliple regression analysis, Syst. Zool. 28:338–344.

Amadon, D., 1943, Bird weights as an aid in taxonomy, Wilson Bull. 55:164–177.

Atchley, W. R., Gaskins, C. T., and Anderson, D., 1976, Statistical properties of ratios. I. Empirical results, Syst. Zool. 25:137–148.

Bairlein, F., 1981, Ökosystemanalyse der Rastplätze von Zugvögeln, Ökol. Vögel 3:7–137.

Baker, M. C., 1979, Morphological correlates of habitat selection in a community of shore birds (Charadriiformes), Oikos 33:121–126.

Beecher, W. J., 1962, The bio-mechanics of the bird skull, Bull. Chicago Acad. Sci. 11:10–33.

Berthold, P., 1973, Relationships between migratory restlessness and migration distance in six Sylvia species, Ibis 115:594–599.

Berthold, P., 1979, Beziehungen zwischen Zugunruhe und Zug bei der Sperbergrasmücke *Sylvia nisoria*: eine ökophysiologische Untersuchung, *Vogelwarte* **30**:77–84.

Bock, W. J., 1964, Kinetics of the avian skull, *J. Morphol.* **114**:1–42.

Bock, W. J., 1966, An approach to the functional analysis of bill shape, *Auk* **83**:10–51.

Bock, W. J., 1977, Toward an ecological morphology, *Vogelwarte* **29**:127–135.

Bock, W. J., and Wahlert, G. V., 1965, Adaptation and the form–function complex, *Evolution* **19**:269–299.

Brown, R. H. J., 1963, The flight of birds, *Biol. Rev.* **38**:460–489.

Cody, M. L., 1978, Habitat selection and interspecific territoriality among the sylviid warblers of England and Sweden, *Ecol. Monogr.* **48**:351–396.

Cody, M. L., and Walter, H., 1976, Habitat selection and interspecific interactions among Mediterranean Sylviid warblers, *Oikos* **27**:210–238.

Conover, M. R., and Miller, D. E., 1980, Rictal bristle function in Willow flycatcher, *Condor* **82**:469–471.

Cornwallis, L., 1975, The comparative ecology of eleven species of wheatear (Genus *Oenanthe*) in S.W. Iran, Ph.D. thesis, Univ. Oxford, Oxford.

Curry-Lindahl, K., 1981, *Bird Migration in Africa*, Volume 1, Academic Press, London.

Dilger, W. C., 1956, Adaptive modifications and ecological isolating mechanisms in the thrush genera *Catharus* and *Hylocichla*, *Wilson Bull.* **68**:171–199.

Dyer, M., 1976, On the function of rictal bristles, with reference to Nigerian birds, *Nigerian Orn. Soc. Bull.* **12 (42)**:45–48.

Eaton, S. W., O'Connor, D. P., Osterhaus, M. B., and Anicete, B. Z., 1963, Some osteological adaptations in Parulidae, Proc. XIII Intern. Orn. Congr., pp. 71–83, Ithaca.

Engels, W. L., 1940, Structural adaptations in thrushes (Mimidae: genus *Toxostoma*) with comments on interspecific relationships, *Univ. Calif. Publ. Zool.* **42**:341–400.

Gaston, A. J., 1974, Adaptation in the genus *Phylloscopus*, *Ibis* **116**:432–450.

Gibson, A. R., Gates, M. A., and Zach, R., 1976, Phenetic affinities of the wood thrush, *Hylocichla mustelina* (Aves, Turdinae), *Can. J. Zool.* **54**:1679–1687.

Grant, P. R., 1966, Further information on the relative length of the tarsus in land birds, *Postilla* **98**:1–13.

Greenewalt, C. H., 1975, The flight of birds, *Trans. Am. Phil. Soc.* **65 (4)**:1–67.

Hoerschelmann, H., 1966, Allometrische Untersuchungen an Rumpf und Flügel von Schnepfenvögeln (Charadriidae und Scolopacidae), *Z. Zool. Syst. Evol. Forsch.* **4**:209–317.

Hummel, D., 1980, The aerodynamic characteristics of slotted wing-tips in soaring birds, Acta XVII Congr. Intern. Orn., Volume 1, pp. 391–396, Berlin.

Hurlbert, S. H., 1981, A gentle depilation of the niche: Dicean resource sets in resource hyperspace, *Evol. Theory* **5**:177–184.

Huxley, J., 1943, *Evolution, the Modern Synthesis*, Harper, New York.

Illies, J., 1970, Die Gattung als ökologische Grundeinheit, *Faunist. ökol. Mitt.* **3**:369–372.

Inger, R. F., 1958, Comments on the definition of genera, *Evolution* **12**:370–384.

James, F. C., 1982, The ecological morphology of birds: A review, *Ann. Zool. Fenni.* **19**:265–275.

Johnston, R. F., 1972, Ecologic differentiation in North American birds, in: *A Symposium on Ecosystematics*, Volume 4 (R. T. Allen and F. C. James, eds.), Univ. Ark. Mus. Occ. Pap. pp. 101–132, Fayetteville.

Johnston, R. F., and Selander, R. K., 1971, Evolution in the house sparrow, II. Adaptive differentiation in North American populations, *Evolution* **25**:1–28.

Karr, J. R., and James, F. C., 1975, Eco-morphological configurations and convergent evolution on species and communities, in: *Ecology and Evolution of Communities* (M. L. Cody and J. M. Diamond, eds.), Belknap Press, Cambridge, pp. 258–291.

Kokshaysky, N. V., 1973, Functional aspects of some details of bird wing configuration, Syst. Zool. 22:442–450.

Lederer, R. J., 1972, The role of avian rictal bristles, Wilson Bull. 84:193–197.

Lederer, R. J., 1975, Bill size, food size, and jaw forces of insectivorous birds, Auk 92:385–387.

Lederer, R. J., 1980, Prey capture by flycatchers and the importance of morphology to behavior, Sociobiology 5 (1):43–46.

Lederer, R. J., 1984, A view of avian ecomorphological hypotheses, Ökol. Vögel 6:119–126.

Leisler, B., 1975, Die Bedeutung der Fussmorphologie für die ökologische Sonderung mitteleuropäischer Rohrsänger (Acrocephalus) und Schwirle (Locustella), J. Ornithol. 116:117–153.

Leisler, B., 1977a, Die ökologische Bedeutung der Lokomotion mitteleuropäischer Schwirle (Locustella), Egretta 20:1–25.

Leisler, B., 1977b, Ökomorphologische Aspekte von Speziation und adaptiver Radiation bei Vögeln, Vogelwarte 29:136–153.

Leisler, B., 1980, Morphological aspects of ecological specialization in bird genera, Okol. Vögel 2:199–220.

Leisler, B., 1981, Die ökologische Einnischung der mitteleuropäischen Rohrsänger (Acrocephalus, Sylviinae), I. Habitattrennung, Vogelwarte 31:45–74.

Leisler, B., and Thaler, E., 1982, Differences in morphology and foraging behaviour in the goldcrest Regulus regulus and firecrest R. ignicapillus, Ann. Zool. Fenni. 19:277–284.

Leisler, B., Heine, G., and Siebenrock, K. H., 1983, Einnischung und interspezifische Territorialität überwinternder Steinschmätzer (Oenanthe isabellina, O. oenanthe, O. pleschanka) in Kenia, J. Ornithol. 124:393–413.

Morrison, D. F., 1976, Multivariate Statistical Methods, 2nd ed., McGraw-Hill, New York.

Mosimann, J. E., 1970, Size allometry: Size and shape variables with characterizations of the lognormal and generalized gamma distributions, J. Am. Stat. Assoc. 65:930–945.

Nachtigall, W., and Kempf, B., 1971, Vergleichende Untersuchungen zur flugbiologischen Funktion des Daumenfittichs (Alula spuria) bei Vögeln, I. Der Daumenfittich als Hochauftriebserzeuger, Z. vergl. Physiol. 71:326–341.

Newton, I., 1967, The adaptive radiation and feeding ecology of some British finches, Ibis 109:33–98.

Olson, E. C., and Miller, R. L., 1958, Morphological Integration, University of Chicago Press, Chicago, Illinois.

Oxnard, C. E., 1973, Form and Pattern in Human Evolution. Some Mathematical, Physical and Engineering Approaches, University of Chicago Press, Chicago, Illinois.

Oxnard, C. E., 1975, Uniqueness and Diversity in Human Evolution: Morphometric Studies of Australopithecines, University of Chicago Press, Chicago, Illinois.

Peters, D. S., Mollenhauer, D., and Gutmann, W. F., 1971, Bau, Konstruktion und Funktion des Organismus, Natur Museum 101:208–218.

Raikow, R. J., 1973, Locomotor mechanisms in North American ducks, Wilson Bull. 85:295–307.

Root, R. B., 1967, The niche exploitation pattern of the blue-gray gnatcatcher, Ecol. Monogr. 37:317–350.

Rüggeberg, T., 1960, Zur funktionellen Anatomie der hinteren Extremität einiger mitteleuropäischer Singvogelarten, Z. Wiss. Zool. 164:1–118.

Stork, H. J., 1968, Morphologische Untersuchungen an Drosseln. Eine Analyse von Anpassungsstrukturen im Körperbau von sechs europäischen Arten der Gattung *Turdus* L., *Z. Wiss. Zool.* **178**:72–185.

Winkler, H., 1971, Die Bedeutung der Organisation angeborenen Verhaltens für das Verständnis der Ökologie der Wirbeltiere, *Sitzber. Österr. Akad. Wiss. Mathem.-naturw Kl.* **179**:110–127.

Winkler, H., and Bock, W. J., 1976, Analyse der Kräfteverhältnisse bei Klettervögeln, *J. Ornithol.* **117**:397–418.

Wright, S., 1941, The "age and area" concept extended, *Ecology* **22**:345–347.

Zbinden, N., and Blondel, J., 1981, Zu Raumnutzung, Territorialität und Legebeginn mediterraner Grasmücken (*Sylvia melanocephala, S. undata, S. cantillans, S. hortensis*) in Südfrankreich, *Orn. Beob.* **78**:217–231.

CHAPTER 6

PROBLEMS IN AVIAN CLASSIFICATION

ROBERT J. RAIKOW

1. INTRODUCTION

That the higher-level classification of birds is in an unsatisfactory state is clearly demonstrated by the continued proliferation of new classifications and by the failure of biologists in general to adopt any one system as standard. Recent years have seen comprehensive classifications by Mayr and Amadon (1951), Stresemann (1959), Wetmore (1960), Storer (1971), Morony et al. (1975), Wolters (1975–82), Cracraft (1981), and others. The Peters' checklist and the American Ornithologists' Union checklist continue to evolve, and many attempts have been made to reclassify avian subgroups. Pizzey (1980, p. 13) stated that "The classification of birds (avian taxonomy) is in some ways like the peace of God—it passeth all understanding." Olson (1981, p. 193) agreed with this sentiment, but also recognized the basic reason: ". . . the present classification of birds amounts to little more than superstition and bears about as much relationship to a true phylogeny of the Class Aves as Greek mythology does to the theory of relativity." In this paper, I will ask the question "why don't we have a satisfactory classification of the birds?" This is necessarily a personal view, depending considerably on what is meant by "satisfactory," and I will begin by discussing some

ROBERT J. RAIKOW • Department of Biological Sciences, University of Pittsburgh, Pittsburgh, Pennsylvania 15260, and Carnegie Museum of Natural History, Pittsburgh, Pennsylvania 15213.

of the characteristics that a satisfactory classification should have. Next I will consider some reasons why we have fallen short of this goal, and will suggest several ways in which future taxonomic studies might be improved. In doing these things, I hope to interpret for the general reader some of the debates and controversies currently of interest in systematic biology. This paper is meant to express a particular point of view rather than to review the subject either impartially or exhaustively. I will take the position that a satisfactory classification must be based on phylogeny, that the traditional eclectic or "evolutionary" method is unable to generate such a classification, and that the most promising approach is some form of cladistic classification in which all taxa are strictly monophyletic and are ranked by a single method.

2. A SATISFACTORY CLASSIFICATION

Most biologists would agree that a classification should reveal the pattern of order that underlies the diversity among species and that the source of this diversity is the evolutionary history of the group. There is, however, much variation in the way that different systematists view the relationship between the nature of this diversity and its expression in a classification. The study of diversity is the province of *systematics*, whereas classification is the goal of the more narrowly defined field of *taxonomy*, although a clear distinction has not always been made between these disciplines. Taxonomy uses the results of systematic studies as the basis for constructing classifications, but classification is not the only use to which systematic findings may be put, and the study of the causes and significance of diversity is the primary interest of many systematists. This paper is concerned mainly with avian classification, but the analysis will require consideration both of the systematic study of diversity and of the application of such work to the classification of birds.

Three general schools are currently active in avian systematics. These are the traditional or *eclectic* school, sometimes imprecisely termed "evolutionary" by its adherents, the *phenetic* or numerical taxonomic school, and the *cladistic* or phylogenetic school. Eclectic taxonomists try to express both anagenetic and cladogenetic aspects of phylogeny in their classifications (*anagenesis* refers to character transformation in an evolving lineage and *cladogenesis* to the splitting of a lineage). Phenetic taxonomists classify on the basis of overall similarity alone, whereas cladists attempt to express only the cladogenetic or branching aspect of phylogeny in their classifications. Both phenetics and cladistics as formally defined methodologies are of exceedingly recent origin in the history of avian systematics. Earlier workers groped toward a classification using a combination of "similarity" and "descent" concepts, varying in proportion and subjective owing to the lack of a technique for deducing descent in any really objective way. A clearly phylogenetic approach was not possible until the importance of identifying shared derived characters was identified, leading to the development of cladistics. Previously, taxa were assessed on the basis of shared characters in general, without thought as to whether they were primitive or derived, and this approach still characterizes the phenetic school. Before considering the relative merits of the different approaches to classification, I will suggest certain characteristics that a satisfactory classification should have.

2.1. Meaningfulness

A meaningful classification will be unambiguous about the information content of both taxa and categories and must therefore be consistent in the methods by which taxa are recognized and ranked. All taxa should be generated in the same way so that all represent the same kind of biological phenomenon. The rank of all taxa should be determined according to one system so that the nature of the relationship between any taxa of the same or different rank will be automatically designated.

Cladistic and phenetic classifications are meaningful in the present sense and eclectic classifications are not. Phenetic taxa represent groups clustered by similarity; the species in a taxon may be assumed to be more similar to each other (at least in the characters analyzed) than to species in other taxa. Likewise, rank is based on relative similarity and the classification forms a nested set of similarity relationships. Cladistic taxa represent clades or groups clustered by shared derived characters and most workers hypothesize that they represent monophyletic groups in the historical sense. Rank is determined by position in the cladogram, so that the ranking pattern mirrors the pattern of nested clades on which it is based. Whatever one's preference, the meaning of the patterns in phenetic and cladistic classifications is not ambiguous.

Eclectic classifications are ambiguous rather than meaningful, because different methods are used within a single classification both for recognizing taxa and for ranking them. One knows that every taxon was formed and ranked by one or more of several possible methods, but seldom knows which. Most of our classification system was constructed in this way and as a result its information content is limited.

The literature on the philosophies and methods of the different schools is vast and a comprehensive review is outside the scope of this essay. For a general introduction to phenetic methods, see Sneath and Sokal (1973). For cladistics, see Cracraft and Eldredge (1979), Eldredge and Cracraft (1980), Wiley (1981), and Patterson (1982). The eclectic school is discussed by Simpson (1961), Mayr (1969, 1981), and Bock (1973, 1982). The central forum for discussion of these questions in the past two decades has been the journal, *Systematic Zoology*. These sources will provide an entry into the literature for those who wish to pursue it.

2.2. Practicality

A satisfactory classification should serve both specialists in various fields and also the general public, the familiar "general reference system" of biology. An important aspect of practicality is simplicity or ease of use. The more complicated a classification is, the harder it will be to remember the groups and their relationships. Ironically, there appears to be an inverse relationship between complexity (information content) and utility. Critics of cladistic classifications have often suggested that they may be too complex to be useful because they contain so many categories and taxa. However, methods have been developed to reduce this problem, as discussed in Sections 3.3.2 and 3.3.3. In addition, it is not always necessary to name all of the clades in a phylogeny. Swierczewski and Raikow (1981) named only the larger (basal) groupings and designated the smaller, unnamed taxa in their detailed cladogram with letters of the alphabet so that they could be referred to in the discussion. This seems to be a workable solution, and if one subsequently wants to discuss such a group more thoroughly, formal names can be added without restructuring the whole classification. Furthermore, a classification need not be used in all of its complexity for every application. Specialists may need the detailed classification in technical works, for example, whereas the writer of a popular field guide can simply use a few major categories.

Another aspect of practicality is stability. Every time a classification is modified, people have to learn the new arrangement, books must be rewritten, and collections reorganized. From one point of view the most practical classification would never change at all, and many taxonomists consider a classification that resists modification desirable. At the other extreme are those, such as Cracraft (1981), who regard stability as an indication that nothing new is being learned about a group and see change as evidence of an increasing understanding of relationships.

At this time we know little about the relative stability of eclectic, phenetic, and cladistic methods of classification. Our traditional eclectic classifications of birds have changed frequently, but we lack a comparable history of phenetic or cladistic classifications. Critics of both schools have argued that they are likely to produce classifications that are easily modified by different analytical techniques or by the addition of small amounts of new data. I believe that cladistic classifications are potentially stable because they are based on phylogeny, and as the true historical phylogeny is approached from various directions, classifications should also converge. This idea must be tested by separate attempts to classify the same groups using cladistic analyses of different kinds of data. As yet there are few such attempts, but the limited results to date are encouraging. Simpson and Cracraft (1981) studied the osteology of the Piciformes with similar results to Swierczewski and Raikow's (1981) myological study. The ongoing studies of Charles Sibley and his colleagues (e.g., Sibley and Ahlquist, 1983) on passerine phylogeny by DNA hybridization are producing results strikingly similar to my analyses based on morphology. Stability is an important aspect of practicality in classification, but there is currently no agreement on what level of stability is desirable or on the relative stability of the classifications produced by different approaches.

A final aspect of practicality is the degree to which a classification is useful in the study of problems such as speciation, biogeography, macroevolutionary analysis, and phenotypic evolution and rates. Cracraft (1983) argues for the superiority of cladistic classification in such efforts.

Thus it would seem that an ideal classification should be firmly based on phylogeny, should be internally consistent in the information content of its taxa and categories, and should be of practical utility. The ongoing debates over both methods of classification and classifications themselves indicate that we are still far from such an ideal. I will now consider some possible reasons for this situation.

3. WHY WE LACK A SATISFACTORY CLASSIFICATION

3.1. Different Concepts of Relationship

One reason that we lack a satisfactory classification is that we don't agree on the meaning of "relationship." Several different concepts exist and until we settle on just one of them as the basis for a theory of relationships, we cannot hope to develop a classification based on such a theory.

3.1.1. Cladistic Relationships

A clade is a group clustered by synapomorphy (shared derived characters), that is, a strictly monophyletic group. A problem sometimes arises because of the distinction between a cladogram, which is simply a synapomorphy distribution scheme, and a phylogenetic tree, which is a hypothesis of genealogical relationships. This distinction was not recognized at first, and cladograms were merely devices for hypothesizing phylogeny. Later it was recognized that when taxa from different geological ages are analyzed together, it may be impossible to decide whether an earlier form is directly ancestral to a later form or shares a common ancestor with it. Because of this, it is sometimes possible for one cladogram to be consistent with several phylogenies. This need not be too much of a problem: if one studies only extant species, then each cladogram specifies only one phylogeny, which has the same branching pattern as the cladogram. The inclusion of fossils, however, raises additional problems of analysis that still require resolution (e.g., Raikow, 1981; Olson, 1983; Raikow and Cracraft, 1983).

3.1.2. Phenetic Relationships

These are relationships of overall similarity. The same characters are compared within a group of taxa, which are then clustered in a hierarchical manner on the basis of their shared similarities. This was done "by hand" for many years as the basic procedure of eclectic taxonomy. When done in this way, character conflicts can be resolved by decisions as to which conflicting characters are more likely attributable to convergence or parallelism. With computerized numerical phenetics, large numbers of characters and taxa can be analyzed quickly, but the judgmental resolution of character conflicts may be lost, and it is usually impossible to tell readily which characters were used in defining each cluster.

3.1.3. Genetic Relationships

Phenotypic similarity is sometimes used as an estimate of genetic similarity (see Wiley, 1981, p. 261), but this is different from direct measures of genetic distance. Not only is there an imperfect correlation between genome and phenotype, but this is not a justified use of the idea of genetic relationship because it does not directly examine the genome. Inasmuch as there are methods for doing the latter, concepts of genetic affinity should be restricted to them and phenotypic relationships limited to phenetic or cladistic categories.

Direct measures of genetic relationships attempt to assess how much of the genome is shared between species. Earlier methods used the electrophoretic separation of mixtures of proteins, whereas DNA hybridization is a more recent approach. It is not clear whether electrophoretic methods are best considered to be phenetic or genetic measures. Electrophoresis assesses certain features of molecular structure, which are a direct product of gene action. It measures only a small part of the genome, and only the products of structural genes at that, whereas the most important evolutionary changes may result from the modification of the regulatory genome, which the electrophoresis of structural gene products bypasses. Electrophoretic studies are probably best considered comparable to morphological studies of single systems (e.g., osteology, myology, pterylography) rather than as comprehensive genome comparisons.

Deoxyribonucleic acid hybridization, on the other hand, examines essentially the entire genome, determining how much of the total genetic information is the same in different species. Various parts of the genome may evolve at different rates at different times in the history of a group, but because the entire genome is sampled, this effect is believed to be swamped by the large number of events, giving long-term constancy to the rate of genome evolution within a group. If this is true, then the degrees of difference between species will be direct measures of their divergence times and the relationships depicted will be cladistic. Deoxyribonucleic acid hybridization makes possible the construction of much more precise and detailed phylogenetic hypotheses than does electrophoresis. This promising new approach is explained by Sibley and Ahlquist (1983). The results are exciting, but a critical appraisal of its possible limitations is needed. Is the assumption of overall rate constancy realistic at all taxonomic levels? How can the results of DNA studies be tested? What will be the meaning of congruence or noncongruence of DNA and morphological studies?

3.2. No Theory of Avian Relationships

Whatever we mean by "relationship," we cannot have a satisfactory classification until we agree on the pattern of relationships among all groups of birds. The extent of our knowledge is revealed by the current classifications. Most of these are similar to each other and divide the class Aves into approximately two dozen orders. Each order contains a group of species considered to be more closely related to each other (in some sense) than to the members of other orders. The orders thus represent the largest groupings for which patterns of affinity have been accepted; they reflect our levels of understanding, which is why they vary so much in number and diversity of species. For example, the order Passeriformes, which contains more than half of all living species of birds, might be considered disproportionately large except that it is recognized as being strictly monophyletic (Raikow, 1982). On the other hand, the order Coliiformes has only six species. It is given ordinal rank simply because taxonomic procedure requires that these species be placed in some order, and their relationships to other birds are so obscure that we cannot determine what other order to put them in. If one were to diagram a current classification, it would show a single node from which 25 to 30 separate lineages (orders) spring. Nobody would suggest that this represents a phylogeny, that birds had arisen by the simultaneous fractionation of a single common ancestor into two dozen separate lineages. Some interordinal relationships have been recognized informally (e.g., "picopasserine assemblage," "higher non-passerines"), and Cracraft (1981) has gone further by designating a superordinal category, the Division, to incorporate hypothesized relationships. However, because he recognized the weakness of the supporting evidences, he did not formally name these taxa.

3.3. Expression of Relationships in a Classification

Even if we agreed on a phylogeny of the birds, we would still not be able to construct a satisfactory classification because we disagree on how relationships should be expressed in a classification. The problem is clarified by recognizing that it involves two separate procedures: the recognition of taxa and the ranking of those taxa into a hierarchical scheme.

3.3.1. Recognition of Taxa

In cladistic classification only monophyletic groups are recognized as taxa. Furthermore, cladists hold a precise definition of monophyly— a monophyletic group consists of an ancestral species (in practice almost always hypothetical) and *all* of its descendants. Earlier definitions of monophyly were more general, requiring only that a monophyletic group share a common ancestor, but not that it necessarily include all descendants. Ashlock (1971) suggested that strictly monophyletic groups be termed *holophyletic* and that groups containing only some descendants of a common ancestor be restricted to the older term *paraphyletic* (See Wiley, 1981, for a thorough discussion of this terminology). "Paraphyletic" has become widely accepted, but cladists tend to

prefer "monophyly" to "holophyly," perhaps because the strict definition is the meaning toward which the earlier definitions were evolving. Cladists argue that monophyletic groups (clades) have a historical
existence; they are evolutionary units. Restriction to one criterion for
the recognition of taxa also gives a classification consistency as previously discussed in Section 2.1.

In phenetic classification, taxa are based on overall similarity without the distinction between primitive and derived characters analyzed
by cladists. This results in taxa that are phenetically coherent, but it is
difficult to know whether they are holophyletic or paraphyletic because
the relationship of the pattern produced to phylogeny is obscure. It is
generally believed that phenetic clustering approximates phylogeny,
but the method is not designed to obtain this end (e.g., Wood 1983, pp.
104–105). As with cladistic classification, phenetic classification uses
a single method of clustering and thus is meaningful.

In eclectic classification, taxa may be recognized on the basis of
various different criteria. Most often they are based on phenetic similarity, usually without regard to primitive/derived character polarity,
although the point is generally unclear because eclectic taxonomists
seldom explain their methods. Other criteria such as geographic distribution, adaptive specializations, and intuition are also employed.
There is generally no way for a reader to tell how specific taxa were
recognized and the method is therefore not meaningful.

3.3.2. Ranking of Taxa

In any Linnaean classification, the taxa are arranged in a nested
hierarchy of progressively more inclusive ranks or categories. In cladistic classifications, the pattern of cladistic relationships, usually taken
to hypothesize genealogy, is the basis for ranking. The clades are recognized as taxa and their rank is determined by their position. More
inclusive groups are ranked at higher category levels than less inclusive
groups. In its simplest form, a cladistic classification places all sister
taxa at the same rank (Fig. 1). This is totally unambiguous; the classification exactly expresses the genealogy. This procedure has been criticized, however, because it produces classifications with so many ranks
and taxa that they may become impractical. Various methods have
therefore been devised to simplify cladistic classifications while retaining the essential feature of consistency (see Eldredge and Cracraft,
1980, Chapter 5; Wiley, 1981, Chapter 6). It is significant that one can
always reconstruct the phylogeny from the classification. Critics have
argued that this is unimportant because the author will usually have

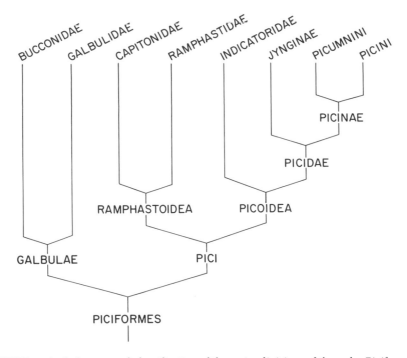

FIGURE 1. A phylogeny and classification of the major divisions of the order Piciformes. The phylogeny was developed first as a cladogram based on morphological characters, and the classification was then superimposed on it. In this classification, all the taxa are holophyletic and all sister taxa have the same rank (from Swierczewski and Raikow, 1981).

presented a diagram along with the classification, but this criticism fails on two counts. First, an author may *not* provide the diagrams; for example, Cracraft (1981) considered them redundant. Second, the significance of being able to reconstruct the phylogeny is not that one may want to do it, but that the capability demonstrates that the information is actually present in the classification. This situation differs fundamentally from eclectic classifications, which often claim to contain genealogical information, but in which that information is not retrievable (see Section 3.3.3 and Fig. 3).

An alternative method of ranking taxa in cladistic classifications is by the absolute age of the taxa, if there is a method for estimating it. Sibley and Ahlquist (1983) estimate divergence by calibrating DNA hybridization measurements against dates based on geological events and the fossil record. These approaches are meaningful because they are applied consistently throughout the analysis.

Phenetic classifications rank taxa on the basis of their relative position in the phenogram. As with cladistic classification, this method is consistent and therefore meaningful.

It is in the ranking of taxa that the eclectic school reaches the height of its arbitrariness. Taxa are ranked on one or more of a variety of criteria, including phenetic distinctiveness, adaptive level (grade), and number of species. Proponents argue that this method is superior because it considers both genealogy and similarity in contrast to the cladistic school, which considers only the first, and the phenetic school, which considers only the second. Thus eclectic taxonomists claim to use more information. However, it is virtually impossible to combine the two different sets of information within a single hierarchy: one can express either genealogy or similarity, but not both, because their patterns are usually not identical. In attempting to combine the two, the eclectic method simply uses each to obscure the revelations of the other. They are combined in inconsistent ways so that the hierarchical structure of an eclectic classification expresses neither a pattern of genealogy nor one of similarity. It expresses some combination of the two and because the nature of their interaction is unknown, one cannot tell what the pattern means.

3.3.3. Example: The New World Nine-Primaried Oscines

Several years ago (Raikow, 1978), I published a phylogeny and classification of the New World nine-primaried oscines in which I used the cladistic method to analyze phylogeny, but rejected it for the purpose of classification. One reason was the idea that a classification should not mirror a phylogeny, but could be used for different purposes. It is significant that I did not specify those purposes, for even now I don't know exactly what I had in mind beyond some vague notion that taxa should represent adaptive grades. I would now like to reconsider that phylogeny in order to show how a cladistic classification of the same group can more accurately express the postulated historical relationships without undue complexity.

Figure 2 shows a cladogram of the New World nine-primaried oscines. It was simplified from Fig. 10 of my 1978 paper by deleting the generic names and other details not important for the present purpose. This cladogram is a nested arrangement of clades, except for a few genera whose positions remain uncertain as indicated by basally dashed lines. Ignoring these for the moment, we can simplify the dia-

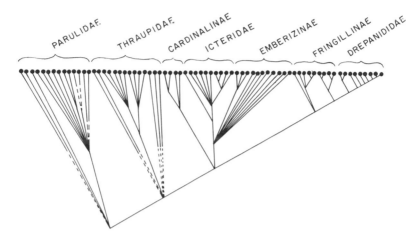

FIGURE 2. A phylogeny of the New World nine-primaried oscines based on a cladogram constructed with morphological characters, and simplified from Fig. 10 of Raikow (1978). The dots at the top indicate genera, and dashed lines indicate forms of uncertain position. The taxa are from the classification proposed in Raikow (1978).

gram to that shown in Fig. 3A, where the families and subfamilies are those that I named in my 1978 classification (Table I). Figure 2 shows that some of these taxa, such as the Cardinalinae, Icteridae, and Drepanididae are holophyletic (consisting of *all* the descendants of a common ancestor), whereas the Parulidae, Thraupidae, Emberizinae, and Fringillinae are paraphyletic (containing only *some* descendants of a common ancestor). The family Fringillidae in this classification is an especially poor taxon (Fig. 3A). It consists of the Cardinalinae, Emberizinae, and Fringillinae and represents the "finch" or seedeater grade. The recognition of this grade-level taxon required a distortion of the hypothesized phylogenetic relationships. For one thing, the Emberizinae are more closely related genealogically to the Icteridae, which are excluded from the family, than to the included Cardinalinae and Fringillinae. Likewise the Fringillinae are more closely related to the excluded Drepanididae than to the included Cardinalinae or Emberizinae. This shows how an eclectic classification cannot simultaneously express two different kinds of information. Indeed, although the family Fringillidae in this example represents an adaptive grade, it does not even include all of the species in the assemblage that exhibit that grade, for the finchlike Hawaiian Honeycreepers (*Psittirostra*, etc.) are ex-

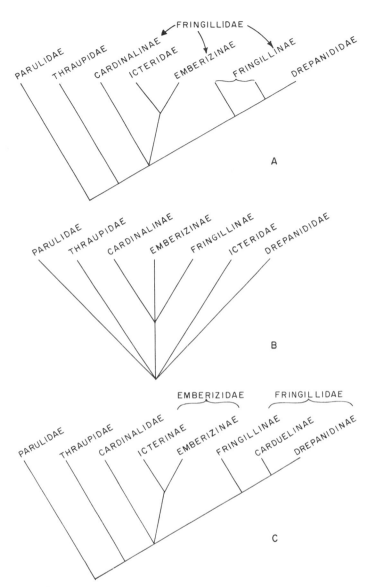

FIGURE 3. Relationships between phylogeny and classification as illustrated by the New World nine-primaried oscines. The classification is given in Table I. (A) Outline phylogeny simplified from Fig. 2, showing taxa named in Raikow (1978). (B) The hierarchical structure of the classification differs from the phylogeny shown in (A) because the phylogenetic information has been distorted in an attempt to express grades. (C) The same phylogeny shown in (A) with a new classification (Table I) based only on the branching pattern of the phylogeny. This aligns the taxa with the evolutionary history of the group and avoids the inconsistencies of (A). See text for discussion.

TABLE I
Classifications of the New World Nine-Primaried Oscines

Raikow (1978)	Raikow (1984)[a]
Parulidae	"Parulidae"
Thraupidae	Thraupidae
Fringillidae	Cardinalidae sedis mutabilis
Cardinalinae	Emberizidae sedis mutabilis
Emberizinae	Icterinae
Fringillinae	"Emberizinae"
Drepanididae	Fringillidae sedis mutabilis
	Fringillinae
Icteridae	Carduelinae
	Drepanidinae

[a]This chapter.

cluded. I retained them in the Drepanididae for arbitrary reasons that cannot be discerned from the classification itself.

The other taxa in my 1978 classification also represent vaguely defined adaptive grades. The hierarchical structure of this classification is shown in Fig. 3B, which should make it clear that in recognizing grades as taxa, my classification completely abandoned the pattern of evolutionary relationships that my study had hypothesized. Because people generally believe that classifications express evolutionary relationships, this classification becomes misleading as well as being inaccurate.

A new classification is shown in Table I, and its relationship to the phylogeny is shown in Fig. 3C. As noted earlier, the basic idea of cladistic classification is that only clades are classified; this takes care of the problem of recognizing taxa. With respect to ranking, the simplest approach, giving all sister taxa equal rank, may lead to excessively complex classifications with groups of any size. One way to reduce this problem is with Nelson's (1973) convention of "phyletic sequencing" (Eldredge and Cracraft, 1980; Wiley, 1981, pp. 206–210). In this technique, a series of holophyletic groups that branch off successively are classified at the same rank, and the sequence in which they are listed in the classification repeats the sequence in which they branch off in the cladogram.

A second useful convention allows one to designate in the classification the occurrence of an unresolved trichotomy or polychotomy. In Fig. 3C there is a node at which three lineages arise, the Cardinalidae, Emberizidae, and Fringillidae. This does not mean that the three lineages evolved by the simultaneous splitting of a common ancestral line, but only that the probable occurrence of two successive divisions was not resolved in my analysis. This uncertainty may be indicated by giving the three groups equal rank and placing them *sedis mutabilis* "at the level of the hierarchy at which their relationships to other taxa are known" (Wiley 1981, p. 211).

Compare the old and new classifications in Table I. Each contains two ranks, family and subfamily. The old classification had eight taxa, whereas the new one has ten. One of the new taxa is the subfamily Carduelinae, which was added because the "Fringillinae" of the old classification actually consists of two separate clades (Figs. 2, 3A) whose separate existence was not expressed therein. The second new taxon is the Family Emberizidae, representing the clade containing the Icterinae and Emberizinae. This was not recognized in the 1978 classification, which separated the two groups in an attempt to express grades. The new classification recognizes groups with an historical existence as postulated in my phylogenetic analysis. It is therefore rooted in the historical process of evolutionary change and divergence that gave rise to the pattern of diversity shown, it is meaningful, and it is sufficiently short and simple to be practical.

Two problems remain. In the more detailed cladogram (Fig. 2), there are a few genera whose positions are uncertain. How should they be dealt with? Classifying only to subfamily as in Table I does not address the problem. If we were to classify the group down to the species level, they would normally be placed, without comment, in the family to which they best appear to belong. I would merely insist that they be listed *incertae sedis* at that level so as to emphasize their problematical position.

The second problem is that two taxa in this revised classification are not strictly monophyletic (can you identify them in Fig. 2?); hence the classification departs from the rigid consistency that I have advocated. They are the Parulidae and Emberizinae. The Parulidae is paraphyletic because two genera (and a possible third) arise from the first node of the cladogram, but that node actually defines the entire assemblage. What is the nature of the problem raised here by the Parulidae? Some might suggest that it reveals a flaw in the method that we cannot include these genera in a holophyletic family Parulidae, where they appear to belong. Are we faced with the dilemma of either excluding

Basileuterus and *Geothlypis* from the Parulidae in order to keep the family holophyletic or else including them at the cost of making the family paraphyletic? Does this intuitively "right" family allocation expose a fatal methodological rigidity?

I would suggest instead that this case reveals one of the strengths of the cladistic method. It is impossible to hide such inconsistencies in a properly delineated cladogram in the way that they can be glossed over in a written narrative or in the vague family trees sometimes presented in eclectic works. The cladogram not only reveals the problem, but by depicting its nature points toward its solution. The real problem here is that the node is insufficiently resolved; it gives rise to four (possibly five) lineages. The solution lies not in redefining the nature of taxa, but in new research that will provide additional characters to permit a further series of branch points, that is, a more detailed phylogeny.

Thus one can stop short of the ultimate goal of a fully dichotomous phylogeny, but one cannot hide the fact that this has been done. There is still a practical problem in classification. If one wants to present a classification for a group not wholly resolved, there is a way to indicate that a taxon is not holophyletic, namely by enclosing its name in quotation marks. Wiley (1981, p. 213) suggests that these taxa not be ranked, but merely listed in the classification so as to reveal their unknown status. The Thraupidae in the new classification could be treated the same way if one decided to include within it those genera with basally dashed lines whose position is unsettled. The point is that this is a temporary expedient indicating that the systematic study on which the classification is based has not been adequately concluded. This example illustrates the value of the cladistic method of classification as an analytical tool to assess the sufficiency of its systematic foundation.

To examine further the relationship between phylogeny and classification, I will consider various approaches to grouping the Hawaiian Honeycreepers and the fringilline and cardueline finches. The phylogenetic relationships among these forms (more specifically among many but not all of the generally recognized genera) were hypothesized in two papers (Raikow, 1977, 1978) and are shown in Fig. 4. Various kinds of groupings are circumscribed on this cladogram. Group A constitutes a commonly recognized taxon, the fringilline/cardueline finches (variously named and ranked), and represents an adaptive grade, the "seedvise" mechanism, an advanced seed-cracking system different from those of the emberizine and cardinaline finches (except perhaps for *Fringilla*; Bock, 1960). However, it does not represent all of the New World nine-primaried oscines that possess this mechanism, which also occurs among the finchlike Hawaiian Honeycreepers. Furthermore, this

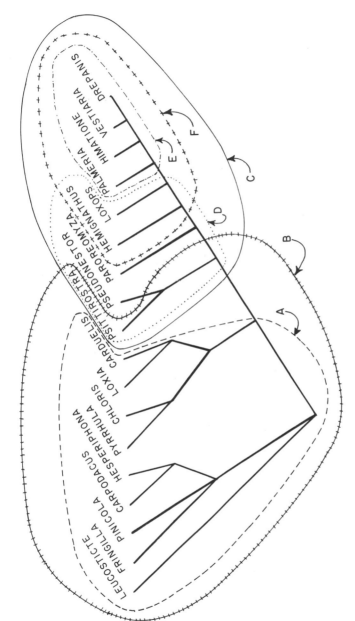

FIGURE 4. A phylogeny of the fringilline and cardueline finches and the Hawaiian honeycreepers after Raikow (1977, 1978). The six assemblages circumscribed on this phylogeny represent groups that have been or could be recognized as taxa under the vague concepts of eclectic classification. The visual confusion evident here is a reflection of the methodological confusion of the eclectic school.

assemblage is paraphyletic. I would therefore regard it as an inappropriate taxon because it has no unique characteristics. Perhaps it is unfair to apply this judgement to a taxon that was created by phenetic clustering, so that its inappropriate characteristics were not at the time detectable. This, however, illustrates the desirability of developing a phylogeny for a group before attempting to classify it, at least if one accepts the concept that classification should be based on genealogy.

Group B constitutes the traditional cardueline finches plus those Hawaiian Honeycreepers that share the same seed-eating specialization. This group is now a phenetically coherent grade assemblage containing all of the species of this grade. However, it is still paraphyletic, and it has never been recognized as a taxon as far as I know.

Group C is a holophyletic group all of whose members live in the Hawaiian islands and which constitutes a single adaptive radiation. It does not represent a grade, however, because several distinct feeding modes have evolved within the group, including insect-gleaning and nectar-feeding in addition to the seed-cracking of the founder species (Raikow, 1977). This group is a commonly recognized taxon, the Hawaiian Honeycreepers, variously classified from family to tribe level.

Group D is a taxon (essentially the subfamily Psittirostrinae of Amadon, 1950) having a certain phenetic coherence based on plumage and so forth and is paraphyletic. It does not represent a grade, but contains the seed-eating and insect-gleaning forms, plus some of the nectar-feeders.

Group E is a taxon (essentially the subfamily Drepaniinae of Amadon, 1950) and is holophyletic. All of its members represent a single grade, nectar-feeding, but it does not include all of the nectar-feeding honeycreepers, some of which are in the preceding subfamily.

Group F is holophyletic, and represents a single adaptive grade, the nectar-feeding drepanids. It would be an appropriate taxon, but to my knowledge has never been so recognized.

Some of these groupings have been designated as taxa, and others have not, but all could be so recognized under the loose concepts of eclectic taxonomy. There is no limit to the number and kinds of groups that can be so defined, especially in the common absence of a detailed phylogenetic hypothesis. The result is that there is no way to determine from an eclectic classification what kind of biological phenomenon a taxon represents.

3.4. Confusion between Phylogeny and Taxonomy

Taxonomic procedure, although designed for practical purposes, may have imposed on the thinking of biologists certain unconscious views about evolution and relationships. The rules require that every species be classified within a genus, family, and order within the class Aves (Mayr, 1969, p. 89). The purpose of this is utilitarian, but the effect may have been to give people the idea, however unconscious, that birds are divided into three levels of relationships between class and species. That is, the categories themselves have taken on personalities, so that one speaks of generic characters or familial characters and debates whether a certain character is sufficiently distinctive or unusual as to be used to classify a form at a lower or higher level. Some taxonomists believe that the cocks-of-the-rock (*Rupicola*) merit family status because their thigh artery is different from that of other cotingas, as though rank were a reward for service or accomplishment.

Consider the debate between Olson (1983) and Raikow and Cracraft (1983) over the question of whether the order Piciformes is monophyletic. If one regards birds as a single class divided into 25 to 30 orders, then each order (especially a large one) constitutes a major group. Given this view, the question of piciform monophyly assumes great importance in the effort to understand avian relationships. On the other hand, ignoring extinct species for the sake of simplicity, one might regard Aves not as a collection of two dozen orders, but as one of 9021 extant species (Bock and Farrand, 1980) having a phylogenetic history in which there is a number of branching points exactly one fewer than the number of species. In that view the question of piciform monophyly is about the validity of one out of 9020 nodes in a phylogeny, and on a purely numerical basis may be seen to be a much less significant question.

This all exemplifies the way in which taxonomic rules have clouded the concept of relationship. Relationship is phylogenetic (historical) not taxonomic (procedural). Concepts of relationship should be based, therefore, on reconstructions (hypotheses) of this history, not on practical rules of classification or nomenclature.

Biologists tend to assume that classifications express phylogeny. Nonphylogenetic classifications come to form unrecognized phylogenetic hypotheses and thereby become misleading in two ways: (1) by giving the idea that phylogenetic relationships are known, and (2) by giving the idea that specific classifications represent those relationships. These are some of the reasons why it is desirable to base classifications solely on phylogeny and to construct phylogenies first and classifications thereafter.

3.5. Shortcomings in the Explanation of Systematic Methods

In many taxonomic works the methodology is not adequately explained and the data are not properly documented. There has been in the past a sense that taxonomy is as much an art as a science, that it cannot be taught formally but must be learned by experience, and that valid taxonomic groups are those recognized by established workers. Many writers have not felt constrained to explain their methods in detail and this could be in part because the methods themselves have often been poorly defined. One sometimes sees papers that lack a materials and methods section, or if one is present, it may only list the specimens used. Given the diversity of current approaches, it is incumbent upon authors to explain clearly what method they are using and exactly how they apply it. Along with this there is a need to document more clearly the actual data on which analyses are based and the way in which the conclusions are derived from those data.

Biologists in general have not required taxonomists to explain their work in a thorough fashion. Many ornithologists appear to accept taxonomic decisions uncritically or to reject them on the basis of inconvenience rather than lack of merit. Cracraft (1983) has suggested that many nonsystematists fail to see the relevance of the debate over classification to their own work in related fields and lack the curiosity to evaluate the controversy. Yet the degree to which many ecological and evolutionary studies require a taxonomic baseline would suggest that such workers would benefit by making the effort to understand the situation. Authoritarianism is a strong tradition in taxonomy and has contributed to deficiencies in the area of methodological rigor and explanation. Instead of a "general reference system" for biology, we have sometimes had a general reverence system. This may be one reason that among scientists in general, taxonomy has not enjoyed a high level of respect for many years.

I would like to suggest some ways in which the authors of taxonomic studies could improve the quality of their work by avoiding certain common shortcomings that have marred many earlier efforts, including my own.

3.5.1. Diagramming the Relationships

There have been many papers with detailed discussions of the comparative data and the proposed relationships of the species, but without a diagram showing the pattern of relationships. The failure to provide a diagram at least raises the suspicion that the author could not make one. It may be instructive for a reader to go through such a paper and try to construct the diagram that the author neglected to provide. In doing this I have sometimes found that the data do not support all the details of the author's hypothesis or that it is impossible, in the absence of an explanation of method, to determine how the author reached his conclusions. A visual presentation of the proposed relationships clarifies the pattern in a way that words alone cannot do, because the whole arrangement can be seen at once. Such a diagram serves also as a framework for the written discussion. If there are uncertainties, they can be illustrated by standard graphic techniques such as dotted lines. Indeed, there is no reason why one should not present alternative hypotheses to illustrate the ambiguities that arise from conflicting characters or different character-weighting schemes.

3.5.2. Correlating Characters with Groups

Characters are sometimes ascribed to groups with which they do not actually correlate. A distinction should be made between attributes that are diagnostic for a group and those that merely occur in it. A character will be descriptively correct but not diagnostic if it occurs in the group but also outside it. Feathers are diagnostic of the class Aves, but the amniote egg is not, although both occur in all species of birds. These points may seem obvious, but it is surprising how often they are overlooked. It is not uncommon to find that characters in one or a few species are attributed to higher taxa that include those species, with relationships being suggested on this basis. A few species of family A share a character with a few in family B, and a relationship is thereby suggested between the two families. There are many possible explanations for such character distributions depending on whether the characters are primitive or derived for the groups in question, on the likelihood of independent origin of derived states, on the conflicts of these characters with others, and so forth. Misleading implications of affinity between higher taxa on this basis are best avoided by providing tables showing the actual occurrence of every character in every species, so that the reader can judge how complete the data really are and therefore how reasonable the hypothesis is. Failure to provide tabulated data is one of the most common shortcomings of taxonomic papers.

3.5.3. Distinguishing between Similarity and Genealogy

The concepts of genealogical and phenetic relationship were discussed above. Phenetic relationship is often used as an estimate of genealogical relationship, for example, as a method of constructing a phylogeny. The general patterns are often similar, but discrepancies arise when one lineage splits into two and one daughter lineage then evolves faster than the second. The result is that forms that are most closely related genealogically may not be as similar to each other as one of them is to a more distant relative. This point has been discussed fully elsewhere (Raikow and Cracraft, 1983).

Related to this problem, however, is a common looseness in writing that may lead to misleading implications. One often reads that some species "has no close relatives" or "no close living relatives." Unless, perhaps, one believes in special creation, such statements are meaningless. Every species has a *closest living relative* in the genealogical sense. The point intended by such statements is that the form in question is sufficiently distinct that this relative cannot be recognized. For example, Holyoak (1978, p. 185) stated that "The two genera *Philepitta* and *Neodrepanis* are not very closely related, although there is ample evidence that they should be regarded as members of a single family." I would interpret this statement as meaning that the two genera are not very similar to each other, but that they nevertheless share a more recent common ancestor with each other than with any other birds. However, this interpretation rests on the concept of taxa as being genealogical units rather than phenetic units, and I do not know if Holyoak shares this view. Therefore his statement remains ambiguous because of the failure to distinguish between recency of common descent and similarity of form.

3.5.4. Evaluating the Strength of the Hypothesis

A phylogenetic hypothesis is a nested set of individual groupings, each of which is a separate hypothesis. Candor requires a frank expression of the confidence in which each node in the phylogeny is held. The cladistic method emphasizes this approach by specifying the characters that support each node and by making it difficult to hide character conflicts, suppositions of parallelism, and evolutionary character reversal. Ideally, these should be indicated graphically on the phylogenetic tree or cladogram illustrated. One may then discuss the reasons that one phylogeny has been preferred over other possibilities.

Besides considering individual nodes, it is desirable to consider the strength of the hypothesis as a whole. There are at least three approaches that may prove useful. First, is the hypothesis concordant with the results of other studies of the same group using different methods? Does pterylography support a phylogeny based on myology, for example? If different systems, which are not obviously coupled in a direct way, give similar results, then there is reason to suppose that both are independently tracing the same history. Bones and muscles would be less independent than bones and DNA. A second consideration is the plausibility of a phylogeny in terms of how we think evolution works. A hypothesis that required the repeated loss and reappearance of some complex structure would be suspect on this basis. Finally, as Crowson (1982) has pointed out, we may consider the credibility of hypothetical common ancestors as living organisms. Cracraft (1982) suggested that loons and grebes are sister groups because they share a number of derived characters, which other workers have considered to be convergent. Several of these characters are associated with the birds' foot-propelled diving habits. Loons have webbed feet and grebes have lobed feet, and the two groups use different methods of propulsion. This difference takes on significance when one tries to reconstruct the common ancestor of loons and grebes that Cracraft's phylogeny hypothesizes. Since the shared derived similarities are mostly associated with their locomotor habit, the sister-group hypothesis implies that the common ancestor was already a foot-propelled diver. Webbing and lobation are very different ways in which various birds have evolved aquatic specialization, and it is difficult to conceive the form that Cracraft's loon–grebe common ancestor would have; was it webbed or lobed? There is no reason to think that a webbed foot would evolve into a lobed one or vice versa, nor is it easy to imagine a foot structurally and functionally intermediate between the two. On this basis the hypothesis is thrown into question.

4. CONCLUSIONS

Systematic biology is in a period of ferment that is exciting for those taking part, but sometimes disturbing or confusing for biologists in other disciplines. In this paper I have attempted to interpret, especially for nonsystematists, some of the current debates about classification, while avoiding those that have not impinged much on ornithology and minimizing use of the endlessly growing terminology. As stated in the introduction, this has been a partisan approach rather than an impartial overview of the subject.

To simplify greatly, the matter in contention seems to boil down to this: relationships between species may be expressed in terms of overall similarity or of genealogy. Whatever one's kind of data or method of analysis and however clearly or uncertainly that methodology is explained or understood, one is ultimately trying to express either a pattern of resemblances or one of descent. Given this, it might be expected that there would be two taxonomic schools, but there are instead three. One is the phenetic school, which uses the methods of numerical taxonomy to construct phenograms or patterns of similarity in morphological characters or biochemical distances. This school, despite a number of important individual studies, has not been greatly active in ornithology and so I have not discussed it at length. The second school is that of cladistics, which aims at reconstructing the pattern of phylogeny. The third school, eclectic or traditional taxonomy, is the approach that has been used for many years in ornithology and on which our standard classifications are mainly based; for this reason I have given it much attention. In this school, classification is based on a combination of similarity and genealogy.

Shall we then classify by similarity, by genealogy, or by a combination of the two? To put it another way, do we want our classifications to be maximally predictive of resemblance or of descent? Inasmuch as the pattern of avian diversity has its origin in evolutionary history, I agree with most avian systematists that purely phenetic classification is not the road to follow and I will say no more about numerical taxonomy. The real debate is between the cladistic and eclectic schools.

A classification is a statement about a hierarchical arrangement of subordinated groups. If those groups are generated by a single method, then there is no ambiguity about their meaning. This is the advantage of cladistic classification: it tells us with certainty what an author has decided about the pattern of genealogical descent uniting the species included.

The eclectic system, in contrast, has attempted to include both relationships of similarity and of descent within a single classification. Because the two give different patterns, they cannot be combined except by some arbitrary and subjective intermixture of the two. As a result, eclectic classifications are ambiguous in that we never know what kind of relationship is being hypothesized. This approach has fostered in avian systematics an atmosphere of authoritarianism that is contrary to the scientific ideal of inquiry and debate.

What then is our prospect for obtaining in the near future a satisfactory classification of the birds? I believe that the prospect is good and that the desired goal will result from the application of cladistic classifying methods to the findings of both traditional (e.g., morphological) and new (e.g., DNA hybridization) data, and the fruitful comparison of their results. Cladistic classification is not without difficulties, however, and will utimately be subject to the test of time and usage. Who would dare to predict what the seventh edition of the A.O.U. checklist will look like? Not I.

ACKNOWLEDGMENTS. I am grateful to my friends and colleagues William Coffman, Joel Cracraft, Mary C. McKitrick, Kenneth C. Parkes, William

Searcy, Charles Sibley, and D. Scott Wood for reading and criticizing various versions of the manuscript. None of them liked the whole paper, but those who can be identified with a particular systematic school tended to dislike parts that adherents of other schools favored, and perhaps this may suggest that the effort, while partisan, was not excessively unbalanced. My research is supported by N.S.F. grant DEB-8010898.

REFERENCES

Amadon, D., 1950, The Hawaiian Honeycreepers (Aves, Drepaniidae), *Bull. Am. Mus. Nat. Hist.* **95(4)**:151–262.

American Ornithologists' Union, 1983, *Check-list of North American birds*, 6th edition.

Ashlock, P. D., 1971, Monophyly and associated terms, *Syst. Zool.* **20**:63–69.

Bock, W. J., 1960, The palatine process of the premaxilla in the passeres, *Bull. Mus. Comp. Zool.* **122(8)**:361–488.

Bock, W. J., 1973, Philosophical foundations of classical evolutionary classification, *Syst. Zool.* **22**:375–392.

Bock, W. J., 1982, Biological classification, in: *Synopsis and Classification of Living Organisms*, Volume 2 (S. P. Parker, ed.), McGraw-Hill, New York, pp. 1067–1071.

Bock, W. J., and Farrand, J., Jr., 1980, The number of species and genera of recent birds: A contribution to comparative systematics, *Am. Mus. Novitates* **2703**:1–29.

Cracraft, J., 1981, Toward a phylogenetic classification of the recent birds of the world (Class Aves), *Auk* **98**:681–714.

Cracraft, J., 1982, Phylogenetic relationships and monophyly of loons, grebes, and hesperornithiform birds, with comments on the early history of birds, *Syst. Zool.* **31**:35–56.

Cracraft, J., 1983, The significance of phylogenetic classifications for systematic and evolutionary biology, in: *Numerical Taxonomy*, NATO ASI Series, Volume G1 (J. Felsenstein, ed.), Springer-Verlag, Berlin, pp. 1–17.

Cracraft, J., and Eldredge, N. (eds.), 1979, *Phylogenetic Analysis and Paleontology*, Columbia University Press, New York.

Crowson, R. A., 1982, Computers versus imagination in the reconstruction of phylogeny, in: *Problems of Phylogenetic Reconstruction* (K. A. Joysey and A. E. Friday, eds.), Academic Press, London, pp. 245–255.

Eldredge, N., and Cracraft, J. (eds.), 1980, *Phylogenetic Patterns and the Evolutionary Process*, Columbia University Press, New York.

Holyoak, D. T., 1978, Asitys, in: *Bird Families of the World* (C. J. O. Harrison, ed.), Harry N. Abrams, New York, pp. 184–185.

Mayr, E., 1969, *Principles of Systematic Zoology*, McGraw-Hill, New York.

Mayr, E., 1981, Biological classification: Toward a synthesis of opposing methodologies, *Science* **214**:510–516.

Mayr, E., and Amadon, D., 1951, A classification of recent birds, *Am. Mus. Novitates* **1496**:1–42.

Morony, J. J., Jr., Bock, W. J., and Farrand, J., Jr., 1975, *Reference List of the Birds of the World*, American Museum of Natural History, New York.

Nelson, G. J., 1973, Classification as an expression of phylogenetic relationships, *Syst. Zool.* **22**:344–359.

Olson, S. L., 1981, The museum tradition in ornithology—A response to Ricklefs, *Auk* **98**:193–195.

Olson, S. L., 1983, Evidence for a polyphyletic origin of the Piciformes, Auk **100**:126–133.

Patterson, C., 1982, Cladistics and classification, New Scientist, **94**:303–306.

Peters, J. L., 1931, Check-list of Birds of the World, Harvard University Press and Museum of Comparative Zoology, Cambridge, Massachusetts.

Pizzey, G., 1980, A Field Guide to the Birds of Australia, Princeton University Press, Princeton, New Jersey.

Raikow, R. J., 1977, The origin and evolution of the Hawaiian Honeycreepers (Drepanididae), Living Bird **15**:95–117.

Raikow, R. J., 1978, Appendicular myology and relationships of the New World nine-primaried oscines (Aves: Passeriformes), Bull. Carnegie Mus. **7**:1–43.

Raikow, R. J., 1981, Old birds and new ideas: Progress and controversy in paleornithology, Wilson Bull. **93**:407–412.

Raikow, R. J., 1982, Monophyly of the Passeriformes: Test of a phylogenetic hypothesis, Auk **99**:431–445.

Raikow, R. J., and Cracraft, J., 1983, Monophyly of the Piciformes: A reply to Olson, Auk **100**:134–138.

Sibley, C. G., and Ahlquist, J. E., 1983, Phylogeny and classification of birds based on the data of DNA–DNA hybridization, Curr. Ornithol. **1**:245–292.

Simpson, G. G., 1961, Principles of Animal Taxonomy, Columbia University Press, New York.

Simpson, S. F., and Cracraft, J., 1981, The phylogenetic relationships of the Piciformes (Class Aves), Auk, **98**:481–494.

Sneath, P. H. A., and Sokal, R. R., 1973, Numerical Taxonomy, W. H. Freeman, San Francisco.

Storer, R. W., 1971, Classification of birds, in: Avian Biology, Volume 1, (D. S. Farner and J. R. King, eds.), Academic Press, New York, London, pp. 1–18.

Stresemann, E., 1959, The status of avian systematics and its unsolved problems, Auk **76**:269–280.

Swierczewski, E. V., and Raikow, R. J., 1981, Hind limb morphology, phylogeny, and classification of the Piciformes, Auk **98**:466–480.

Wetmore, A., 1960, A classification for the birds of the world, Smithsonian Misc. Coll. **139(11)**:1–37.

Wiley, E. O., 1981, Phylogenetics, John Wiley and Sons, New York.

Wolters, H. E., 1975–82, Die Vogelarten der Erde, Paul Parey, Hamburg.

Wood, D. S., 1983, Phenetic relationships within the Ciconiidae, (Aves), Ann. Carnegie Mus. **52(5)**:79–112.

CHAPTER 7

SYRINGEAL STRUCTURE AND AVIAN PHONATION

ABBOT S. GAUNT and SANDRA L. L. GAUNT

1. INTRODUCTION

Studies of syringeal function have historically been hampered by two difficulties, one technical and one perceptual. The technical difficulty is that because the syrinx is at the base of a long trachea and because its functioning is distorted if the surrounding interclavicular airsac is ruptured, direct observation of natural syringeal function has so far proved impossible. Hence, all analyses of syringeal function are based on indirect evidence. Such evidence may be obtained from dissections, manipulations of extracted syrinxes, models, analyses of physiological events associated with phonation, or analyses of the sounds produced.

From a perceptual standpoint, many physiological studies still suffer from a form of typology that would embarrass most ornithologists if used in any other context. Data gathered from experiments with one species are often generalized across the entire range of syringeal structures. "*The* syrinx" is a mythical, albeit popular, organ. One of the goals of this review is to indicate the range of variability in syringeal function.

1.1. Syringeal Structure

The structural differences among the syrinxes of, for example, chickens, parrots, doves, and songbirds, promote various techniques of

ABBOT S. GAUNT and SANDRA L. L. GAUNT • Department of Zoology, The Ohio State University, Columbus, Ohio 43210.

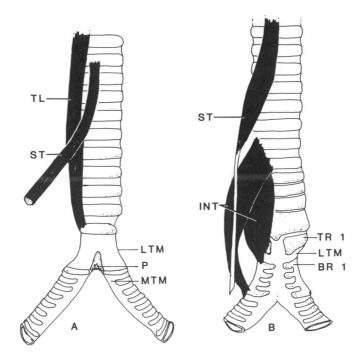

sound production. Even within a closely related group, such as parrots, different species may use different techniques. Differences may exist even between sexes, for example, the syrinxes of male and female Mallard Ducks (*Anas platyrhynchos*) act quite differently (Lockner and Youngren, 1976). Indeed, we have evidence (Gaunt and Gaunt, 1977) that individual chickens (*Gallus gallus*) use radically different techniques to produce similar sounds.

Traditionally, syrinxes are classified either by position (tracheal, bronchial, or tracheobronchial) or by the arrangement and number of associated muscles. In neither classification are the categories completely distinct. Further, both position and muscular arrangement have functional implications. A tracheal syrinx, such as a parrot's (Fig. 1B), has only one air passage, and that can be constricted by only a single set of membranes in its lateral walls. On the other hand, the tracheobronchial syrinx of oscines (Figs. 1C, 4) contains both lateral and medial membranes in each bronchus, and thereby forms a duplex organ. A dove's syrinx is intermediate, with medial membranes in each bronchus and lateral membranes in the trachea.

A syrinx may have only extrinsic muscles or both extrinsic and intrinsic musculature of varying composition. We consider as "extrinsic" those muscles that affect syringeal configuration by changing the position of the trachea and "intrinsic" those muscles that can affect

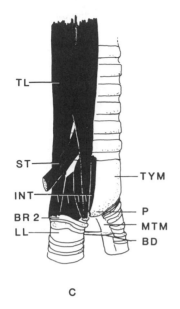

FIGURE 1. Ventral views of idealized syrinxes. Muscles are shown on right sides only. (A) A simple syrinx, which has only extrinsic muscles. This idealization is an amalgamation of several species. (B) Psittacid syrinx with one extrinsic muscle (ST, sternotrachealis) and two intrinsic muscles (INT; a lateral tracheobronchialis and a medial syringeus). This syrinx is basically tracheal, the other two are tracheobronchial. (C) Oscine syrinx with two extrinsic and several intrinsic muscles. Abbreviations: BD, bronchidesmus (a tissue of unknown function); BR, bronchial semiring (both bronchial and tracheal elements are numbered from the syrinx); LTM, lateral tympanic membrane; P, pessulus; TL, tracheolateralis muscle; TR, tracheal ring; TYM, tympanum (common to many birds, it consists of the first few, fused tracheal elements). Other abbreviations as in following figure.

the configuration of the syrinx directly, but even this distinction is not always clear. Nor is the number of muscles present always obvious. Many authors have applied names of every slip of tissue they could distinguish in a dead specimen, but how many of those function as independent entities is difficult to say. Ames (1971) provides an excellent review of structural variation in the passerine syrinx. The best general review of syringeal structure remains Beddard (1898).

The traditional classifications of syrinxes are widely known among ornithologists because they are reasonably easy to observe and occasionally have been used for taxonomic purposes. However, other variables, for example, the number, shape, and thickness of the flexible portions (membranes or labia), or the presence of skeletal inclusions in the membranes, are as important to syringeal function. Few of these attributes have been examined, and those only superficially. Perhaps a major reason for the lack of attention is that most of the information requires difficult and tedious histological studies. An excellent compilation of histological material is Warner's (1969)*doctoral thesis, some of which has been published (Warner, 1971, 1972 a,b).

Although a typical syrinx may be a myth, certain morphologic features are common to all syrinxes, and variants can be discussed with reference to a few generalized syringeal types (Fig. 1). First, all syrinxes

*Available in toto on microfilm for loan from Center for Research Libraries, Chicago.

are portions of the respiratory tract (Fig. 2). Thus, the basic source of power for phonation is the ventilatory musculature of the body wall.

Second, all syrinxes contain flexible portions that not only permit changes of shape, but also may act as sources of sound. An oddity of syrinxes vis-à-vis the vocal organs of other terrestrial vertebrates is that the flexible portions are so arranged that under conditions of high respiratory flow as may occur during excitement or heavy exercise, they will collapse into the lumen unless restrained. Indeed, a major role of the tracheal musculature appears to be maintaining the patency of the airway (Nottebohm, 1971; Youngren et al, 1974). Such controlling musculature, then, represents a third general component of syrinxes.

Finally, all syrinxes are enclosed within the interclavicular airsac and are, therefore, exposed directly to the pressures of that airsac on all except the dorsal side, which is usually in direct contact with, and frequently adherent to, the esophagus.

2. SOUND GENERATION

The generation of sound requires both a source of energy and a sound producing device. As the energy for vocal sounds is derived from

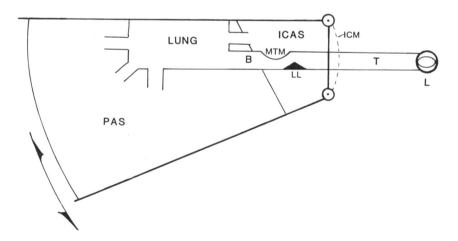

FIGURE 2. Model of an idealized, avian vocal system. The bird's body is depicted as a large bellows with three rigid sides: the dorsal spinal column and rib cage, the anterior coracoids, and the ventral sternum. Abbreviations: B, bronchus; ICAS, interclavicular airsac (including, in oscines, the fused anterior thoracic airsacs); ICM, interclavicular membrane; L, larynx; LL, lateral labium (replaced in many birds by a heavy, lateral tympanic membrane); MTM, medial tympanic membrane; PAS, posterior airsacs; T, trachea.

airflow in the ventilatory system, studies of pressures and flow in that system are vital to the understanding of phonation.

2.1. The Hérissant Effect

The earliest experiments in syringeal physiology were performed by Hérrissant (1753), who discovered that opening the interclavicular airsac (ICAS) to ambient pressure greatly impaired a bird's ability to call. Closing the airsac restored vocal ability. Rüppell (1933) undertook an experimental examination of this "Hérissant effect" by placing a gull's syrinx inside a glass chamber and passing air through the syrinx while varying pressure in the chamber. He determined that the surrounding pressure must exceed that in the syrinx before sound could be produced and proposed that high pressure in the chamber moved the syringeal membranes into the lumen. Unfortunately, this interpretation was flawed because Rüppell assumed, reasonably for that time, that valves in the ventilatory system directed airflow and that pressure in the airsacs could be maintained independent of the pressure in the windpipe. We now know that such valves do not exist (Fig. 2; Duncker, 1971). Therefore, under static conditions, pressures must be equal throughout the ventilatory system. However, a pressure differential across the syringeal membranes can develop under dynamic conditions when flow induces a Bernoulli effect that lowers the pressure on the inner tracheal walls.

Gross (1964a) performed similar experiments using chicken syrinxes and an apparatus that more accurately reflected avian anatomy. He also tested syrinxes set at different lengths, hence different membrane tensions. His experiments showed that, for a given membrane tension, higher pressures in the chamber require higher bronchial pressures to produce sound. As the syrinx was shortened and the membranes relaxed, sound production required lower pressures in both bronchus and chamber. The importance of relaxing the membranes was further illustrated by the fact that sounds can be elicited from the syrinxes of tinamous (Beebe, 1925), owls (Miller, 1934), geese (Paulsen, 1967), and ducks (Lockner and Youngren, 1976) in the absence of high internal pressures if the syringeal membranes are manually relaxed into the syringeal lumen. However, when a chicken's syrinx was maximally shortened, the pressure required to elicit sound increased 40-fold (Gross, 1964a). These findings indicate that sound production depends in part on a relationship between airflow through a syrinx and the pressure in the surrounding airsac, and that the relationship is modified by the degree of relaxation, or inward folding, of the membranes.

Some of our first experiments (Gaunt *et al*, 1973) examined pressure and flow in a songbird. When a Starling (*Sturnus vulgaris*) utters a distress call, pressure in the airsacs rises to 20 to 30 times ventilatory levels, but pressure in the trachea cranial to the syrinx remains well below ventilatory levels and scarcely exceeds atmospheric (Fig. 3). If a large bore cannula is inserted into a Starling's ICAS, the bird can call loudly as long as the external opening of the cannula is closed. When the cannula is opened, the bird is nearly silenced, and flow through the syrinx almost stops. In contrast, tracheal pressures in a crowing rooster are much higher than ventilatory pressures and fluctuate in parallel with those of the airsacs (Fig. 3; Gaunt *et al*, 1976; Brackenbury, 1977, 1978a). Taken together, these facts indicate that chickens and Starlings do something rather different when uttering loud calls. A Starling's syrinx imposes a higher resistance than that of a chicken. We hypothesized that opening the ICAS provides a low-resistance exhaust port, and Starlings are rendered silent because air bypasses the high-resistance syrinx. However, rupture of the ICAS silences adult chickens

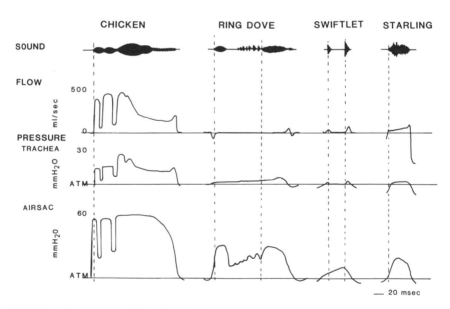

FIGURE 3. Pressure and flow during phonation in several species. All values to the same scale. Data for rooster's crow from Brackenbury (1978a) and Gaunt *et al*. (1976); for dove's coo, Gaunt *et al*. (1982); for Grey Swiftlet's clicks, Suthers and Hector (1982), and for Starling' distress call, Gaunt *et al*. (1973).

even if bypass of the syrinx is prevented by sealing the secondary bronchi from the respiratory system into the airsac (Youngren et al, 1974). If the ICAS in a chicken is opened to atmospheric pressure, high exhalatory pressure in the bronchi forces the membranes to balloon outward rather than move into the airstream (Gross, 1964a; Brackenbury, 1977; Gaunt and Gaunt, 1977).

Chicks and ducklings cannot be silenced simply by opening the ICAS (Gottlieb and Vandenberg, 1968). The effects of rupturing a duck's ICAS differ not only with a bird's age but also with its sex (Lockner and Murrish, 1975). Young birds and females show marked changes in vocal quality as well as intensity, whereas drakes show only some reduction in intensity. The reduced intensity can be attributed to a reduced flow through the syrinx. The qualitative changes in the ducklings are similar to maturational changes, which are due to thickening of the membranes (Abs, 1980). Thickening a membrane lowers its natural frequency of vibration. Reduced flow reduces the Bernoulli effect, thereby reducing transmural pressure and membrane tension (Gaunt and Wells, 1973), which in turn changes vocal quality in the adult females. Thus, the Hérissant effect is but one example of an apparently single phenomenon that can involve several mechanisms.

We have coined the term "passive closure" (Gaunt et al, 1973) to describe most of the above situations, in which constriction of the syringeal lumen depends solely on the inward movement, sometimes aided by a Bernoulli effect, of relaxed membranes. Passive closure appears to be a technique used by all syrinxes possessed of only extrinsic muscles and may occasionally occur in more complex syrinxes (Fig. 4A). However, the diameter of the lumen of syrinxes with intrinsic muscles can be directly constricted by rotating a skeletal element and associated tissues into the lumen (Fig. 4B). Such "active closure" appears to be important for parrots and songbirds. The lateral borders of the caudal end of a psittacine syrinx contain a series of fused bronchial elements that form a pair of rigid bars (Fig. 1B). Contraction of the tracheobronchialis muscles rotates the bars into the lumen. The cranial ends of the bronchial bars are attached to the last tracheal elements by thick syringeal membranes that also serve in part as the tendons of the syringeus muscles. The syringeal membranes can be thought of as springs, the tension of which can be varied by the contraction of the syringeus. By varying the degrees of contraction of the syringeus and trachiobronchialis muscles, a parrot can precisely control the position of the anterior tips of the bronchial bars, thereby controlling both the diameter of the lumen and the tension of the supporting membrane (Nottebohm,

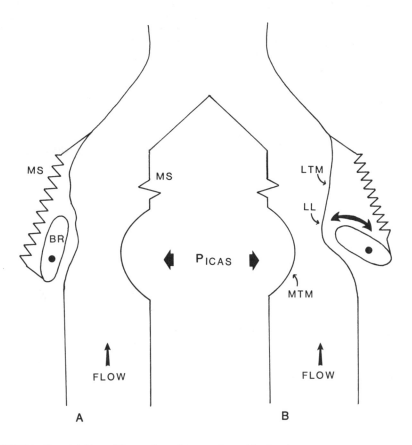

FIGURE 4. Constriction of the syringeal lumen in an idealized, oscine syrinx. (A) Passive Closure, in which constriction arises from the inward movement of a membrane. (B) Active Closure, in which constriction depends, at least partially, from the rotation of the lateral labium. Abbreviations: MS, syringeal muscles; P_{ICAS}, pressure in the interclavicular airsac. Other abbreviations as in preceding figures. Redrawn from Gaunt and Wells, 1973.

1976; Gaunt and Gaunt, unpublished). In oscines, the rotating elements are the lateral labia (Figs. 1C, 4B), which are controlled by several intrinsic muscles (Chamberlain *et al*, 1968; Gaunt *et al.*, 1973).

2.2. Air Consumption versus Intensity and Duration of Sound

Gross changes in the loudness of a call depend on the driving pressure. Just as humans, birds must blow harder in order to call more loudly. Airflow is high during most chicken calls. Roosters appear to use more than 90% of their ventilatory capacity during crowing (Brackenbury, 1978a), but a Starling's distress call moves about the same volume of air as normal respiration (Gaunt et al., 1973). On the other hand, some calls of some roosters do not use much air (Gaunt et al., 1976). The proper distinction may not be between passive and active models of closure, but between high and low resistance syrinxes. The resistance of a passively constricted syrinx depends on the degree to which the airway is occluded. If the lumen is only partially occluded, then the imposed resistance is low and flow can be high. However, if the syrinx is so completely contracted that the infolded membranes press against each other or against the opposite wall, then resistance will be high and flow reduced, which could explain the sudden rise in pressure necessary to activate a chicken's syrinx at full contraction (Gross, 1964a).

It is not possible to predict which syringeal configuration a bird may use to produce a given call. An artificially manipulated rooster's syrinx can produce a wailing sound in either maximally contracted or stretched configurations (Fig. 5). Records from different roosters pro-

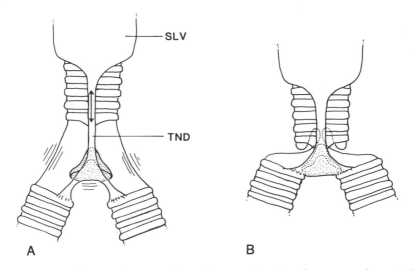

FIGURE 5. Dorsal view of two position of a rooster's syrinx that can produce similar sounds. (A) "Stretched" configuration in which the syrinx is drawn relatively taught. Tension from the dorsal tendon draws the pessulus craniad, thereby slightly relaxing the membranes and allowing them to interact with the airflow. (B) Fully contracted configuration in which the lateral membranes are folded into the lumen and form a pulse generator. Abbreviations: SLV, syringeal sleeve; TND, tenden.

ducing such sounds showed that tracheal pressures are either low or high. In the latter case, activity was absent in at least one muscle responsible for tracheal shortening (Gaunt et al., 1976; Gaunt and Gaunt, 1977), which fits well with an hypothesis that the syrinx was not fully contracted and resistance was low. Similarly, the different techniques of the syrinxes of hen and drake Mallards involve considerable differences in resistance to airflow (Lockner and Murrish, 1975).

The duration of a call or song depends on the rate at which a bird's store of air is consumed. Some birds can produce apparently uninterrupted sound for long periods, for example, the Grasshopper Warbler (Locustella naevia) can sing continuously for at least 57 sec (Brackenbury, 1978b), which represents a prodigious feat given an air supply of less than 10 ml. How such feats are accomplished is one of the few aspects of avian phonation about which investigators seriously disagree. A high-resistance syrinx that extracts a maximum amount of sound from a minimum airflow is one possible solution. An alternative solution proposed by Calder (1970) is that birds engage in partial inhalations, or "minibreaths" between notes, including the individual notes of trills. We have long disagreed with Calder's interpretation of his data. In order for any inhalation to occur, airflow must reverse direction, and for that to happen the internal pressure must drop below atmospheric. Calder obtained his data from an impedance pneumograph, which measures changes in electrical impedance of an animal's body as its volume changes during ventilation. However, impedance also changes with every shift in the animal's posture. Calder's data clearly show changes of thoracic impedance that synchronize with the notes of a trill. Because a pneumograph provides no reference to atmospheric pressure, we cannot determine whether the changes in impedance signify oscillations in volume sufficient to reverse the flow. In two other passerines in which flow was measured directly, a Starling (Gaunt et al., 1973) and an Evening Grosbeak (Hesperiphona vespertina; Berger and Hart, 1968), the flow during trilling sounds consists of a series of pulses of air, but flow does not reverse. We have postulated that such trilling sounds depend on a flow pattern that is constant in direction but pulsatile in velocity, and that this flow would be engendered by rapid pulsations of the body wall (Gaunt et al., 1976). That hypothesis was supported by studies of the tremolo in a dove's coo (Gaunt et al., 1982) and the fear trills of chicks (Phillips and Youngren, 1981). The initial interpretation of the chick's fear trill was based on only electromyographic data, but we have cannulated a chick's airsacs and confirmed that pressure remains well above atmospheric.

A possible expansion of the minibreath model is that sound is not produced continuously, but is interrupted by rapid but normal inhalations. A bird the size of a Starling can achieve full inhalation by normal ventilation in something less than 0.1 sec (Gaunt et al., 1973); small birds may require even less time. Moreover, if ventilation takes the form of panting, then the shallow inhalations can be very rapid, perhaps sufficiently rapid to render detection difficult. The songs of Skylarks (Alauda arvensis) and Grasshopper Warblers may incorporate rapid inhalations, perhaps coupled to the resonant frequency of the ventilatory system. A Skylark's song is broken into phrases by complete inhalations. The warbler's song consists of a series of double pulses that can continue uninterrupted for nearly a minute. Each double pulse may represent the exhalation portion of a minibreath cycle and the interval between the double pulses the inhalatory portion, with the cycles resembling panting. The break between the pulses within an exhalatory phase is presumed to be caused by activity of the syringeal muscles (Brackenbury, 1978b,c).

Another mechanism for prolonging the continuity of a song is for phonation to continue during normal inhalation. Continuous phonation during inhalation has been shown for the European Nightjar (Caprimulgus europaeus; Hunter, 1980), which may sing continuously for as long as eight minutes. The chickens, doves, and parrots that we have studied all produce occasional sounds during inhalation. The incidence of such "accidents" increases with excitement. Clearly, any behavior that occurs consistently in a specific situation can be selected as a signal of that situation. Thus, "accidents" can be preserved and incorporated as a portion of the repertoire.

2.3. Sound Generators

Three kinds of sound generators are commonly used by tetrapods (Fig. 6). The first two of these are mechanical, involving the oscillation of a flexible portion of the vocal tract in an airstream, the third is aerodynamic, that is, a whistle. Numerous investigations implicate the flexible components of a syrinx as mechanical oscillators, but which component varies with author and species. For instance, Miskimen (1951) attributed sound production in Starlings to the medial tympanic membranes, but Chamberlain et al., 1968)designated the lateral tympanic membranes of a Crow (Corvus brachyrhynchos) as the source of sound. Klatt and Stefanski (1974) assigned the lateral labium a primary sound-generating role in the Indian Hill Mynah (Gracula religiosa, but a modulating role in the Chaffinch (Fringilla coelebs). If the membranes are so injured that they lose flexibility, phonation is no longer possible

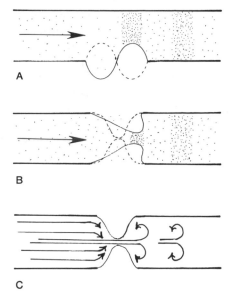

FIGURE 6. Models of sound generators. (A) The "compression" model. In the simplest case, oscillation is normal to airflow, but waves or "ripples" travelling in the direction of flow, as shown here, are likely. These waves are presumed to be responsible for the generation of harmonics (see text). In the extreme case, they can become: (B) a pulse generator. The version depicted here consists of two membranes that act in the manner of human vocal folds. Alternatively, the valve could consist of a single membrane or labium that extended completely across the lumen. (C) A whistle, in which a more-or-less rigid constriction of the airway forms a slot. Again, the system could comprise a single, mobile element brought against the opposite wall.

(Gross, 1964b, 1979). Such results are evidence for the necessity of flexilibity, but do not determine whether that flexibility is to permit oscillation or simply to reconfigure the syrinx.

The first kind of generator is that described by Greenewalt (1968). As the membrane oscillates, its movements toward and away from the opposing syringeal wall alternately compress and rarify the intervening air (Fig. 6A). In the simplest case, the membrane oscillates normal to the airflow. However, ripples or waves traveling in the direction of flow are likely to be induced in the surface of the membrane (Greenewalt, 1968; Dürrwang, 1974, Klatt and Stefanski, 1974).

A second sound generator involves the occlusion of the air passageways by the oscillator (Fig. 6B). In this system, the flexible portions of a syrinx form a valve. Pressure builds on the lung side of the valve until it is opened and a burst of air escapes, releasing the pressure. The Bernoulli force generated by flowing air, elastic-restoring forces in the tissues of the membranes, and in some cases muscular contraction restore the valve to its original position. The system generates a series of pulses of air that are responsible for a sound with a rich frequency spectrum and a fundamental frequency equal to the rate of pulsation. This technique is that of most mammals and amphibians. Many early workers, for example, Rüppell (1933), attributed avian phonation to such a pulse generator, but that mechanism has been largely discounted in recent considerations of syringeal function in favor of the compression–rarification model. However, we judge that pulse generators are fairly common among birds. Paulsen (1967) has provided high-speed cine pictures of the syringeal membranes of a Domestic Goose (*Anser anser*) behaving in a manner appropriate for a pulse generator, and Beebe's (1925) observations of the syrinx of a Variegated Tinamou (*Crypturus variagatus*) also suggest a similar motion. At least one of the techniques that we suspect chickens use to wail involves such a mechanism (Fig. 5; Gaunt *et al.*, 1976), and our recent studies of the action of the psittacine syrinx suggest that pulse generation must be one of the mechanisms used by parrots (Gaunt and Gaunt, unpublished). A mynah's external labium may oscillate similar to human vocal cords (Klatt and Stefanski, 1974).

The third kind of generator is a whistle that requires no mechanical oscillator, but depends on the periodic formation of vortices (Fig. 6C). Whistled sounds are produced in a variety of ways, but one common technique is a "hole-tone" that develops when a constriction to a column of flowing air induces shearing forces that generate vortices (Chanaud, 1970). A slot can be formed by any syrinx that can be actively closed and perhaps as well by many syrinxes with heavy membranes that could be shaped as we shape our lips (Gaunt *et al.*, 1982). A slot would also be formed by closely opposed, folded membranes in which the folded edges have been drawn extremely taut. Indeed, this type of whistled sound can be obtained from freshly dissected human larynges that are mechanically manipulated, and even some living humans can produce "glottal whistles" (van den Berg, 1968). Whistles are capable of producing sounds composed primarily of a single frequency with little or no overtone structure. Sounds devoid of overtones cannot be produced by freely vibrating membranes, but are found in many bird calls.

These three types of generators are not mutually exclusive and, indeed, tend to blend into one another. A compression system generates pulses whenever it sufficiently occludes the lumen, but forms a whistling slot if its internal margin becomes sufficiently rigid. A single species could use all three mechanisms at different times.

3. SOUND MODULATION

A sound is a wave that can be characterized in terms of three basic parameters: frequency, for which the subjective impression is pitch; amplitude, for which the subjective measurement is loudness; and phase, which is generally ignored in analysis. Frequency modulation (FM) is easily visualized as changes along the vertical axis of a sonogram. Amplitude modulation (AM) is more easily seen on an oscilloscope as fluctuations in the wave envelope. However, the wave envelope can be modified by "pseudomodulations" that arise from the interactions of more than one tone, for example, beat frequencies (Greenewalt, 1968; Stein, 1968).

Early workers assumed that because the avian system was wind powered, it could be analogized to other wind-powered systems such as musical instruments or the human vocal system, for which the techniques of modulation were better understood. In wind instruments, a sound generator is coupled to a resonating tube the effective length of which can be varied to control the pitch of the elicited sound. The human voice is modulated by coupling a sound generator in the larynx to a series of chambers in the upper respiratory tract. Those chambers act as resonant filters to emphasize some frequencies and diminish others. The interactions of different combinations of emphasized or "formant" frequencies produce the different vowel sounds. The resonant properties of chambers can be changed by movements of the tongue, lips, soft palate, and so forth, thus producing different combinations of formant frequencies (Lieberman, 1975).

A clear problem with an analogy to the human vocal system is that birds do not have such complex anterior chambers in the vocal tract, and, except for parrots, do not have a fleshy, mobile tongue. Birds do, however, have a trachea that, at least in theory, could be lengthened and shortened by action of the extrinsic muscles.

3.1. The Two-Voice Model

In the late 1940s, a number of relatively inexpensive, electronic devices for visualizing sound, chief among them the oscillograph and the sound spectrograph, became available. These instruments provided an entirely new way to explore syringeal function, for it became possible to "dissect" a sound, then extrapolate back to the kinds of physical phenomena that could produce that sound. The first published sonographic examination of a bird call provided evidence that modulation of avian phonations was radically different from that of human speech. Potter et al. (1947) noted that the song of a Brown Thrasher (Toxistoma rufum) contained two harmonically unrelated tones. Similar reports soon appeared in the general ornithological literature (e.g., Borror and Reese, 1956). Greenewalt's (1968) investigations showed that two harmonically unrelated and independently modulated tones are present in many avian calls. Because a single resonator such as a trachea cannot simultaneously effect more than one modulatory pattern, Greenewalt hypothesized that the sound sources of the syrinx are not strongly coupled to tracheal resonance, but are source modulated. Thus, the two-voice phenomenon is a product of a duplex syrinx in which each side produces and modulates its sounds independently.

Stein (1968) compared the avian vocal system with electronic devices such as radios rather than with wind instruments. He showed that many of the characteristics of avian phonations could be explained as the products of a carrier wave subjected to a modulating frequency. He also noted that independent use of the two sides of a syrinx could explain the two-voice phenomenon.

If the two-voice model is to be accepted, then three things must be shown. First, we need confirming evidence that tracheal resonance does not play a prominent role in the modulation of avian sounds. Second, we must show that the two sides of a duplex syrinx can be independently controlled, and third, that those controls are capable of producing the modulations necessary to produce the kinds of sounds birds make.

One approach to the first question is to examine the relationship between tracheal length and the fundamental frequency of the broadcast sound. Several investigators have tried a direct approach to this question. Rüppell (1933) found that the pitch of a crane's vocal system rose as he shortened the trachea. Myers (1917) shortened and externalized the trachea of a living rooster, with a resulting increase in pitch of the call.

A second approach to the same question is to modify the resonating properties of a system without changing its anatomy. For instance, the density of the enclosed medium affects a system's resonant properties. Thus, a human breathing a helium/oxygen mixture, which has a lower density than normal air, has a distinctly higher pitched voice. Hersh (1966) tested several species of birds in He/O_2. Birds with whistled songs showed little or no coupling of vocal frequencies to tracheal resonance, but some birds with a richer frequency spectrum showed at least weak coupling. No bird showed the strong coupling associated with wind instruments or human vocal systems.

We have recently applied both of these approaches to the same group of birds. In a few species of cranes, the trachea is of normal shape and length, but in others it is elongated and complexly coiled within the sternum. We subjected birds of both tracheal types to He/O_2 and to surgical alteration of tracheal length. Preliminary results are inconclusive but nevertheless intriguing. Regardless of tracheal type, He/O_2 raises the pitch of calls that have distinct overtones. The shift of pitch is obvious in sound spectrographs, but is not auditorially as dramatic as that in a human voice. Calls in which the frequencies are broadly distributed but not grouped into overtones are seemingly less affected. Shortening the trachea distinctly raises the pitch of the calls of a bird with a short, straight trachea. Shortening a coiled trachea seems to have little effect on pitch, but produces a severe reduction in sound intensity.

Indirect evidence concerning the role of tracheal resonance is also ambiguous. The calls of many birds show enormous discrepancies between the actual pitch and that predicted from tracheal length. Further, forbidden frequencies, that is, frequencies that would be diminished by a resonating filter, are absent (Greenewalt, 1968). During their development, the young of many birds suddenly shift from high-pitched peeping sounds to the adult calls of lower frequency. This "breaking of the voice" is associated with a hormonally induced change in membrane thickness, not tracheal elongation, and can be triggered in even small chicks by application of testosterone (Abs, 1970, 1980). These data speak against strong tracheal coupling. However, in some instances, the syrinx may be sufficiently coupled to tracheal resonance to influence, but not to determine, a bird's frequency range. The calls of Ross' Goose (*Chen rossii*) are similar to those of the larger Snow Goose (*Chen caerulecens*), but are higher pitched, as would be expected for a shorter trachea (Sutherland and McChesney, 1965). A similar relationship of pitch to size obtains among North American gulls (Russe and Fairchild, unpublished data).

Although the role of tracheal resonance remains ambiguous, the independence of the two sides of the syrinx has been firmly established. A Chaffinch's song shows the two-voice phenomenon in the form of overlapping, independently modulated notes. As part of a series of studies on vocal learning in Chaffinchs, Nottebohm (1971) severed the branches of the hypoglossal nerves that control the syrinx. Severing the nerve on one side of an adult bird caused an irreversible loss of a portion of the song, and one note of any pair of overlapping notes was always lost. Damage to the left side produced the more drastic effect. If the operation was performed on a young bird that had not yet learned its song, that bird learned to sing an entire song from the intact side of the syrinx. However, temporal relationships in the song were adjusted to eliminate overlapping notes. Thus, either side of the syrinx has the mechanical ability to produce all of the modulations necessary for a song, but both sides are required to produce two independent notes simultaneously.

Several authors (e.g., Borror and Reese, 1956; Thorpe, 1961; Miller, 1977) have reported instances of three to four voices in birds. These reports are usually based on the presence of more than two nonharmonically related tones in a sound spectrogram. A possible explanation for the source of such tones is the formation of formant frequencies (see Section 3, p. 226). That explanation requires a primary source and one or more responding resonators. The system works best if the source produces a rich frequency spectrum. In a mynah bird imitating human speech, the lateral labia could serve as pulse generators responsible for the first formant and the medial tympanic membranes as resonators responsible for additional formants (Klatt and Stefanski, 1974). Because one of the oscillators is driven by the activity of another, formants are not independent voices. To demonstrate the presence of multiple voices, one must show either that each of the voices can be independently modulated (à la Greenewalt and Stein) or, preferably, that it is possible to eliminate voices without disturbing the others (à la Nottebohm).

3.2. Linked Modulations

An interesting feature of the modulation of avian phonations is that AM and FM are often linked (Greenewalt, 1968). In the lower reaches of a bird's vocal range, amplitude and frequency increase together, but at some point the relationship reverses and as frequency further increases the amplitude drops. Greenewalt supposed that the effect was due to a combination of the position of the membrane and its tension. If a relaxed membrane, which produces a low frequency, extends sufficiently into the lumen, its vibrations are constrained by proximity of the opposing wall and amplitude is restricted. If the membrane's tension is increased, it withdraws from the opposing wall and the amplitude of its vibrations can increase. As tension continues to increase, the membrane eventually stiffens and the amplitude of its oscillations again decreases. Greenewalt further proposed that for those cases in which the AM–FM linkage is absent, the space available for oscillation might be controlled independently by rotating the lateral labium into the lumen opposite the membranes.

Although ingenious, this particular solution has certain difficulties. Among other things, the model works best if the tension of the membrane and the diameter of the lumen are linked and if the pressure in the ICAS, airflow, and tension are independent. In fact, these relationships do not necessarily hold and in many cases appear to be exactly the opposite.

Gaunt and Wells (1973) contended that the linkage between AM and FM is more easily explained as a classic case of resonant coupling.

The syringeal membranes act as resonators responding to periodic perturbations of the flowing air column. Given a perturbation of constantly increasing frequency, the amplitude of the elicited sound increases, then decreases as the frequency of the perturbation approaches and passes through the natural frequency of the membrane.

3.3. Modulation and Airflow

Although even subtle changes of amplitude can be effected by varying the flow rate, some AM does not involve changes of flow rate. Therefore, a source for these modulations must be sought elsewhere. Stein (1968) proposed that the lateral labium might be used as an amplitude modulator. He suggested that it could be inserted into the syringeal lumen where it would oscillate in the same airflow that stimulates the medial tympaniform membranes. Its oscillation interacts with a carrier frequency generated by the membranes to produce an amplitude-modulated sound.

We have suggested that some of the AM in the calls of Streptopelia could be explained if the infolding lateral membranes form a flap valve (Gaunt et al, 1982). Whistled tones are produced at slots opposite the medial membranes. When the tracheal valve is open, sound passes freely and when the valve is closed, sound is muffled.

Klatt and Stefanski (1974) reversed the roles postulated by Stein for the lateral labium and medial membranes. We, too, were able to construct a plausible model for the action of a dove's syrinx with the roles of the lateral and medial membranes reversed. A whistled tone formed between the infolded lateral membranes could be amplitude modulated if oscillations of the medial membranes sufficiently altered airflow. However, we reasoned, as did Stein, that the slow modulating frequency was better attributed to oscillations of a relatively massive structure such as labia or lateral membranes. Hence, we chose the model in which the lateral membranes serve as modulators.

The use of such acoustic mufflers appears to be widespread. The aerial display call of a Dunlin (Caladris alpina) is buzzy, whereas those of some other caladrine sandpipers are not (Kroodsma and Miller, 1982). The buzziness arises from a repetitive AM (Miller, 1983). A Dunlin's syrinx is notable for the fact that the lateral labium forms a large bronchial valve that extends almost half-way across the lumen. The valve is not present in two other sandpipers, including the Knot (Caladris canuta; Warner, 1969).

In doves, vibration of the lateral membranes by the airstream is sufficient to effect a modulation. Beyond setting the configuration of the trachea, no further activity of the syringeal muscles is necessary. We presume that a Dunlin's syrinx behaves in a similar manner. However, in other species the intervention of the lateral labium or similar structures may involve the active participation of syringeal muscles (see Section 3.5).

Brackenbury (1978b) suggested that rapid pulses in the calls of the Grasshopper Warbler and Sedge Warbler (*Acrocephalus schoenobaenus*) are produced by alternate contractions of the extrinsic muscles. He further suggested that higher frequency pulses within the song are derived from the activity of intrinsic muscles and that the tempo, rhythm, and duration of the pulses are controlled by the abdominal muscles. Although we have no electromyographic data from oscine syringeal muscles, this rather intriguing model, with its division of labor, fits the general pattern of electromyographic findings.

On an *a priori* basis, FM appears to be easier to explain than AM, because an obvious way to modulate the frequency of a membrane oscillator is to adjust the membrane's tension by changing the configuration of the syrinx through the action of syringeal muscles. However, this hypothesis is largely without experimental verification, perhaps because the animals that have so far been chosen for experiments do not show significant FM in their calls. We have one set of records that may show a case in which FM depends on the activity of intrinsic muscles. One of the calls of the Peach-faced Lovebird (*Agapornis rosiecollis*) is a disyllabic "chee-yeep," in which the second portion is distinctly higher in pitch. Activity of intrinsic muscles during this call shows two peaks, the first occurring just before the onset of sound production, the second as the pitch shifts (Gaunt and Gaunt, 1982, unpublished). However, additional work is necessary to determine whether the fundamental frequency of the call changes, as would be expected if membrane tension changes, or whether sound energy simply is shifted into higher overtones.

Other studies of techniques for FM have produced a consistent, but curious result. Beebe (1925) found that the pitch of sound elicited from a tinamou's syrinx was highest when the syrinx was fully contracted with the membranes pressed together and decreased as the membranes were drawn apart. But drawing the membranes apart stretches them and increases tension. Thus, Beebe's observations are exactly the opposite of what we would expect if the tension of the membranes controls pitch. Hersh (1966) also reported that tensing the tympaniform membranes reduced the frequency of a bantam chicken's calls, but Gross (1964a) found the opposite. However, a rooster's syrinx evidently can make similar sounds in either contracted or extended configurations (Gaunt and Gaunt, 1977).

The observations of Beebe and Hersh can be explained in two ways. If the source of the sound is aerodynamic rather than a membrane vibration, then the larger the aperture between the membranes, the slower the flow and the lower the pitch. If the membranes act as a pulse generator, then the tighter they are pressed together, the higher the resulting pitch. Further, the oscillatory frequencies of a pulse generator are sensitive to the driving pressure (Klatt and Stefanski, 1974; Liberman, 1975).

Phillips and Youngren (1981) found that denervation of the sternotrachealis muscles in chicks eliminated low-intensity calls and reduced the high-frequency components of other calls. Contraction of the sternotrachealis relaxes the membranes and allows them to move into the lumen. If the muscle is inactive, then the low flow rates may not be sufficient to stimulate the stiffened membrane. Again, however, we are faced with the counterintuitive fact that maintaining the membranes in a tensed condition eliminates high-frequency components.

Suthers and Hector (1982) surgically removed both pairs of extrinsic muscles from Grey Swiftlets (*Collacalia spodiopygia*). The swiftlets then produced calls of somewhat longer duration than usual showing distinct FM. Frequency rose, then fell so that a sonographic figure of the call was an inverted "U". This situation makes a little more sense. In the absence of muscle contraction, the tension of the membrane depends solely on the transmural pressure differential, which increases and decreases with flow. Thus, frequency increases during the initial part of the call, then drops as the air supply is exhausted or is terminated. In this case, membrane tension does seem to be important, but again flow rate plays a key role in the modulation of frequency.

The FM of a whistle depends to some extent on the kind of whistle. The frequency of many whistles increases linearly with the flow rate. Since rate depends on resistance, there is a potential linkage to the size of the slot with the pitch decreasing as the size of the opening increases. Some whistles tend to show relatively small increases in frequency until flow rate reaches a threshold at which the tone jumps markedly and again becomes reasonably constant (Chanaud, 1970; Wilson et al., 1971). This phenomenon has been noted in a number of avian calls, particularly those of both real and modeled dove syrinxes (Abs, 1980).

3.4. Modulation and Muscle Action

Modulation at source requires the ability to control and change syringeal configuration, presumably by using the syringeal muscles.

Several authors (Brockway, 1967; Gross, 1964b; Miskimen, 1951; Smith, 1977; Phillips and Youngren, 1981) have attempted to discern the functions of extrinsic muscles by severing or denervating them. Such procedures generally may produce aberrant vocalizations, but rarely silence birds. Indeed, Gross found that the only way to effect long-term silencing of chickens or peafowls (*Pavo* sp.) was to destroy the lateral membranes. Suthers and Hector (1982) determined that denervating the sternotrachealis did not silence Grey Swiftlets but removing the muscle did. They concluded that an "intact, if flaccid sternotrachealis" is necessary for click production. Chaffinches denervated by Nottebohm (1971) experienced severe respiratory difficulties when breathing rapidly. These results suggest that in general the extrinsic muscles, at least, are used to control the position of the trachea. If that position is not adequatly controlled, the resulting sounds are abnormal. However, chickens, which use very high rates of airflow, retain the ability to cluck even if the syrinx is denervated (Phillips *et al.*, 1972). If the membranes are sufficiently flaccid, high airflow may draw them into the lumen, where they can make a sound or, on inhalation, restrict ventilation.

Ablation and denervation experiments suffer from two major drawbacks. They permanently disable an animal and they attempt to determine normal behavior by inducing the abnormal. Both of these problems can be reduced and often completely eliminated by the use of electromyography.

Electromyography (EMG) is a technique based on the fact that contracting muscles produce an electronic discharge that can be detected by implanting fine wire electrodes (0.02 to 0.05 mm OD) in the muscle and connecting them to proper amplification and display instruments. The kinds of electrodes used in syringeal studies provide information about the timing and relative strength of contractions, but do not permit quantitative comparisons of the force of contraction in different muscles. Much of our work, therefore, is devoted to determining the timing of muscular activity and whether or not the electromyograms (also called EMGs) form repeatable patterns that can be correlated with features of the sounds investigated. The technique is extremely useful for investigating obligatory, repeated behaviors (e.g., breathing or feeding) that can provide repeated sets of data, but it encounters some difficulties when dealing with variable activities such as phonation. We record EMGs and sound on a multichannel tape recorder so that events can be replayed for detailed analysis.

Three groups of investigators have used two different approaches to EMG for syringeal analysis. Phillips, Youngren, and their co-workers have coupled EMG with electrical stimulation of the brain (ESB) (Youngren *et al.*, 1974; Peek *et al.*, 1975; Lockner and Youngren, 1976) of anesthetized birds. This technique has distinct advantages. The EMG electrodes can be precisely placed and even moved during the process of the experiment, thereby permitting many tests in the course of a single operation. Because ESB can elicit specific calls, each experiment can include many trials and replication of experiments is simplified. On the other hand, the full repertoire of the animal is not displayed. Birds tested in this manner are usually positioned on their backs, a position of convenience for the experimenter but one in which birds do not normally vocalize and in which the respiratory system may behave in a somewhat aberrant manner. Further, most general anesthetics induce slow, deep ventilations, the effects of which are probably trivial during phonation but do raise some questions about the role of syringeal musculature in respiration. Investigators using this technique have all reported respiratory-related activity in the extrinsic muscles, but we find such activity only during rapid or excited respiration.

We, and Suthers and Hector, prefer to study *ad libitum* phonations by bringing the implanted electrodes under the skin to a backpack from where they can be connected to display instruments when desired. The ICAS is resealed and the bird allowed to recover. This technique eliminates most of the charges of artificiality, but pays a heavy price in repeatability. The investigator is essentially at the mercy of an animal and can only urge its cooperation. Thus, most studies of *ad libitum* calls have been done with birds that have limited vocabularies and predictable behavior.

Even the earliest EMG tests showed that simple and obvious solutions are not always correct. The syrinx of a chicken appears to be controlled by two pairs of muscles. According to classic notions (e.g., Myers, 1917), the sternotrachealis ("on" muscle) draws the trachea caudad, thus relaxing the vocal membranes and activating the syrinx; the tracheolateralis ("off" muscle) restores the system to the nonvocalizing configuration (Fig. 5). These muscles, however, contract simultaneously during phonation (Fig. 7). Moreover, the tracheolateralis of a crowing rooster activates first and remains active throughout the call, although it tends to reduce its activity as activity in the sternotrachealis peaks. Further, a third tracheal muscle, the cleidotrachealis (also known as cleidohyoideus, sternohyoideus, ypsilotrachealis, tracheohyoideus) is also active during phonation. The caudal end of the tracheolateralis and the tracheal insertion of the sternotrachealis are surrounded by a sleeve of connective tissue that extends caudad on the dorsal surface of the syrinx as a tendon to the pessulus (Figs. 5, 8). When the tracheolateralis contracts, it prevents caudal motion of the trachea. Because the contracting sternotrachealis cannot move the tra-

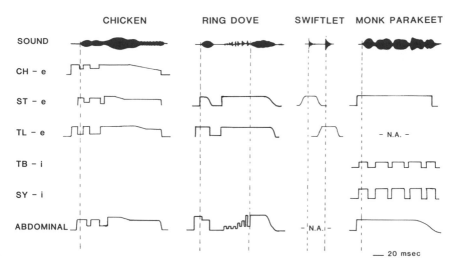

FIGURE 7. Summary of electromyographic activity versus sound in several species. Data for chicken, Gaunt and Gaunt, (1977); for dove, Gaunt et al. (1982); for Swiftlet, Suthers and Hector (1982). CH, cleidohyoideus muscle; SY, syringeus muscle; TB, tracheobronchialis muscle. Abbreviations for extrinsic muscles followed by "e", for intrinsic by "i". Other abbreviations as in preceding figures.

chea, it presses laterad against the sleeve, thereby pulling on the tendon and moving the pessulus craniad, which allows the membranes to vibrate. This system seems to be used during repetitive clucking and the "stretched" version of the wail. Whether or not a similar method is used in crowing is not certain, but crowing definitely depends on a balancing of the forces of the two extrinsic muscles, which do not act as simple antagonists in an on/off fashion (Gaunt and Gaunt, 1977; Peek et al, 1975; Youngren et al, 1974).

The simultaneous action of both extrinsic muscles is also seen during the phonation of doves (Gaunt et al, 1982). In ducks and swiftlets, on the other hand, the two extrinsic muscles behave in a manner more consonant with the notion of on/off antagonists (Lockner and Youngren, 1976; Suthers and Hector, 1982).

The role of the cleidotrachealis remains mysterious. Its action clearly draws the trachea caudad. It may serve as a buffer that isolates the trachea from other motions of the neck (Gaunt and Gaunt, 1977), or it may help to disseminate sound by enlarging the pharynx (White, 1968).

Suthers and Hector (1982) have combined EMG with pressure and flow data in an analysis of phonations of the Grey Swiftlet (Figs. 3, 7). The Swiftlet's echo-locating call consists of two brief, sharp clicks sep-

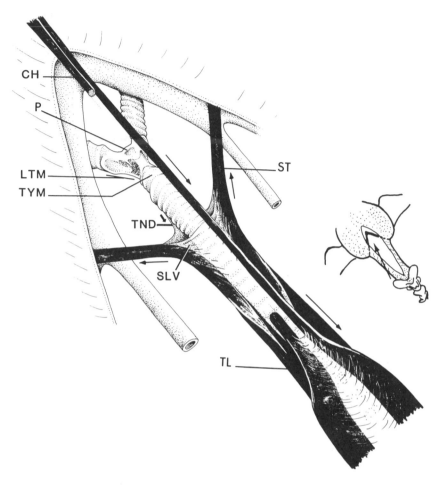

FIGURE 8. Ventral view of a rooster's syrinx showing the arrangement and motion (arrows) of the elements during phonation. Abbreviations as in preceding figures. Redrawn and modified from Gaunt and Gaunt, 1977.

arated by a brief interval. Electromyograms show the sternotrachealis contracting from just before the onset of the first click into the interclick interval, where its activity is replaced by that of the tracheolateralis. Flow through the trachea increases until the production of the first click, whereupon it drops to zero, then sharply rises again with the onset of the second click. Pressure in the airsacs increases as sternotrachealis begins its contractions, remains high during the interclick interval, and drops with the production of the second click, which lasts slightly longer than the first. The call can be produced by either side of the syrinx. Suthers and Hector suggested that the sternotrachealis draws the trachea caudad, relaxing the lateral tympaniform membrane and allowing the first bronchial cartilage to rotate into the lumen. As the gap in the lumen narrows, Bernoulli forces vibrate the medial tympanic membrane, thereby producing the first click, which is terminated as the inwardly moving bronchial cartilage comes in contact with the internal membrane. Contraction of the tracheolateralis then removes the cartilage and as the lumen again opens, a second click is produced. In this species, the lateral membrane is used to interfere physically with sound production to produce a pulse. A sonogram of a Grasshopper Warbler's pulsatile calls shows a pattern similar to that of Grey Swiftlets, and a similar syringeal mechanism, perhaps using the lateral labium as the damper, seems likely (Brackenbury, 1978b).

The syrinx of the Monk Parakeet (*Myopsitta monachus*) contains an unusual flap valve associated with the first tracheal elements. Contraction of the syringeus muscle rotates that element and withdraws the flap. Relaxation of the syringeus allows the flap to move into the airstream where it can be blown closed. Contraction of the tracheobronchealis moves the flexible portions into the lumen where they are positioned for phonation. Simultaneous contraction of these two *intrinsic* muscles, therefore, both sets the syrinx into a vocalizing configuration and removes any impediment to the flow of sound. These muscles can be shown to contract in a one-for-one sequence with pulses of sound (Fig. 7). The mechanism suggested here differs from the mechanism previously suggested for AM in doves and Dunlins only in that the parrot's flap valve is under direct muscular control.

The pulsatile calls of the Monk Parakeet do not seem to involve pulsations of the abdominal musculature, but depend completely on modifications of syringeal configuration (Gaunt and Gaunt, 1982, unpublished). However, in *Streptopelia*, pulsatile contractions of the ventilatory muscles effect a pulsatile flow that interacts with a tracheal muffler, and possibly with a beat frequency, to impart a rolling quality to a portion of the coo (Gaunt et al, 1982). In chicks, pulsatile abdominal contractions can produce a trill with notes coming as rapidly as 50 Hz (Phillips and Youngren, 1981). Here again, similar effects have disparate causes.

3.5. Simple versus Complex Syrinxes

It is easy to suppose that the complex intrinsic musculature of a songbird's syrinx is an adaptation (i.e., has evolved under selective pressure) for vocal virtuosity. However, ornithologists, for example

Beddard (1898), have long recognized that the relationship between the number of muscles in a syrinx and its vocal ability is not simple. Indeed, the correlations are so weak that some investigators (e.g., Nottebohm, 1975) have attributed vocal plasticity and learning to behavioral (neurological?) shifts rather than to structural elaboration.

What then is the role of intrinsic musculature in avian phonation? A clue can be found in Greenewalt's observation that although birds with complex syrinxes do not produce more complex modulations than those with simple syrinxes, they can produce a greater variety of modulations. Perhaps, then, simple structure imposes limitations on the plasticity of syringeal performance, and those restrictions can be released by the actions of intrinsic muscles (Gaunt, 1983a).

We have tried to show that avian phonation involves many interactions between syringeal components, flow, and pressure. The exact nature of these interactions depends on the configuration of the syrinx. Extrinsic muscles adjust the position of the trachea and must change syringeal configuration in toto. Therefore, any change in their activity affects all interactions within the syrinx simultaneously. A bird with a simple syrinx cannot adjust the position of the lateral membranes without changing the tension of the medial membranes and the flow through the syrinx. If the bird has a duplex syrinx, both left and right sides will be equally affected. Such a system can produce a variety of modulations, but changing from one specific modulatory pattern to another may be difficult. Intrinsic muscles provide a means to adjust different structural components of the syrinx independently, thereby reducing interactions. Precise, predictable control becomes possible and neurological shifts promoting vocal plasticity become practical.

The temporal characteristics of song, that is, the duration of sounds and the intervals between them, can be controlled by varying the airflow. Important information is encoded into the temporal features of many bird songs (Emlen, 1972; Bailey and Baker, 1982), and varying those features appears to be a major form of non-genetic plasticity in the vocalizations of birds with simple syrinxes (Gaunt, 1983a).

If this scheme is true, then it is unlikely that intrinsic musculature evolved as an adaptation for vocal plasticity. Present speculation concerning selective pressures that might have led to the evolution of intrinsic muscles center on the possibility that intrinsic musculature increases the effectiveness of a syrinx by allowing it to produce as much or more sound for a smaller use of energy or consumption of air than a simple syrinx (Gaunt et al., 1973, 1976; Brackenbury, 1977). Roosters, for instance, are remarkably inefficient (1.6%) at converting airflow to sound (Brackenbury, 1977). Further, there is an apparent allometric relationship in the effectiveness of conversion of mechanical energy to sound by a syrinx. In general, larger birds produce louder sounds, but many oscines sing much more loudly than would be expected for their size, especially if they are compared with roosters (Brackenbury, 1979a). Some tiny oscines are capable of truly spectacular performances, for example, the Winter Wren (*Troglodytes troglodytes*). Brackenbury suggests that intrinsic musculature may compensate for scaling effects.

3.6. Sound Amplification

A topic that could profit from additional attention is how birds, especially small ones, can sing loudly, that is, vocal amplification. The amount of sound radiated from a system depends in part on the amount of air moved. If the generator involves a small, mechanical oscillator, then relatively small amounts of air are moved. The sound can be enhanced by coupling the generator to some sort of resonating structure with a larger surface. For example, the bridge of a violin passes the vibration of the strings to the violin's body, which is shaped to resonate at many frequencies. We have already suggested that membranes may be coupled with an airstream in a way that amplifies the sound by resonance. Some reptiles and amphibians use vocal sacs to increase the effectiveness of sound radiation. Several species of birds also possess tracheal or esophageal pouches that are inflated during phonation. The acoustic properties of these structures have not been examined, but the cooing of an injured dove that could not inflate its esophagus was notably soft (Gaunt et al., 1982).

In many species of birds, the trachea is remarkably elongated and sometimes coiled or expanded in odd ways. Some of these species can utter loud, raucous calls and the bizarre tracheae are commonly assumed to play a role in that performance. Yet little is known of the direct effects of such tracheae. We have already mentioned that bypassing the coils of a crane's trachea reduces sound. The coils of a crane's trachea are imbedded in and firmly attached to the hollow sternum, which has direct communication with large airsacs that have expanded around the anterior viscera. This arrangement suggests the possibility that the trachea serves to communicate sound vibrations to the sternum, and that the sternum-cum-airsacs serves as a resonator to amplify the sound. Interestingly, the anterior airsacs of oscines are also much enlarged and completely line the inner surface of the sternum. In manucodes, the trachea lies directly under the skin and may itself serve as a sound radiator. The bullae associated with the syrinxes of many male anatids influence sound production, but, oddly, they act as filters that reduce sound (Lockner and Youngren, 1976).

Another way to amplify sound while also adding directionality is to shape the orifice of the vocal tract to minimize impedance between the tube and the outside world. A megaphone is a familiar example of such an impedance-matching device. Both Gross (1964a) and White (1964) have suggested that a crowing rooster enlarges its pharynx to help disseminate sound.

Brackenbury (1979b) has proposed a somewhat different possibility. He divides phonation into an alternating component, sound, and a direct component, airflow. He suggested that birds with exceptionally high velocities of airflow may be able to couple the two components to enhance sound production. He also noted that the radiant power of a sound increases as the fourth power of its frequency, which could provide a significant benefit for the syrinxes of small birds.

4. HARMONICS

The final topic we will consider is the formation of harmonic tones, that is, sound with overtones at whole number multiples of a fundamental frequency. A freely oscillating, edge-clamped membrane should always produce overtones, but not at harmonic intervals (Casey, 1981; Rossing, 1982). Except in special circumstances that do not apply to most syrinxes, such membranes tend to produce sounds without distinct pitch. Some broadband bird calls may be considered pitchless, but most avian phonations are either whistled sounds without overtones or have harmonic overtones. In our judgement, a convincing mechanism for generating the latter remains to be proposed.

Explanations for the formation of harmonics by syringeal membranes have generally included hidden assumptions. A common assumption is that membranes behave like strings, which produce harmonic overtones (Gaunt, 1983b). A string can be treated as a one-dimensional, linear object, the length of which is a major factor in determining its fundamental frequency. Overtones arise when the oscillation breaks up into more than one wave. The additional waves occupy equal portions of the string's length, so the resulting frequencies must be at harmonic intervals. Most models for the generation of overtones by syrinxes postulate a more-or-less plausible method for the induction of several waves on the membrane's surface, but assume that the resulting overtones will be at harmonic intervals (e.g., Greenewalt, 1968; Dürrwang, 1974). However, the distribution of waves on a two-dimensional surface is a complex procedure that does not produce frequencies at whole number multiples of the fundamental. The problem is exacerbated because published versions of the models are often accompanied by line drawings that, as the equivalent to a view of the edge of a section through a membrane, look like drawings of strings.

Indeed, the problem of harmonics as we have posed it contains the assumption that syringeal membranes conform to the criteria of edge-clamped membranes, an assumption open to challenge at several points. It certainly does not apply to those situations in which membranes fold along one axis and the oscillation moves toward a free edge. Even if the membranes do not form a fold, tension is almost certainly unevenly distributed, being greater along the lateromedial axis than in the caudocranial direction. The edge conditions of a syringeal membrane are not uniform. The cranial margin is usually affixed to a rigid pessulus, the lateral and medial margins to a series of flexible, cartilaginous bars, and the caudal margin blends into the connective tissues of the bronchus. Finally, some syringeal membranes contain skeletal inclusions, and center-weighting a membrane is one way to bring its overtones closer to harmonic values (Rossing, 1982). Evidently the problem of harmonics in avian phonations is likely to remain unresolved until we have a far better understanding of the mode(s) of vibration of the syringeal membranes.

5. SUMMARY

The Greenewalt–Stein model remains the basis of our understanding of the process of avian phonation. Since its inception in 1968, it has been reshaped rather than replaced. Perhaps the major trend in its evolution has been a heightened awareness of the potential for variation. Not only is the mode of operation of an oscine syrinx different from that of a simple syrinx, but not all simple syrinxes necessarily behave the same way. Membranes may serve as primary oscillators, responding resonators, or simply as flexible structures that permit reconfiguring a syrinx to exploit aerodynamic processes. Modulations may derive from the activities of syringeal muscles, from the interactions of different oscillators stimulated by the same driving airflow, or from contractions of the ventilatory musculature. The syrinx itself is but one element in a vocal system.

The invitation to prepare a review of syringeal physiology for *Current Ornithology* included the suggestion that we emphasize our personal experiences. Hence, this review has been biased toward our interests and toward a biomechanical approach. For those who may wish to consider other approaches, we recommend a recent review by Brackenbury (1982),whose interests lie more with fluid mechanics and energetics. We will end this review by joining Brackenbury in concluding that in regard to syringeal physiology, much that many ornithologists may consider as firmly established is, in fact, informed supposition, and that the hard questions remain.

ACKNOWLEDGMENTS. We are grateful to Robert Carrington Stein who first directed us toward a path that we have followed for now 15 years. The path has been shared by many companions. Among those whose encouragement, criticism, and enthusiasm have made the journey more rewarding, we count especially Drs. Michal Abs, John H. Brackenbury and Lincoln Fairchild and our former students, Dwight H. Hector and Richard M. Casey. This review has benefitted from the critical suggestions of Drs. Lincoln Fairchild and Lewis Greenwald. Our work has been supported by grants from the Frank M. Chapman Fund of the American Museum of Natural History, the Graduate School and College of Biological Sciences of The Ohio State University, the National Science Foundation (GB-40069, DEB-7911774, and PCM-8302203), and the National Geographic Society (2636-83). The Borror Laboratory of Bioacoustics of The Ohio State University has generously provided access to its equipment and recordings.

REFERENCES

Abs, M., 1970, Uber Hormonwirkungen auf Lautäusserungen von Haustauben, J. Ornithol. 111:227–229.

Abs, M., 1980, Zur Bioakustik des Stimmbruchs bei Vogeln, Zool. Jb. Physiol. 84:289–382.

Ames, P. L., 1971, The morphology of the syrinx in passerine birds, Peabody Mus. Nat. His. Bull. 37:1–195.

Bailey, E. D., and Baker, J. A., 1982, Recognition characteristics in covey dialects of Bobwhite Quail, Condor 84:317–320.

Beddard, F. E., 1898, The Structure and Classification of Birds, Longmans, Green & Co., New York.

Beebe, [C] W., 1925, The Variegated Tinamou, Crypturus variegatus variegatus (Gmelin), Zoologica 6:195–227.

Berger, M. and Hart, J. S., 1968, Ein Beitrag zum Zusammenhang zwischen Stimme und Atmung bei Vögeln, J. Ornithol. 109:421–424.

Borror, D. J. and Reese, C. R., 1956, Vocal gymnastics in Wood Thrush songs, Ohio J. Sci. 56:177–182.

Brackenbury, J. H., 1977, Physiological energetics of cock-crow, Nature 270: 433–435.

Brackenbury, J. H., 1978a, Respiratory mechanics of sound production in chickens and geese, J. Exp. Biol. 72:229–250.

Brackenbury, J. H., 1978b, A comparison of the origin and temporal arrangement of pulsed sounds in the songs of the Grasshopper and Sedge Warblers, Locustella naevia and Acrocephalus schoenobaenus, J. Zool. Lond. 184:187–206.

Brackenbury, J. H., 1978c, A possible relationship between respiratory movements, syringeal movements, and the production of song by Skylarks Alauda arvensis, Ibis, 120:526–528.

Brackenbury, J. H., 1979a, Power capabilities of the avian sound-producing system, *J. Exp. Biol.* **78**:163–166.

Brackenbury, J. H., 1979b, Aeroacoustics of the vocal organ of birds, *J. Theor. Biol.* **81**:341–349.

Brackenbury, J. H., 1982, The structural basis of voice production and its relationship to sound characteristics, in: *Acoustic Communication in Birds*, Volume 1, (D. E. Kroodsma and E. H. Miller, eds), Academic Press, New York, pp.53–73.

Brockway, B. F., 1967, The influence of vocal behavior on the performer's testicular activity in Budgerigars (*Melopsittacus undulatus*), *Wilson Bull.* **79**:328–334.

Calder, W. A., 1970, Respiration during song in the Canary (*Serinius canaria*), *Comp. Biochem. Physiol.* **32**:251–258.

Casey, R. M., 1981, Theoretical analysis of tympanic membranes in avian syrinx, M. Sc. Dissertation, The Ohio State University, Columbus.

Chamberlain, D. R., Gross, W. B., Cornwell, G. W., and Mosby, H. S., 1968, Syringeal anatomy in the Common Crow, *Auk* **85**:244–252.

Chanaud, R. C., 1970, Aerodynamic whistles, *Sci. Am.* **222** (l):40–46.

Duncker, H.-R., 1971, *The Lung Air Sac System of Birds: A Contribution to the Functional Anatomy of the Respiratory Apparatus*, Springer-Verlag, Berlin.

Dürrwang, R. F., 1974, Functionelle Biologie, Anatomie und Physiologie der Vogelstimme, Ph. D. Dissertation, University Basil, Basil, Switzerland.

Emlen, S. T., 1972, An experimental analysis of the parameters of bird song soliciting species recognition, *Behavior* **41**:130–171.

Gaunt, A. S., 1983a, An hypothesis concerning the relationship of syringeal structure to vocal abilities, *Auk* **100**: 853–862.

Gaunt, A. S., 1983b, On sonograms, harmonics, and assumptions, *Condor* **85**:259–261.

Gaunt, A. S., and Gaunt, S. L. L., 1977, Mechanics of the syrinx in *Gallus gallus* II. Electromyographic studies of *ad libitum* vocalizations, *J. Morphol.* **152**:1–19.

Gaunt, A. S., and Gaunt, S. L. L., 1982, Electromyography of the syringeal muscles in parrots, *Am. Zool.* **22**:918.

Gaunt, A. S., and Wells, M. K., 1973, Models of syringeal mechanisms, *Am. Zool.* **13**: 1227–1247.

Gaunt, A. S., Gaunt, S. L. L., and Casey, R. M., 1982, Syringeal mechanics reassessed: Evidence from *Streptopelia*, *Auk* **99**:474–494.

Gaunt, A. S., Gaunt, S. L. L., and Hector, D. H., 1976, Mechanics of the syrinx in *Gallus gallus*. l. A comparison of pressure events in chickens to those in oscines, *Condor* **73**:208–223.

Gaunt, A. S., Stein, R. C., and Gaunt, S. L. L., 1973, Pressure and air flow during distress calls of the Starling, *Sturnus vulgaris* (Aves: Passeriformes), *J. Exp. Zool.* **183**:241–262.

Gottlieb, G., and Vandenbergh, J. G., 1968, Ontogeny of vocalization in duck and chick embryos, *J. Exp. Zool.* **168**:307–326.

Greenewalt, C. H., 1968, *Bird Song: Acoustics and Physiology*. Smithsonian Institution Press, Washington, D. C.

Gross, W. B., 1964a, Voice production by the chicken, *Poult. Sci.* **43**:1005–1008.

Gross, W. G., 1964b, Devoicing the chicken, *Poult. Sci.* **43**:1143–1144.

Gross, W. B., 1979, An operation for reducing the vocal intensity of peafowl, *Avian Dis.* **23**:1031–1036.

Hérissant, [F.-D.], 1753, Recherches sur les organes de voix des quadrupédes et de celle des oiseaux, *Acad. Roy. Sci. Mem.*(Paris), pp.279–295.

Hersh, G. L., 1966, Bird voices and resonant tuning in helium-air mixtures, PH. D. Dissertation, University of California, Berkeley.

Hunter, M. L., Jr., 1980. Vocalization during inhalation in a nightjar, *Condor* **82**:101–103.

Klatt, D. H., and Stefanski, R. A., 1974, How does a mynah bird imitate human speech? *J. Acoust. Soc. Am.* **55**:822–832.

Kroodsma, D. E., and Miller, E. H., 1982, Introduction, in: *Acoustic Communication in Birds*, Volume 1, *Production, Perception and Design Features of Sound* (D. E. Kroodsma and E. H. Miller eds), Academic Press, New York, pp. xxi–xxxi.

Lieberman, P., 1975, *On the Origins of Language: An Introduction to the Evolution of Human Speech*, Macmillan, New York.

Lockner, F. R. and Murrish, D. E., 1975, Interclavicular air sac pressures and vocalization in Mallard Ducks *Anas platyrhynchos*, *Comp. Biochem. Physiol.* **52A:** 183–187.

Lockner F. R. and Youngren, O. M., Functional syringeal anatomy of the Mallard, I. *In situ* electromyograms during ESB elicited calling, *Auk* **93:**324–342.

Miller, A. H., 1934, The vocal apparatus in some North American owls, *Condor* **36**:204–213.

Miller, D. B., 1977, Two-voice phenomenon in birds: Further evidence, *Auk* **94**:567–572.

Miller, E. H., 1983, The structure of aerial displays in three species of Calidridinae (Scolopacidae), *Auk* **100:**440–451.

Miskimen, M., 1951, Sound production in passerine birds, *Auk* **68:**493–504.

Myers, J. A., 1917, Studies on the syrinx of *Gallus domesticus*, *J. Morphol.* **29:**165–214.

Nottebohm, F., 1971, Neural lateralization of vocal control in a passerine bird. 1. Song, *J. Exp. Zool.* **177:**229–262.

Nottebohm, F., 1975, Vocal behavior in birds, in: *Avian Biology*, Volume 5 (D. S. Farner and J. S. King eds), Academic Press, New York, pp. 289–332.

Nottebohm, F., 1976, Phonation in the Orange-winged Amazon Parrot, *Amazona amazonica*, *J. Comp. Physiol.* **108(A):**157–170.

Paulsen, K., 1967, *Das Prinzip der Stimmbildung in der Wirbeltierreihe und beim Menschen* Akad. Verlag., Frankfurt am Main.

Peek, F. E., Youngren, O. M., and Phillips, R. E., 1975, Repetitive vocalizations evoked by electrical stimulation of avian brains. IV. Evoked and spontaneous activity in expiratory and inspiratory nerves and muscles of the chicken (*Gallus gallus*), *Brain Behav. Evol.* **12:**1–42.

Phillips, R. E. and Youngren, O. M., 1981, Effects of denervation of the tracheo-syringeal muscles on frequency control in vocalizations of chicks, *Auk* **98:**299–306.

Phillips, R. E., Youngren, O. M., and Peek, F. W., 1972, Repetitive vocalizations evoked by local electrical stimulation of avian brains. I. Awake chickens (*Gallus gallus*), *Anim. Behav.* **20:**689–705.

Potter, R. K., Kopp, G. A., and Green, H. C., 1947, *Visible Speech*, D. van Nostrand Co., Princeton, New Jersey.

Rossing, T. D., 1982, *The Science of Sound*, Addison-Wesley, Reading, Massachusetts.

Rüppell, W., 1933, Physiologie und Akustik der Vogelstimme, *J. Ornithol.* **81:**433–542.

Smith, D. G., 1977, The role of the sternotrachealis muscles in bird song production, *Auk* **94:**152–155.

Stein, R. S., 1968, Modulation in bird sounds *Auk* **85:**229–243.

Sutherland, C. A. and McChesney, D. S., 1965, Sound production in two species of geese, *Living Bird* **4:**99–106.

Suthers, R. A. and Hector, D. G., 1982, Mechanism of the production of echolocating clicks by the Grey Swiftlet, *Collocalia spodiopygia*, *J. Comp. Physiol.* **148:**457–470.

Thorpe, W. H., 1961, *Bird Song*, Cambridge University Press, London.

van den Berg, Jw., 1968, Sound production in isolated human larynges, *Ann. N. Y. Acad. Sci.* **155**:18–26.

Warner, R. W., 1969, The anatomy of the avian syrinx, Ph.D. Dissertation, University of London.

Warner, R. W., 1971, The structural basis of the organ of voice in the genera *Anas* and *Aythya* (Aves)., *J. Zool.* **164**:197–207.

Warner, R. W., 1972a, The syrinx in family Columbidae, *J. Zool.* **166**:385–390.

Warner, R. W., 1972b, The anatomy of the syrinx in passerine birds, *J. Zool.* **168**:381–393.

White, S. S., 1968. Movement of the larynx during crowing in the domestic cock, *J. Anat* **103**:390–392.

Wilson, T. A., Beavers, G. S., DeCoster, M. A., Holger, D. K., and Regenfuss, M. D., 1971, Experiments on the fluid mechanics of whistling, *J. Acoust. Soc. Am.* **50**:366–372.

Youngren, O. M., Peek, F. W., and Phillips, R. E., 1974, Repetitive vocalizations evoked by local electrical stimulation of avian brains. III. Evoked activity in the tracheal muscles of the chicken (*Gallus gallus*). *Brain Behav. Evol.* **9**:393–421.

CHAPTER 8

ASSESSMENT OF COUNTING TECHNIQUES

JARED VERNER

1. INTRODUCTION

Counts of birds are used for many purposes in a variety of field studies. Simple presence-or-absence information suffices to study avian biogeography, but indices of abundance are needed to track the changing mix of species' populations associated with plant succession. Still other studies, as of trophic dynamics, require knowledge of population densities. This chapter critically assesses the suitability of various counting methods for each of these purposes. It is not a thorough review of the topic. These have been provided already by Kendeigh (1944), Blondel (1969), Berthold (1976), and Dawson and Verner (in preparation); Robbins (1978) and Shields (1979) should also be consulted, although the scope of their reviews was less extensive.

Most standard methods for counting birds include computations purported to deliver estimates of density, and researchers using these methods typically compute density estimates, even when something less would answer their question(s). This results in more costly studies than needed. Even worse, however, is the fact that most methods used to deliver density estimates are inadequate for the purpose.

The objectives of this chapter are (1) to identify the limitations of standard methods used to count birds, and (2) to suggest ways to begin

JARED VERNER • Forestry Sciences Laboratory, Fresno, California 93710 to cope with those limitations. The chapter is organized into six sections, beginning with a discussion of the different scales of abundance obtainable by various counting methods. This is followed by a brief survey of broad study objectives and selection of appropriate scales of measurement for each objective. The next section briefly summarizes biases affecting counts, as their understanding is critical for development of accurate counting methods. The fourth section evaluates the most commonly used of the standard counting methods. Finally, section five gives conclusions and section six offers some recommendations for improving the ways we count birds.

This assessment emphasizes use of existing counting methods to estimate densities of birds, as chapter length precluded much else. The general theme is that most practitioners have used unwarranted confidence when estimating densities of birds by the common counting methods. Although the situation is far from hopeless, much more research is needed to develop efficient, reliable methods for estimating densities of birds. In certain sections I have relied heavily on an extensive review of counting methods by Dr. David G. Dawson and myself (Dawson and Verner, in preparation). In one sense, then, this is a condensed version of that review, although some of my views expressed here do not quite line up with those of Dr. Dawson, as expressed in our review. The manuscript has been markedly improved by technical suggestions of Drs. Louis B. Best, David G. Dawson, John T. Emlen, Jr., Douglas H. Johnson, and David A. Manuwal, although responsibility for errors of fact or interpretation remains mine alone.

2. SCALES OF ABUNDANCE

Proper study design depends on selecting an appropriate scale at which to measure the abundance of birds, defined as follows (some of these definitions depart from those used by statisticians):

1. A *nominal* scale records only the presence or absence of species.
2. An *ordinal* scale depends on measurements sufficiently accurate to rank species in the correct order of abundance. An ordinal scale is suitable at a coarse level, as in placing species in a hierarchy of classes (e.g., abundant, common, uncommon, rare). A fully accurate ordinal scale is essentially the same as a ratio scale, however, because most communities have two or more species nearly equal in abundance.
3. A *ratio* scale assumes that all target species are sampled equally (the same percentage of each species is detected, so one can

legitimately say that species A is twice as abundant as B, or that B is 70% as abundant as C).

4. An *absolute* scale of abundance assumes a total count, or at least accurate coefficients for adjusting a biased count to be equivalent to a total count, permitting one to compute density if the area sampled is known. Total counts are possible only in special cases, and in no case known to me is it possible to achieve total counts of all bird species present even in a relatively simple ecosystem. Some mixed-species colonies of seabirds may permit one to approach this ideal.

3. DEFINING OBJECTIVES AND SELECTING AN APPROPRIATE SCALE OF MEASUREMENT

Every study should begin with a statement of its objectives, which then guide selection of appropriate methods to be used in the study. The literature shows a marked tendency by researchers to use more detailed information than needed to answer many questions. In this section, I discuss the levels of detail in sample data (lists, counts, and censuses), their associated scales of measurement, and the sorts of biological questions appropriately addressed by each (Table I).

3.1. Lists

Lists simply record all species detected in an area; they are most useful when standardized by effort and area sampled. They give information at a nominal scale, useful in three basic areas of interest— biogeography, species richness, and frequency of occurrence.

3.1.1. Biogeography

Lists of bird species by region are used to build our understanding of the distribution of birds, as in developing bird atlases. Root (1983) persuasively argues, however, that abundance information would considerably improve our interpretation of avian biogeography.

3.1.2. Species Richness

Most studies of factors influencing the dynamics of avian communities have addressed species diversity [normally expressed as an index such as H' (Shannon and Weaver, 1963) that integrates the num-

TABLE I

Study Objectives, Measurement Scales Needed, and Applicable Levels of Sampling Detail

Study objective	Measurement scale needed			Applicable level of sampling detail			
	Nominal	Ratio	Absolute	List	Unadjusted Count	Adjusted[a] Count	Census
Biogeography	X			XX[b]	X	X	X
Species richness	X			XX	X	X	X
Frequency	X			XX	X	X	X
Annual population trends		X			XX	X	X
Seasonal population trends		X				XX	X
Successional trends (intra- and interspecific)		X				XX	X
Habitat suitability or preference		X				XX	X
Density-dependent and density-independent effects		X				XX	X
Species diversity		X				XX	X
Population fluctuation			X				XX
Trophic dynamics			X				XX

[a]Adjusted counts compensate for differences in area sampled and for differences in detectability of birds in different habitats and seasons. They consequently result in an estimate of density. Unadjusted counts do not permit an estimate of density, but properly designed counts that control for observer differences, phenological changes, and other such variables should deliver abundance measures on the same ratio scale for the same species in the same or structurally comparable habitats.

[b]XX shows the least costly (hence preferred) sampling level for the indicated study objective. An X in this section shows a more costly way to the same study objective.

ber of species and the number of individuals in an assemblage of or-ganisms]. Species richness (number of species) and species diversity are highly intercorrelated variables (e.g., Tramer, 1969; Larson, 1980), however, and I believe that we may learn as much about the dynamics of avian communities by using species richness as by using species diversity as the dependent variable (Larson and Verner, in preparation). Species richness can be measured more accurately and cheaply than species diversity and it is more easily interpreted.

3.1.3. Frequency

Frequency is simply the percentage of sites at which a species was detected, irrespective of its abundance at each. Within limits, frequency scores can give a crude index of relative abundance (Yapp, 1956; Daw-son and Verner, in preparation), but this advantage is lost for species that are detected on all areas (frequency = 1.0). Frequency scores should be used to compare studies only when large numbers of sites (preferr-ably 100 or more) are sampled, giving reasonably fine resolution of frequency intervals. Because the number of frequency intervals equals the number of sites sampled, studies with few sites permit only a crude differentiation among species by frequency. Frequencies are also un-suitable for comparing studies differing in time spent sampling each site or differing in plot sizes (more species should be detected with more time spent on a site and on larger plots).

3.2. Counts

Counts as used here refer to any systematic, standardized effort to enumerate birds, as by moving along established lines of prescribed length (transect counts) or remaining a prescribed period at one location (station or point counts). Counts can give information about the relative abundance of bird species, and some observers believe that they can be used to estimate density by adjusting for detectability and area sam-pled. Although I am not optimistic about using counts to estimate den-sities, with further research we will probably be able to adjust counts accurately enough to answer most questions of interest to ornithologists. I believe that at least a ratio scale of abundance is needed to address all study objectives listed below. Little need is suggested for information at an ordinal scale (Table I).

3.2.1. Annual Population Trends

Detecting trends in a species' population requires repeated mea-sures on a ratio scale. Because such studies are intraspecific and confined to the same population (therefore the same habitat), simple counts will suffice for annual trends if properly designed to control phenology, observer differences, and other variables.

3.2.2. Seasonal Population Trends

Study of seasonal trends assumes that detectability is constant from one season to another or that one can correct for seasonal changes in detectability.

3.2.3. Successional Trends and Habitat Suitability or Preference

These studies can involve one or more species. Analyses are normally interspecific, however, and they require comparisons of relative abundance between sites differing in vegetation structure, composition, or both. The basic assumption is that abundances of target species are measured accurately on a ratio scale. Therefore, either each species must be equally detectable in all habitats (unlikely), or we must have accurate ways to adjust for differing detectabilities in different habitats. Although one can make inferences about successional trends from lists of species in the various stages, assuming species turnover with advancing succession, lack of abundance information limits one's ability to evaluate the relative importance of each stage for any species in question.

3.2.4. Density-dependent and Density-independent Effects

Murray (1979, p. 11) defines a density-dependent factor as "any component of the environment whose intensity is correlated with population density and whose action affects survival and reproduction." Conversely, a density-independent factor is "any component of the environment that affects survival and reproduction but whose occurrence or intensity is not correlated with population density." In spite of the use of "density" in this concept, true density values are not essential for its study. Relative abundance at a ratio scale is sufficient, and because studies are normally intraspecific and within a given habitat, simple counts adjusted for area and detectability should suffice.

3.2.5. Species Diversity

Indices of species diversity have been used widely in studies of environmental factors influencing avian community dynamics. Such studies can compare communities in the same area at different times (temporal variation), in different areas at the same time (spatial variation), or in different areas at different times (temporal and spatial variation). It is not necessary that correct densities be determined for all species, but calculation of a species diversity index does assume that the abundances of all species in target communities are measured on the same ratio scale.

3.3 Censuses

A census is a total count, so it measures abundance on an absolute scale. Total counts are possible only in special circumstances, as with colonially nesting birds whose nests are relatively conspicuous or with colonially roosting birds that can be counted as they enter or leave the roost. Otherwise, intensive color-banding studies or other such involved efforts are required to approach an estimate of the total number of birds in an area. Thus true censuses are rare in ornithological studies, although many forms of counting are wrongly called censuses.

An accurate total count of a population (a census) and accurate determination of the area it occupies are required to determine population density. Density is simply computed as $D = N/A$, where D = density, N = number of birds counted, and A = area sampled. A common standard for expressing density is the number of birds/40 hectares, because it approximates 100 acres. The sections on "lists" and "counts" show that most questions in ornithology can be answered with information less detailed than population densities. Only two general areas of ornithological research require measurements of bird populations at a scale capable of delivering true densities.

3.3.1. Population Fluctuation

Study of the growth or decline of a population in terms of its absolute size, as opposed to trends, requires knowledge of the true population size. Such information may be sought in certain studies of population dynamics.

3.3.2. Trophic Dynamics

Studies of energy flow in ecosystems typically attempt to measure the amount of energy passing through various components of the trophic system. Often birds are grouped in a single pathway, but sometimes they are subdivided. In any case, accurate assessment of energy consumption by all birds in a community assumes an absolute scale of abundance.

4. BIAS IN COUNTING BIRDS

Bias is a measure of average deviation from the real value of a paramater by estimators of that parameter. Any factor that biases an estimate contributes to its inaccuracy. It does not necessarily affect its precision, however. Precision refers to the degree of repeatability of a measure, irrespective of its accuracy. It has several standards, the most common being variance and standard deviation.

Sources of bias in bird counts can be grouped in five broad categories—observer, habitat, birds, study design, and weather (Table II). Although many studies have addressed sources and magnitudes of bias, it still remains among the factors in bird counting most in need of intensive quantitative study. Space precludes a thorough review of bias. Instead, in this section I itemize the types of bias within each broad category, provide some key references to guide the reader to more detailed discussions, and highlight a few quantitative studies of bias that affect two or more different counting methods. Biases shown to violate specific assumptions of various counting methods are discussed again in the appropriate sections evaluating those methods.

4.1. Effects of Observer

Differences in (1) acuity (visual, auditory), (2) alertness (fatigue, motivation), (3) experience, (4) knowledge, and (5) number of observers can bias bird counts. Emlen and DeJong (1981) showed that observers with slight to moderate hearing losses in high-frequency ranges detected various species at only 25% to 90% of the distances they were detected by observers with normal hearing. Hearing tests of 274 people attending a symposium on bird counting showed that differences in hearing ability can change the sampled area by an order of magnitude (Ramsey and Scott, 1981). Ramsey and Scott also found that average hearing thresholds of persons over 40 years of age failed to meet the 20-dB criterion recommended by Emlen and DeJong (1981) in the frequency ranges 3 to 6 kHz, corresponding to frequencies typical of the songs of many passerine birds.

4.2. Effects of Habitat

Bird counts can be biased by (1) species composition of vegetation, (2) structure of vegetation (e.g., height profile, patchiness), (3) noise

TABLE II
Some Sources of Bias in Bird Counts[a]

| | | Coping with bias | | |
| | | Standardize by | | |
Source of bias	Research priority	Study design	Testing and training	Study and correct
Observer				
Acuity	3	1	1	
Alertness	3	1	1	
Experience	3	1	1	
Knowledge	3	1	1	
Number of observers	2	2		1
Habitat				
Composition of vegetation	1			1
Structure of vegetation	1			1
Noise	2	1		2
Terrain	2	2		1
Birds				
Density	1			1
Flocking behavior	1	2		1
Intraspecific detectability				
Age	1	1		1
Sex	1	1		1
Season	2	1		2
Time of day	2	1		2
Interspecific detectability	1	1		1
Movement	1	1		1
Social or breeding system	1	2		1
Weather				
Fog	3	1		2
Precipitation	3	1		2
Relative humidity	3	1		2
Snow cover	3	1		2
Temperature	3	1		2
Wind	3	1		2
Study Design				
Distance between sites	3	1		2

(Continued)

TABLE II (Continued)

Source of bias	Research priority	Coping with bias		
		Standardize by		
		Study design	Testing and training	Study and correct
Duration of sampling periods	3	1		2
Frequency of sampling	3	1		2
Plot size or transect length	3	1		2
Site selection	3	1		2
Timing	3	1		2

[a]My subjective priority ranking of research needs (1 is high), and an estimated priority ranking of means to cope with bias (1 is high).

(e.g., streams, vehicles), and (4) terrain associated with different habitats. Observers commonly acknowledge the likelihood that habitats bias bird counts, but most simply assume that standard adjustments for differences in detectability adequately compensate for the differences. Surprisingly few studies have quantitatively assessed biases attributable to habitats (but see Yui, 1977; Gill, 1980; Rodgers, 1981).

Bias associated with noise (Dawson, 1981a; Karr, 1981) or from terrain (Rodgers, 1981; D. K. Dawson, 1981) can probably be reasonably well controlled by study design. Bias resulting from the composition or structure of vegetation is more serious, however, because much research in ornithology endeavors to determine relationships of bird populations in habitats that differ in vegetation. Yui (1977) found marked and consistent differences in detectabilities of most species depending on the height of vegetation in studies comparing transects with spot mapping in Japan.

4.3. Effects of Birds

Many attributes of birds bias counts, including (1) detectability, (2) movement (normal or in response to an observer), (3) density, (4) flocking behavior, and (5) social or breeding system. Many factors contribute to differences in detectability among birds, both within and among species. Males are typically more detectable than females. Age differences, too, may bias counts. For example, among Willow Tits

(*Parus montanus*) in Sweden, known juveniles were detected significantly more often than known adults. This resulted directly from differences in foraging heights among the age and sex classes (Ekman,
1981).

Diurnal variation in detectability is widely acknowledged because
most observers confine counts to early morning hours, when most birds
are more detectable (e.g., Robbins, 1981a: Skirvin, 1981; see review by
Conner and Dickson, 1980). Quantifying diurnal changes in detectability would permit correction for them in studies that include a wide
range of detectabilities. Alternatively, one could confine counts to periods later in the day when the variance among counts is lower, as
recommended by Yapp (1956) and Dawson and Bull (1975). Although
this might improve the precision of counts, it might not improve one's
ability to estimate densities (e.g., Skirvin, 1981), because total counts
would be lower than if counting were done earlier in the day.

Seasonal changes also influence detectability of birds. For example,
males of most species are most conspicuous during portions of the
breeding season when they sing most actively. A detailed study of three
passerine species in Poland showed that detectability was generally
highest during the prelaying period, lowest during incubation, and
intermediate during the nestling and fledgling periods (Diehl, 1981).
Diehl also found that successful breeders were more frequently detected
than unsuccessful ones. In some cases these biases may be controllable
by appropriate scheduling of sample periods. However, multiple factors
contributing to seasonal variations in detectability can be extraordinarily complex (see Best, 1981; Best and Petersen, 1982). Hopefully
their effects can be corrected for with enough information (Table II).

One class of birds—the nonbreeding "floaters" (see reviews by
Brown, 1969; von Haartman, 1971)—is acknowledged as a source of
bias in bird counts, but is then ignored in most studies attempting to
estimate densities. A study of color-marked birds of 25 species during
the breeding season in Minnesota gave ratios between total marked
birds and those with territories from $4:1$ to $16:1$ among males and
$2.7:1$ to $8:1$ among females (Rappole *et al.*, 1977). These may not be
accurate ratios for a fixed area, however, because movement of territorial birds from nearby areas was not ruled out. The most thorough
study of floaters involved Rufous-collared Sparrows (*Zonotrichia capensis*) in Costa Rica, and included detailed observations of color-banded
individuals over a three-year period (Smith, 1978). The floaters com-
prised at least 50% of the total population in that study. One cannot
judge how general this result is because other populations have not
been studied so thoroughly.

The above sources of bias resulting in intraspecific differences in detectability apply interspecifically as well, but they are multiplied an unknown amount because different species are unlikely to be identical in their sources of intraspecific bias. For example, the fact that species differ in their breeding schedules precludes sampling all of them on an equivalent basis in any study of short duration. Sampling would need to continue with equal intensity throughout the period including the breeding cycles of all target species, so that a phenologically comparable period could be used for each species.

Perhaps the most significant interspecific differences in detectability result from birds' responses to an observer. It is well known that some birds respond by moving away from an observer; others move toward the observer; and some appear to remain concealed in the presence of an observer. Examples of the effects of these behaviors on density estimates are given later in this paper in Section 5.2.2.

Even the density of birds—the object of study—has been shown to bias counts. The detectability of known Red-backed Shrikes (*Lanius collurio*) in Poland during the prelaying period was significantly higher on a plot with low density than on one with high density (Diehl, 1981). Similarly, Duke (1966) reported a significant decrease in the rate of flight calling by American Woodcocks (*Scolopax minor*) in Michigan with increasing density of calling males. The reverse situation has been reported for Ring-necked Pheasants (*Phasianus colchicus*) (Gates, 1966) and Ruffed Grouse (*Bonasa umbellus*) (Rodgers, 1981), and I recorded marked reductions in song rates and the duration of song bouts by male Marsh Wrens (*Cistothorus palustris*) in a low-density population compared with a high-density one in another year on the same plot in central Washington (Verner, unpublished data). The influence of these biases on density estimates has not been systematically studied.

"Saturation" or "swamping" is another density-related source of bias in bird counts (Järvinen and Väisänen, 1976; Frochot *et al.*, 1977; Walenkiewicz, 1977). This occurs when so many birds are detectable at one time that an observer is unable to distinguish or record them all. The effect can be either interspecific or intraspecific. Intraspecific saturation may also result from an observer's inability to differentiate between a new bird and another detected earlier in about the same location (Enemar, 1962). A study in Hawaii designed specifically to examine saturation bias compared counts of four observers, each within 3 m of a central point, who independently counted birds during the same time interval (Scott and Ramsey, 1981a). One observer (the generalist) recorded all species. The other observers (specialists) counted (1) only the most abundant species, (2) only the three rarest species, and (3) three of the more common species. Counts of the specialists were higher than those of the generalist for five of seven species tested and significantly higher for two—the Apapane (*Himatione sanguinea*) and the Hawaiian Thrush (*Phaeornis obscurus*). The density estimate of the Apapane, using collective results of the specialists, was significantly higher ($P < 0.001$) than that of the generalist.

Flocks of birds are generally much more detectable than solitary birds. Consequently, flocking habits can bias counts by increasing the likelihood that an otherwise inconspicuous bird will be detected because it is a member of a flock. I am unaware of any systematic, quantitative study of this source of bias on counts of any bird species.

4.4. Effects of Weather

Weather factors that may bias bird counts include (1) precipitation, (2) wind, (3) temperature, (4) fog, (5) snow cover, and (6) relative humidity. These biases may be mediated through their effects on cues produced by birds, their effects on birds themselves, or their effects on observers. Although all methods for estimating numbers of birds include caveats about sampling during periods with extreme weather conditions, surprisingly little research has addressed the magnitude of bias introduced by weather. To summarize, these biases can be reasonably controlled by avoiding periods of (1) strong wind (> 11 km/hr: Dawson, 1981a; Emlen and DeJong, 1981; Robbins, 1981b; but see Anderson and Ohmart, 1977, for a recommended threshold of 20 km/hr), (2) light to heavy precipitation (O'Connor and Hicks, 1980; Dawson, 1981a; Robbins, 1981b), (3) extreme temperatures ($< 7°C$ or $> 24°C$) (Robbins, 1981b), and (4) fog (Robbins, 1981b). Some of these recommended levels were not specifically identified in the literature cited, but represent my interpretation of results therein. Obviously they need further study.

4.5. Effects of Study Design

Factors associated with study design that may bias counts include (1) site selection (random, stratified random, purposive, regular), (2) distance between sites, (3) plot size or transect length, (4) timing (diurnal, seasonal), (5) duration of sampling periods, and (6) sampling frequency. Nonrandom selection of sites precludes many sorts of data analysis. If the distance between sites is short enough that some individuals are counted at two or more sites, positive biases result in the counts of those species when data are pooled to characterize the bird assemblage of the study area.

4.6. Coping with Bias

The many sources of bias described in this and following sections result in a composite bias of unknown direction and magnitude on the density estimates of most or all species by most or all methods. Many researchers seem to forget that biases do not go away simply because they are acknowledged, because many freely acknowledge a variety of biases in their counts, but then analyze and interpret their data without regard to those biases. Many sources of bias can be controlled (Johnson, 1981) by appropriate standardization of the study design (e.g., site selection, details of technique, observer variables, and noise) (Table II). Others require careful study to allow correction for them (e.g., structure and composition of vegetation, interspecific differences in detectability, movement, flocking habits) (Table II). Much bias can be reduced by training observers (Atkinson-Willes, 1963; Sinclair, 1973; Kepler and Scott, 1981) who should first pass criteria that test their visual and auditory acuity and their skill as observers.

Large sample sizes improve the precision of count data, and I agree with Bock and Root (1981) that confidence in observed trends or patterns increases with sample size. If bias is unknown but constant, increasing sample size for greater precision is a desirable approach for intraspecific comparisons, both temporally (same habitat during the same phenological stage in different years) and spatially (same time period in different but comparable habitats). It does not, however, solve the problem for intraspecific comparisons between different habitats or seasons or for interspecific comparisons of any sort. Greater precision should always be a goal, but it does not eliminate bias. I agree with Johnson (1981) that taking large samples with the expectation that biasing variables will average out is overly optimistic. Although the art of bird counting has progressed sufficiently for many sources of bias to be dealt with, the sources most difficult to cope with still remain (Table II).

5. THE METHODS

Many general methods, most with variants, have been used to estimate the densities of birds. Space precludes thorough discussion of each method and its variants. I have chosen, instead, to emphasize the three methods most often used (mapping, transects, and point counts), with a brief discussion of other methods. This section identifies important assumptions underlying methods to estimate densities and presents case studies about how well data meet those assumptions.

5.1. Mapping

5.1.1. Spot Mapping

Also known as territory mapping, this is the most widely used of the mapping methods. Its essential features were described by Kendeigh (1944), and it was later selected as the method of preference by the International Bird Census Committee, which provided guidelines for its use (Anonymous, 1970; Robbins, 1970). Useful improvements were proposed by Svensson (1980) and Tomialojć (1980). These and other modifications were integrated by Dawson and Verner (in preparation) into a complete set of instructions that, if rigorously applied, should increase the accuracy of results from the mapping method.

Briefly, spot mapping involves plotting the locations of individual birds on maps of gridded areas during each of several visits to a plot, later transferring those locations to separate maps for each species, and finally identifying clusters of locations assumed to represent centers of activity by individual territory holders. Consequently one maps clusters rather than territories, so the method is misnamed. The total number of clusters (territories) = (number of complete clusters) + (sum of fractional clusters overlapping plot boundaries). The total number of birds on the plot is then estimated by multiplying the number of clusters by the mean number of birds represented by a cluster (normally two, assuming an even sex ratio). Principal assumptions of the method are:

1. Populations of target species are stable, and birds remain within exclusive spaces during the sampling period.
2. At least one bird per territory produces cues frequently enough to permit repeated location on successive visits to the plot.
3. Estimates of the proportions of boundary territories (actually clusters cut by plot boundaries) assigned to the plot are unbiased.
4. An accurate estimate is made of the mean number of birds represented by each cluster.
5. Birds are correctly identified.

Assumptions 1 and 2 nearly limit spot mapping to the set of territorial birds during the breeding season. The method is unsuited to (1) birds that regularly produce cues within the boundaries of conspecifics' territories (assumption 1), (2) unusually quiet and secretive species (assumption 2), (3) nonterritorial birds (floaters) in the population (assumptions 1 and 2), and (4) species with territories that are large relative to the size of the study plot (assumption 3). The method works best with species that frequently produce audible cues from within their territories (e.g., most passerines), so that birds on two or more territories can be detected simultaneously. This eliminates much ambiguity in delineating clusters, especially for species that tend to violate territorial boundaries.

Failure of assumption 1 causes difficulty in identifying clusters. For example, mean home range size (by capture–recapture of banded birds) of five forest passerines in France ranged from 2 to 12 times the size of territories, as determined by spot mapping (Ferry et al., 1981). Ovenbirds (Seiurus aurocapillus) studied in eastern Canada occasionally used song posts overlapping adjacent territories, and birds foraged extensively on one another's territories (Zach and Falls, 1979). Enemar et al. (1979) found that multiple song perches of Willow Warblers (Phylloscopus trochilus) sometimes caused interpreters of maps to erroneously divide single territories into two. Jensen (1974) found many cases in which assumption 1 was violated by species in marshes in Denmark. Marsh Warblers (Acrocephalus palustris), Whitethroats (Sylvia communis), and Reed Buntings (Emberiza schoeniclus) held multiple territories. He concluded that the usual view of the territory as a well-delimited area is invalid for certain species or individuals.

Variation among observers in the interpretation of spot-mapping results is unacceptably large for some species, giving further evidence that clusters are not easily distinguished (Svensson, 1974; Best, 1975). The most carefully executed study of variation among observers and map interpreters was reported by O'Connor and Marchant (1981). Four observers, differing in experience with the method, each independently completed ten visits to the same study plot in England. Only one of the observers was also a map interpreter, but the study design was such that no interpreter knew which observer was responsible for any map. A Freidman two-way analysis of variance showed significant evidence of differences between observers that could be related to the observer's prior experience both with the method and with the study plot. Because a low number of clusters was identified for most species, the percentage difference between observers was sometimes very large. Similar conclusions resulted from several other studies (Enemar, 1962; Snow, 1965; Chessex and Ribaut, 1966; Hogstad, 1967; Jensen, 1974; Enemar et al., 1978, 1979).

Such results give me reason to doubt whether acceptable consistency is generally obtained by spot mapping as it is commonly applied by practitioners. O'Connor and Marchant (1981) assert that differences in results among observers could be reduced to acceptably low levels if only highly experienced observers were used. My experience with spot mapping leads me to believe that this may be true, if plots larger than usual are used and more effort is given to recording movements

of individual birds and to obtaining simultaneous registrations of singing males (see Tomialojć, 1980). Consistency of results among different
observers, however, does not confirm spot mapping to be an *accurate*
method for estimating the densities of breeding birds, as all density
estimates could be biased.

Failure of assumption 2 can result in failure to detect locations of
some territories. The International Standard recommends detections on
at least three visits to the plot when the total sample includes eight to
ten visits. If the probability (P) of detecting a bird is at least 0.5 during
any visit to the plot, 90% or more of the clusters should be identified
(Svensson, 1979; O'Connor and Marchant, 1981). The number drops to
about 80% with ten visits at $P = 0.4$ and to about 55% at $P = 0.3$
(O'Connor and Marchant, 1981). The disadvantage of a low probability
of detection may be compensated for by additional visits to the plot,
but this leads to identification of more chance clusters unrelated to
territories by increasing detections of territorial birds off their territories
and of transients. The balance between these effects has been modeled
by O'Connor and Marchant (1981) and elaborated by Dawson and Ver-
ncr (in preparation). Results indicate that 8 to 14 visits may give the
best compromise (Fig. 1), but it is still desirable to know the value of
P. The difficulty is that P is usually estimated from results of the mapping study itself, introducing unacceptable circularity into the process.
Proper use of Fig. 1 to design a spot-mapping study depends on independent estimates of P.

Failure of assumption 3 results either in overestimation or underestimation of the portion of a boundary cluster actually within the plot.
The International Standard recommends that edge clusters "be counted
as belonging to the plot only if *more than half* of the registrations lie
within the plot or on the boundary. Otherwise they will not be counted"
(Anonymous, 1970). The National Audubon Society specifies for the
U.S. Breeding Bird Census that portions of edge territories within the
plot be estimated to the nearest 0.1 territory and summed by species
to the nearest 0.5 territory. Unless one samples the full extent of each
edge territory, both procedures tend to overestimate the proportion of
edge territories within the plot. Marchant (1981), for example, found
that 10% to 25% of edge territories were incorrectly included within
the plot when he followed the International Standard of including territories that were at least half within the plot. Most studies fail to report
whether any effort was made to sample beyond plot boundaries, and I
know of no study that sampled the full extent of edge territories. Bias
inherent in assigning edge clusters to the plot can be reduced by sampling a plot large enough to include at least three complete territories

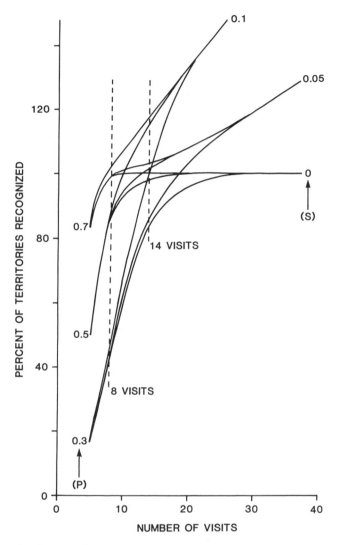

FIGURE 1. The theoretical number of territories recognized with spot mapping (100% is the correct number). The nine lines represent different combinations of two compensating sources of error: (1) the probability (P) of detecting a bird on each visit, and (2) the probability (S) that a registration contributes to identification of a cluster not truly representing a territory. For example, with ten visits, P = 0.5, and S = 0.05, about 95% of the true number of territories would be recognized (from Dawson and Verner, in preparation, based on results of O'Connor and Marchant, 1981).

of each target species, determining the mean number of detections per complete territory, and dividing that into the total number of within-plot detections from all edge clusters. Alternatively, on very large plots, one could simply assign half of the edge clusters to the plot on the assumption that plot boundaries divide the average territory into two equal parts. Manuwal (1983) used mean territory size from clusters completely mapped to adjust for within-plot portions of edge clusters. This is probably better than standard methods, but it assumes that one truly maps territories when all that is mapped are clusters of detections. These may or may not give an accurate estimate of territory size.

Bias from edge territories is greatest for species with territories large in proportion to the plot itself; it declines with increasing plot size because larger plots have a lower edge : area ratio. Density estimates of birds in oak woodlands of California, from reports of the U.S. Breeding Bird Census (*Audubon Field Notes* and *American Birds*), showed a significant negative correlation with plot size (Verner, 1981a). The combined density estimate for all birds on plots of 3.2 hectares was about 2.7 times that on plots of 13 hectares. Results of this and other studies, including consideration of the mean territory size of several breeding species, indicate that a plot size of at least 40 hectares is needed in oak woodlands of California (Verner and Ritter, 1982, in press). Scherner (1981, p. 147) reported the same relationship when comparing results of spot mapping in the German literature. He concluded that "Direct comparisons of species-numbers or -densities obtained on areas having different size are not meaningful. Such area-dependences (statistical errors) exist even if no error in census work occurs!" The International Standard recommends 40 to 100 hectares in open habitats and 10 to 30 hectares in closed habitats. I believe that 30 hectares is too small for shrublands, woodlands, and forests, but a large percentage of the U.S. Breeding Bird Censuses in such habitats were done on plots smaller than 10 hectares. Scherner's (1981) analysis led him to recommend a standard based on numbers of pairs (territories) of a species detected, concluding that at least four to six pairs must be included in the plot to give a reliable estimate of density.

Failure of assumption 4 leads either to overestimates or underestimates of the total count. Usually the number of clusters is simply doubled, assuming that all clusters include a mated pair. Typically no adjustment is made for young in or out of the nest, for floaters, or for differences in mating status (e.g., bachelors, bigamists). Floaters generally go undetected because they violate assumptions 1 and 2. If Smith's (1978) results are any indication, however, we may need to quadruple the number of clusters to get a better estimate of the real density of adults on mapping plots.

Independent assessments of the accuracy of spot mapping for estimating densities of birds give mixed results. Some comparative studies using marked birds (see Section 5.1.2) suggest that mapping gives poor results (Haukioja, 1968; Bell et al., 1973; Diehl, 1974; Jensen, 1974; Mannes and Alpers, 1975; Mackowicz, 1977; Nilsson, 1977; Osborne, 1982). Enemar et al. (1979) found territories of 15 Willow Warblers on a plot in Sweden by both mapping and intensive study of song playback to banded birds. Comparison of the territories delimited by these two methods, however, showed a poorer fit than the total counts suggest. For example, two mapping clusters were distinguished in a single territory, two boundary clusters were erroneously excluded, and a third was erroneously included. These errors coincidentally balanced. A study comparing three species in each of five plots by spot mapping in Poland (Tomialojć, 1980) gave estimates averaging 104 ± 0.5% (mean ± standard error) of the number determined by color banding to be present on the plots, but the possibility of compensating errors in this study cannot be evaluated because maps were not published. Similar results were found in oak woodlands of California, where the territory of a bigamous Bewick's Wren (*Thryomanes bewickii*) was identified by mapping as two territories by two observers independently and those of two monogamists were erroneously identified by mapping to represent one territory (Fig. 2) (Verner and Ritter, 1982). These various results suggest that *well-designed and executed spot mapping* may deliver reasonably accurate estimates of the densities of breeding birds of many species. Before this observation is blindly accepted, however, it needs further testing with many species in a wide variety of habitats.

5.1.2. Total Mapping

This term is applied here to mapping efforts that intensively study the movements of color-banded birds to identify their territories or home ranges. The literature includes a wealth of such studies, but most involve one species in a limited area. Total mapping probably gives the most accurate estimates of the densities of target species, especially when thoroughly executed and, when done during the breeding season, including an exhaustive search for nests. Maintenance of an extensive trapping and netting operation during a total mapping study even offers a chance to learn something about floaters. Indeed, I believe that total mapping should be used as a standard for evaluating the accuracy of other methods of estimating the densities of birds. Unfortunately, how-

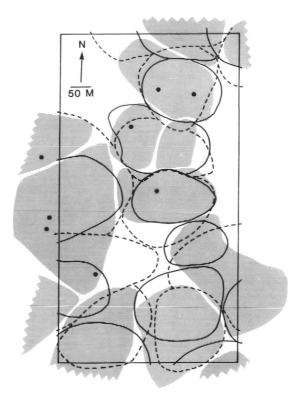

FIGURE 2. A comparison of territorial delineations of Bewick's Wrens in a 32.8-hectare plot of oak-pine woodlands in central California, using total mapping (shaded) and spot mapping by two different observers (solid and dashed lines). Spot mapping and total mapping were done independently by different observers. Points designate locations of nests found (from Verner and Ritter, 1982).

ever, studies involving total mapping rarely include other methods, probably because density estimation is not often the primary objective of such studies. Most comparisons of methods have used spot mapping as the standard of accuracy, even though it is subject to numerous biases.

Total mapping eliminates assumptions 2 and 3 of spot mapping, but it still depends on assumptions 1 (stable populations on territories or home ranges) and 4 (accurate assessment of the mean number of birds per territory). Assumption 4 may be relaxed, however, if the banding operation is extensive and intensive enough to include young birds and floaters. Playback of recorded songs can be used as an aid to total mapping, markedly increasing the detectability of target species and even giving the opportunity to estimate densities of species traditionally avoided because they are so difficult to detect (e.g., owls and rails, Johnson et al., 1981). Jensen (1974, p. 379) is among the harshest critics of spot mapping. Using total mapping, he showed that following rules of the International Standard gave only 36% of all territories with ten visits to the plot. However, he considered "the basic rules of the mapping method infallible if combined with colour ringing and time."

5.1.3. Consecutive Flush

This is basically a version of mapping without repeated visits to an established grid. First described by Wiens (1969) as a method used with grassland species, it is probably more accurate than spot mapping for certain species in certain habitats. An observer simply approaches a singing bird until it flushes. Its initial position, line of flight, and landing position are noted on a field map. The same bird is approached and flushed again, and the same data are recorded. This procedure is repeated until at least 20 consecutive flushes have been mapped, usually requiring 5 to 10 min per bird for the grassland species studied by Wiens (1969). L. B. Best (personal communication) found that it sometimes required considerably more time than that, especially for individuals that tend to avoid portions of their territory when flushed. Wiens believed the method delineated actual territories, because boundaries were stable and nonoverlapping when densities were high. He also observed many territorial disputes at points identified as boundaries by consecutive flush. As with total mapping, consecutive flush requires assumptions 1 and 4 of spot mapping, but not assumptions 2 and 3. Similar to spot mapping, it fails to account for young birds or floaters.

Consecutive flush is clearly most applicable in habitats permitting an observer to keep a bird's location in view continuously, such as grasslands, marshes, and perhaps habitats with only sparse cover of low shrubs. But I have used it with success while spot mapping in sparsely wooded habitats where even one or two consecutive flushes of a singing male added significantly to my ability to interpret maps of clusters. Indeed, information of this sort is sometimes as useful as simultaneous records of singing males.

5.2. Transects

Terminology used in the literature to identify various forms of transect sampling is variable and confusing. I suggest the following terminology, based on field methods:

1. *Line transects without distance estimates.*
2. *Variable-distance line transects.* Perpendicular distances from the line to birds detected are estimated either directly or indirectly (by recording distance from observer to bird and the sighting angle between the transect line and the bird). Species-by-species analyses can use the data either in ungrouped form or grouped into distance intervals. Analytic methods such as proposed by Emlen (1971, 1977) and Burnham *et al.* (1980) result in variable-distance treatments by species.
3. *Fixed-distance line transects.* All birds out to a fixed distance (delimiting the "main belt") are recorded separately from all birds (the "supplementary records") recorded beyond the main belt. This is the standardized Finnish line transect method (Järvinen and Väisänen, 1975).
4. *Strip transects.* The same fixed boundaries are used for all species, as in the main-belt records of the Finnish system.

Transect methods offer greater flexibility in their use than mapping methods. For example, they may be used outside of the breeding season because birds need not be territorial and they have some potential for detecting juveniles and floaters.

5.2.1. Line Transects without Distance Estimates

The observer simply walks a preset line and records all birds detected, without measuring or estimating distances to the birds. This is an efficient method for quickly generating a list of species and for obtaining total counts useful in studies with objectives that can be answered by counts unadjusted for detectability or area (Table I). Results cannot be used to estimate densities, however, as the area sampled is unknown. Its usefulness in giving measures at ordinal or ratio scales depends on the following assumptions:

1. For intraspecific comparisons (i.e., between seasons, years, or habitats), all individuals are equally detectable in all samples, or individual differences in detectability remain in constant proportion in the population in different samples.
2. For interspecific comparisons (i.e., between seasons, years, or habitats, and within habitats), in addition to assumption 1, all *species* are equally detectable in all samples.

The validity of these assumptions is discussed in the next section.

5.2.2. Variable-distance Line Transects

This is the most commonly used of the transect methods, and several models have been developed to approximate distributions of count data with distance from the observer (Emlen, 1971, 1977; Ramsey and Scott, 1978; Burnham *et al.*, 1980; also see Robinette *et al.*, 1974, and Tilghman and Rusch, 1981, for reviews and comparisons of other models). The most rigorous and comprehensive assessment of line transects is that of Burnham *et al.*, (1980). They developed both theoretical and practical aspects of the method, thoroughly reviewed and evaluated earlier models, and tested a number of new models to fit transect data. Their monograph should be read by anyone using transect data to estimate densities of birds. The models developed in the monograph are available in program TRANSECT (Laake *et al.*, 1979). Major assumptions of line transect models are:

1. No bird exactly on the transect line is missed for models using ungrouped data (i.e, not grouped into bands of varying width with distance from the line). None is missed in the first grouping interval for models using grouped data.
2. All birds are correctly identified.
3. With ungrouped data, no bird moves toward or away from the transect line in response to the observer before it is detected. With grouped data, no bird moves to another interval before being detected.
4. No bird is counted more than once.
5. Distances and angles to detected birds are measured exactly for ungrouped data. With grouped data, measurements accurately place all birds within the correct grouping interval.
6. Detections are independent events. This is especially important when results from different transects are pooled.

Burnham *et al.* (1980, p. 14) correctly pointed out that random placement of transects is necessary to justify "extrapolating results to an area of interest larger than just the immediate vicinity of the transect line." However, they disagree with the common assumption that birds must be distributed randomly to satisfy models, because random location of transects satisfies that assumption.

Jolly (1981) observed that failure to meet assumption 1 may be the major reason for underestimating densities if birds near the observer are not detected. Dawson (1981b, p. 556) reported having "plenty of anecdotal evidence that birds may easily be missed even when overhead." On many occasions while conducting transect counts, I have detected birds less than 5 m from the transect line that remained silent in shrubs and appeared to be intentionally moving inconspicuously in a manner that kept them as concealed from my view as possible. Some, such as Scrub Jays (*Aphelocoma coerulescens*), were relatively large birds of species that appear to make no attempt to remain concealed when farther from an observer. Such concealment behavior in response to a nearby observer, if this is a correct interpretation, probably succeeds occasionally, especially in areas with dense vegetation. The fact that it involves near birds could seriously bias density estimates from models that assume 100% detection of birds in the first grouping interval (e.g., Emlen, 1971, 1977; Ramsey and Scott, 1978), but it is less likely to result in serious bias in models that assume 100% detection only on the transect line itself (e.g., Burnham *et al.*, 1980).

I used the Ramsey and Scott (1978) method to calculate two density estimates of five species with a minimum count of 35 in each of two plots in oak-pine woodlands of central California. The first estimate used actual counts; tbe second used actual counts plus 0.5 bird/transect in the first grouping interval (0 to 10 m) to simulate the effect on density estimatcs of missing one bird in the first interval on half of the transects. All simulated density estimates were higher by 12% to 80%; the mean increase on site 1 was 36%, on site 2 it was 44% (Table III). The results thus indicate that birds missed near the observer can cause substantial negative bias in density estimates when data are grouped. The bias is less with higher counts in the first grouping interval, but increasing counts by increasing the width of the first interval increases the likelihood of missing birds. This problem needs further study, especially in habitats with dense vegetation or with high canopies where birds may remain undisturbed and quiet as an observer passes beneath them.

By definition, assumption 1 cannot be evaluated by a single observer. Preston (1979) presented evidence that the number of birds detected on a transect increased with the number of observers involved. He concluded that results of two or more observers could be used to correct for the numbers of birds missed, making transect counts more accurate. Preston's results may be challenged, because some of his assumptions are unverified, but his suggestion to use two observers to improve the accuracy of transects is worth exploring. Hutto and Mosconi (1981) specifically studied this question, showing that two observers detected more total individuals than a single observer. Curiously, however, two observers detected fewer birds than one observer in the first grouping interval (< 15 m). The overall advantage of two observers resulted from their detecting more birds in more distant intervals (15 to 30 m, and > 30 m).

TABLE III

Density Estimates by Line Transects of Oak-woodland Birds in Central California[a]

Species	N^b	Site #1 Density from			N	Site #2 Density from		
		Actual counts	Adjusted counts	Percent difference		Actual counts	Adjusted counts	Percent difference
Ash-throated Flycatcher (*Myiarchus cinerascens*)	50	10.1	11.8	+17	40	7.4	11.4	+54
Scrub Jay (*Aphelocoma coerulescens*)	49	9.7	17.5	+80	50	17.0	22.1	+30
Plain Titmouse (*Parus inornatus*)	80	26.1	29.2	+12	76	19.3	31.2	+62
Bushtit (*Psaltriparus minimus*)	48	28.4	33.5	+18	35	20.8	28.4	+37
Bewick's Wren (*Thryomanes bewickii*)	44	11.4	17.5	+54	46	8.8	11.9	+35

[a]Densities were estimated by using actual counts and actual counts adjusted by adding 0.5 bird per transect to the first grouping interval (0 to 10 m) to simulate the effects of failing to detect birds. Sixteen transects 660 m in length were sampled at each site from April 14 through May 5, 1980.
[b]N = Total count of individuals used to estimate density.

As with assumption 1, assumption 2 may be studied by using more than one observer. I know of no study that specifically addresses this issue and shows effects on density estimates. Certainly the problem should be minimized by using only experienced observers with demonstrated ability to identify all target species by sight and sound.

The possibility of bias resulting from movement of birds in response to a moving observer (assumption 3) was recognized by Yapp (1956), Eberhardt (1968), Emlen (1971), and others. Yapp (1956), Schweder (1977), and Burnham et al. (1980) developed some theoretical aspects of the problem, but it has not been studied empirically for transects. Density will be overestimated for species that tend to be attracted to a moving observer and underestimated for those that tend to withdraw. The effect is less with models using grouped data than with models using ungrouped data, because birds that move within an interval are still assigned to the correct one. Nonetheless, some may move from one interval to another, biasing the density estimate. Generally, movement is not a problem in line transect sampling "if it is both random with respect to the observer and slow relative to the observer's progression down the line" (Burnham et al., 1980, p. 21).

Conant et al. (1981) concluded that Nihoa Millerbirds (Acrocephalus familiaris kingi) on Nihoa Island, Hawaii, were attracted to a moving observer (not a stationary one), possibly to feed on insects disturbed by the observer. But Nihoa Finches (Psittirostra ultima) were apparently attracted to a stationary observer (not a moving one) to prey on seabird eggs while they were left uncovered during the count. Results suggest an 89% overestimate of the density of millerbirds using variable-distance line transects and a 43% overestimate of the density of finches by variable-radius point counts.

I agree with Burnham et al. (1980, p. 120) that "for most species, the movement will be away from the observer." Supporting evidence is found in the donut pattern so commonly seen in histograms of recorded perpendicular distances from transect lines to birds (Fig. 3). Burnham et al. (1980, p. 124) concluded that "it is unforeseen that a method of deriving a reliable estimator of density will be developed for animal movement away from the line without making critical and untestable assumptions." They therefore tested the robustness of existing estimators to undetected movement. The exponential polynomial and isotonic regression models were clearly more robust to movement than the Fourier series (Table IV). Burnham et al. recommended the exponential polynomial model for cases with some movement; they recommended against the use of line transects when considerable movement occurs.

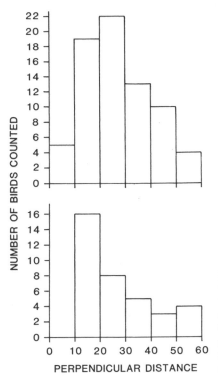

FIGURE 3. Sample histograms of the numbers of detections of Plain Titmice (above) and Bewick's Wrens (below) in oak-pine woodlands of central California with perpendicular distance (meters) from the transect line (16 transects), showing probable movement of birds away from the line in response to a moving observer.

    Assumption 4 (double counting) is likely to be violated when sampling bird populations. Most detections with line transects come from vocal cues, because birds are more often heard than seen, especially during the breeding season. This makes it difficult to study the extent of duplicate counting of individuals that change locations while still detectable by a moving observer. A single observer cannot measure duplicate counting, however, which probably explains why it has not been studied.

    Density estimates can be seriously biased by systematic errors in measures of perpendicular distances from the transect line to birds detected (assumption 5), especially for birds close to the line (compare Burnham et al., 1980). Perpendicular distances should be measured, or at least paced. Such procedures are impractical in studies of birds, however, because (1) they are so mobile, (2) most studies attempt to sample many species, resulting in a large number of detections, and (3) measurement or pacing of perpendicular distances would increase the movement by birds in response to the observer. Most observers

TABLE IV

Mean Percentage Relative Bias in Density Estimates Resulting from Four Levels of Movement away from the Observer in Simulated Transect Counts [a]

Movement ratio[b]	Treatment	Half-normal detection function			Negative exponential detection function		
		Exponential polynomial	Isotonic regression	Fourier series	Exponential polynomial	Isotonic regression	Fourier series
0.05	1	−8.1	−10.9	−19.8	−29.1	−32.3	−37.2
	2	−7.9	−16.0	−25.3	−37.3	−41.8	−45.1
	3	−10.5	−16.7	−24.6	−32.2	−37.9	−41.6
0.10	1	−14.3	−23.1	−28.9	−41.2	−45.1	−48.8
	2	−17.6	−27.7	−75.3	−55.1	−56.7	−59.0
	3	−19.8	−28.1	−62.0	−51.5	−53.3	−55.7
0.20	1	−27.9	−35.8	−70.6	−56.4	−58.0	−59.3
	2	−34.5	−42.8	−90.4	−66.6	−67.3	−69.3
	3	−31.4	−40.3	−87.9	−63.5	−64.9	−66.1
0.40	1	−48.5	−54.1	−87.9	−70.1	−70.4	−72.4

[a]Three models (exponential polynomial, isotonic regression, and Fourier series) and two assumed detection functions (half-normal and negative exponential) were used; N = 100 in each case (adapted from Burnham et al., 1980, Tables 14 and 15).
[b]A movement ratio of 0.05 represents very little movement; one of 0.40 represent extreme movement.

simply estimate perpendicular distances, both from visual and auditory cues, without leaving the transect line. This process is sometimes aided by placement of markers at known distances from the line, by using areas with numerous natural landmarks at known distances, by using gridded study areas with which observers have extensive experience, or by using range finders to measure distances.

Few studies report attempts to measure systematic bias in distance estimation by observers. Dawson (1981b) found significant differences among four observers in the proportions of several species estimated as farther away than 50 m. Emlen (1977) believed that with some practice he could almost invariably judge within 10% to 15% of the correct distance to selected fixed objects at distances up to 60 m. He later abandoned this procedure when he found that variability among briefly trained birders was too great (J. T. Emlen, personal communication). A brief test (unpublished) of distance estimation in oak-pine woodlands of central California by six observers from our laboratory in Fresno, California, showed that five observers without training consistently underestimated distances (n = 100) to visible objects and one consistently overestimated them. The mean deviation from the true distance (all estimates of all six observers) was 29%. After the initial test, three observers spent 3.5 hr each, over a two-day period, training to estimate distances. The following day they repeated the initial trial. Estimates of these three observers before training averaged 69% of the actual distances; after training they averaged 92%. Finally, two of the observers repeated the test three months after training, with no intervening training or practice. They averaged 70% of actual before training, 91% immediately after, and 93% three months after, indicating that the effects of training did not decay during the longer waiting period after training. Surprisingly, in all of these trials, the *percentage* deviation of the estimated distance from the true distance showed no trend with increasing distance; observers were relatively as accurate (inaccurate) in estimating near distances as far distances.

These results have important implications for users of line transects to estimate densities of birds. Most studies fail to report training of observers to estimate distances; I assume that none was done. Without distance markers to serve as guides, untrained observers probably estimate distances no better than our test results suggest and most probably underestimate them. This would introduce a systematic positive bias in density estimates that would be most serious when distances of birds near the observer were underestimated (Burnham et al., 1980).

Estimating the distance to a bird from an aural cue should be even more difficult than from a visual one. Scott et al. (1981) found that 81%

of all detections of forest birds in Hawaii were from aural cues. They tested the ability of observers trained to estimate distances to singing or calling birds (Kepler and Scott, 1981); 95% of the estimates were from 57% to 175% of the measured distances. As with estimated distances to objects seen, those to objects heard were biased toward underestimation. "Observer mean effects ranged from 9.1% below measured to 6.3% above measured" (Scott et al, 1981, p. 336). Effects of these errors on density estimates were simulated for point counts (see Section 5.3), but not transects.

Recent studies by Emlen and DeJong (1981) and DeJong and Emlen (in preparation) show some promise of avoiding the need to estimate distances to singing birds in some cases. Measured maximum distances at which singing males of a given species were audible to trained observers with normal hearing, the "detection threshold distances," were surprisingly constant (coefficients of variation ranged from 9% to 25%). The authors believe it may be possible to prepare tables (see Järvinen and Väisänen, 1977; Järvinen, 1978) of detection threshold distances for various species of songbirds in various habitats, which could then be used as a standard reference for computing density, given (1) transect length, (2) a total count of singing males, (3) an adjustment, independently derived, for frequency with which males sing during the phenological stage when counting is done, and (4) an adjustment for sex ratio. Such a standard, if shown to be valid, could be invaluable in many situations. It would not, however, eliminate the need for distance estimation by observers with impaired hearing. It also would not include floaters or juveniles in the final estimate of density.

Assumption 6, that detections are independent events, is necessary for certain kinds of data analysis. The assumption is unavoidably violated in transect counts of birds. Their mobility and vocal behavior make it likely that the same individual will be detected on two or more transects, unless transects are separated by considerable distances (adequate quantative studies unavailable). The vocal activity of one individual is likely to stimulate vocal activity or movement by another, making it detectable when otherwise it may have been missed. Nonrandom distribution of birds with respect to one another (pairs, families, flocks) often results in detection of more than one at a time.

When birds are detected in groups, the usual procedure is to estimate the distance to the presumed center of the group, record the number of individuals in the group, and use mean distance to groups of average size to estimate density. In this case, groups are treated as individuals in the analysis, so they must meet all assumptions of the models for individual birds and also satisfy the assumption that average group size is correctly estimated. When the probability of sighting a group is related to group size, as in porpoises, it complicates getting a valid estimate of mean group size (Burnham et al., 1980). Such cases require independent information to weight group sizes estimated from transect data.

5.2.3. Fixed-distance Line Transects

This is the standard method used in Finland since the pioneering work of Merikallio (1946, 1958). Density estimates were initially based only on birds detected in the main belt (usually 50 m wide—25 m on either side of the transect line). Järvinen and Väisänen (1975) proposed models to use the supplementary records beyond the main belt when estimating densities, increasing sample size an average of five-fold (20% of records are within the main belt). This can be important when estimating densities of rarer species and when monitoring long-term fluctuations. Assumptions of their models are:

1. No bird exactly on the transect line is missed.
2. All birds are correctly identified.
3. No bird moves into or out of the main belt in response to the observer before being detected.
4. No bird is counted more than once.
5. Distance estimations needed to assign birds to the main belt are made without error.
6. Detections are independent events, especially when results from different transects are pooled.

Järvinen and Väisänen (1975) tested three models using different assumptions about the rate at which the detectability of birds declines with distance from the observer: (1)negative exponential, (2) linear, and (3) normal. With the exception of species for which about 55% or more (judging from Fig. 1, Järvinen and Väisänen, 1975) of the records were within the main belt, the three models gave similar estimates of density. These models consistently gave higher density estimates than those based only on main belt records. The linear model has been used in subsequent work by these investigators, "because it is intermediate between the other two" (Järvinen, 1976). Järvinen recommended a minimum count of 20 birds when estimating the density of a species by the linear model. Counts of this size gave standard deviations of no more than 0.0005 in the constant (k) used to approximate the maximum distance from which a species is detectable. Density estimates should be similarly precise, because D (density in pairs/km^2) = 1000 [(Total count × k)/Transect length in km].

Järvinen and Väisänen (1975) believe that the most serious obstacle to their method is that not all birds are detected at zero distance (assumption 1). It also suffers from a lack of unbiased estimates of k for a given species in different habitats, regions, times of the year, and so on (see Järvinen, 1976). The constant, k, is similar in concept to the detection threshold distance of Emlen and DeJong (1981). Indeed, Järvinen and Väisänen (1977) have assembled a table of empirically estimated k values for 170 species of European birds.

5.2.4. Strip Transects

Strip transects may be regarded as line transects with fixed boundaries that are applied to all species counted. Boundaries are usually set from 25 to 50 m on each side of the center line. They are simpler to use than variable-distance line transects, because one records only the birds detected within the strip, and the total for each species is divided by the area of the strip to obtain a density estimate. The observer must be trained to estimate only one distance accurately—that to the limit of the strip. Assumptions of this method are:

1. All birds within the strip are detected.
2. All birds are correctly identified.
3. No bird moves into or out of the strip in response to the moving observer.
4. No bird is counted more than once.
5. No errors are made in determining whether a bird is within the strip.
6. Detections are independent events, especially when results from more than one transect are pooled.

Discussion of most assumptions pertaining to variable-distance line transects apply here as well. Violations of assumptions 1 and 3 are probably the most damaging to this method. The density estimate of any species is negatively biased in proportion to the percent of undetected birds. The effect should be greater for secretive than for conspicuous species, and it should increase in proportion to the density of vegetation on the strip. Similarly, species differ in their responses to a moving observer (e.g., Burnham et al., 1980; Conant et al., 1981). Some are attracted toward the observer, but probably most are repelled. This can result in some individuals moving beyond strip boundaries before they are detected, lowering the estimated density. Violation of assumption 3 can be reduced by increasing the width of the strip, but this increases the likelihood of violating assumption 1. The result is that density estimates of different species are likely to be differentially biased, so interspecific comparisons using strip transects are inadvisable unless independent evidence shows that the method equivalently samples all species.

5.2.5. Accuracy of Transects

Accuracy of density estimates from transect methods has not been adequately tested against an absolute standard. Several workers (Emlen, 1971, 1977; Franzreb, 1976, 1981; Dickson, 1978; Järvinen et al., 1978a, b; Tiainen et al., 1980; Hildén, 1981; O'Meara, 1981; Redmond et al., 1981) have compared transect results with those of spot mapping, although spot mapping has not been proved to be an accurate method either (Tomialojć, 1980). Results of fixed-distance line transects are believed to be crudely rather than absolutely correct (Järvinen and Väisänen, 1975). Reports of close agreement between spot mapping and fixed-distance line transects (Järvinen et al., 1978a,b; Tiainen et al., 1980) have been challenged by Tomialojć (1981) because comparisons lumped all species. Reanalysis of two of the data sets by species showed that about half of the species in each case had 50% or greater departure of the transect estimates from the mapping estimates. Emlen (1971, p.340) concluded that properly adjusted transect estimates will "resemble those obtained directly by territory mapping in general order of magnitude." Franzreb (1976) observed significant differences in density derived by the two methods and suggested proper caution when applying results of either. With her 1981 data set, however, she reached the surprising conclusion that "for the majority of species, density values using the two methods were similar" (Franzreb 1981, p.165). I cannot reach the same conclusion from examination of her results.

When density estimates by the two methods were summed for all species in each of the six data sets reported by Emlen (1971, 1977) and Franzreb (1976, 1981), the lower estimate averaged 81.3% of the higher (range 59.4% to 96.5%); mapping estimates were higher in five of the six cases (Table V). Summing across species obscured many real differences between methods, because higher estimates for some species were offset by lower estimates for others. If the two methods were exactly equivalent, density estimates by both should be the same for all species. If they were only relatively equivalent, one method should give comparably higher density estimates than the other for all species.

TABLE V

Summary Statistics of Comparisons of Density Estimates by Spot Mapping and Variable-distance Line Transects[a]

| | Method with higher estimate | | | | | | | |
| | Spot mapping | | | | Transect | | | |
	N	Mean	S.D.	C.V.	N	Mean	S.D.	C.V.
Emlen, 1971[b]	11	61.5	21.69	0.35	15	67.8	20.96	0.31
Emlen, 1971[c]	27	42.1	25.95	0.62	2	70.0	29.70	0.42
Emlen, 1977[d]	23	58.7	21.39	0.36	3	81.7	15.63	0.19
Emlen, 1977[e]	10	83.8	7.60	0.09	8	89.6	8.00	0.09
Franzreb, 1976[c]	20	57.5	24.23	0.42	5	75.0	21.02	0.28
Franzreb, 1981[c]	22	59.3	21.99	0.37	10	78.9	13.49	0.17

[a]Calculated from results of Emlen (1971, 1977) and Franzreb (1976, 1981). Tabled means are of percentages [100(lower estimate/higher estimate)]. Species undetected by either method have been omitted.
[b]Transect estimtes by Method "D": 2[(2 × singing males within 200' of line) + nonsong detections)] (consult Emlen, 1971).
[c]Transect estimates by Method "G", the higher of (1) all detections, adjusted for coefficient of detectability; or (2) songs × 2, adjusted for coefficient of detectability (consult Emlen, 1971).
[d]Transect estimates based on all cues detected.
[e]Transect estimates based on song cues, adjusted for song frequency and for undetected females.

Neither situation occurred. Comparison of each species separately showed that density estimates from spot mapping exceeded those from transects in 113 cases; transect estimates exceeded mapping estimates in 43 cases. When mapping estimates were higher, mean transect estimates were less than 62% of those from mapping in five of six comparisons. Conversely, when transect estimates were higher, mean mapping estimates were at least 70% of those from transects in five of six comparisons. The two methods did not give even relatively equivalent density estimates in most cases. This is further supported by the typically large coefficients of variation found when comparing the percentage agreement between the two density estimates (Table V), showing that the two methods did not give density estimates even on the same ordinal scale. One cannot expect density estimates on equivalent ratio or absolute scales from data not on the same ordinal scale.

Emlen's (1977) transect estimates based on song cues, adjusted for song frequency and for undetected females (Table V), were exceptional by comparison with the other five data sets. Nearly the same number of species gave higher density estimates by each method, the percentage difference of the lower estimates averaged more than 80% of the higher in each comparison, and standard deviations and coefficients of variation were only 0.09. Work is continuing on this approach (Emlen and DeJong, 1981), since it promises to make these two methods more equivalent. An added requirement of the method, however, will be to derive estimates of song frequency from a source independent of both the transect and mapping data sets. At best, one of the methods could be accurate and the other inaccurate. It is more likely, however, that both methods are inaccurate.

Various models used to fit transect data have been compared (Robinette et al., 1974; Burnham et al., 1980; Tilghman and Rusch, 1981). Robinette et al. (1974) used data from 18 transect samples of known populations of inanimate objects and two of animals. Results showed that the best estimates of density were given by the estimators of Kelker (1945) and Anderson and Pospahala (1970), using measured perpendicular distances from the transect line. Tilghman and Rusch (1981) also used several estimators to derive densities from line transect data, using spot-mapping estimates as the standard for comparison. They concluded that no method or group of methods was better than others and that "line transect methods provided reasonable indices to the density of most bird species" (Tilghman and Rusch, 1981, p. 207). I disagree. Their density estimates by transects and spot mapping differed by a factor of two or more in 30% of the cases; relative bias was substantial in many cases; the smallest mean coefficient of variation in density estimates for any estimator was 0.33 (White *et al.*, 1982, state that values of 0.10 or less are considered good); and 5 of the 12 estimators gave density estimates that failed to rank species in the same order of abundance as the mapping estimates (Spearman's *rho* > 0.05).

Burnham *et al.* (1980) evaluated the various estimators tested by Robinette *et al.*, (1974), which were later tested by Tilghman and Rusch (1981), and recommended against using any of them. Tests of the Fourier series model (Burnham *et al.*, 1980, pp. 57–63) gave good results with simulated data and data from stakes placed in the ground (known density) and later sampled by line transects by students unfamiliar with stake locations (Laake, 1978). In each case, density estimates were within 4% of known densities, suggesting excellent performance of the model under these conditions. Further study of the model led Buckland (1982) to conclude that it provides a powerful tool for estimating densities from line transect data, with estimates being reliable across a wide range of detection functions. On the other hand, Buckland found problems with nominal confidence intervals given by the Fourier series model, recommending a Monte Carlo interval as the most reliable solution to the problem. The Fourier series model and alternatives examined by Burnham *et al.* (1980) appear to be the best available for analyzing line transect data. They are not, however, sufficiently robust to violation of certain assumptions so often seen with counts of birds. For example, as Burnham *et al.* (1980, p. 21) point out, "If the subject of the study is a highly mobile animal (such as a passerine bird), serious problems due to movement can arise, often to the extent of rendering line transect sampling useless for such species." Wrong identifications or failure to detect birds on or near the transect line can bias density estimates at least as much as movement. Consequently, we should not assume that any of these models gives accurate estimates of bird densities from line transect data, even though they may be the best models available.

5.3. Point Counts

As with transects, point counts can be classified into four types based on field methods:

1. *Point counts without distance estimation.* These cannot be used to estimate density, but they provide an efficient means for measuring species richness and for accumulating total counts of birds (Verner and Ritter, in press).
2. *Variable-radius point counts.* Distances from the observer to birds detected are estimated. Analyses, by species, can use the

data either in ungrouped form or grouped into bands encircling the counting station (Reynolds *et al.*, 1980).

3. *Fixed-radius point counts.* These are analogous to the fixed-distance line transects used in Finland, the analytic procedures of which could be appropriately modified to treat data from point counts in which detections in an inner circle of fixed radius are recorded separately from all detections beyond. I have no literature reference to studies using this method.

4. *Circular plots.* The same fixed radius is used for all species.

Burnham *et al.* (1981) correctly pointed out that point counts are basically transects done at zero speed, so that various models applied to one method can be adapted to the other, and assumptions that apply to transects (see above lists) apply to point counts as well. For example, the equation used to estimate densities from fixed-distance line transects has been modified to do so with point counts, assuming that each species has a definable maximum detection distance, but not requiring that detections in an inner circle be recorded separately from detections beyond the circle (see Järvinen, 1978, and references therein). Variations and refinements of some assumptions are discussed by Ramsey and Scott (1978) in reference to their Poisson model to fit data from point counts.

Because the area sampled with point counts increases geometrically with distance from the observer but only linearly with transects, violation of certain assumptions leads to much larger errors in density estimation with point counts. The assumption that all birds are detected in the first grouping interval cannot be evaluated by a single observer using point counts. It seems reasonable to assume that birds are more likely to be missed in areas with denser vegetation, higher canopies, or both, but I know of no study testing this. Frequent vocal behavior reduces the likelihood of missing a nearby bird, especially with counts of longer duration. (But longer counts increase the chance that birds initially beyond detectability will move near enough to be detected, either naturally or in response to the observer.) I have anecdotal information that birds in shrubs as near as 5 m from the observer occasionally remain concealed and quiet throughout point counts lasting as long as 10 min. W. J. Sydeman (personal communication) has often observed Pygmy Nuthatches (*Sitta pygmaea*) remaining motionless and silent in a tree for periods long enough to avoid detection by an observer conducting point counts of normal duration (5 to 10 min); he believes such birds could be missed even when directly overhead.

As with transects, I used the Ramsey and Scott (1978) method to compute two density estimates of five species each with a mimimum count of 43, based on 80 8-min point counts in each of two plots in oak-pine woodlands of central California. The first estimate used actual counts; the second used actual counts plus 0.25 bird/count in the first grouping interval (0- to 10-m radius) to simulate the effect on density estimates of failing to detect birds. Adjusted density estimates were higher by 253% to 3083% (Table VI); the mean increase on site 1 was 1112%; on site 2 it was 1624%. These results show that birds missed in the first grouping interval can seriously bias density estimates; the magnitude of the effect (Table VI) leads me to believe that fewer than 0.25 birds were missed per count, on the average.

Natural movement of birds throughout their home ranges makes it likely that an otherwise undetectable bird will move within detection range of the observer during a count. "The results of simulation studies...show that bird mobility may seriously bias density estimates derived from variable circular plot surveys, especially for counts of longer duration" (Scott and Ramsey, 1981b, pp. 412–413). The effect was relatively minor for slower-moving species, but it was substantial for those moving at least one effective detection radius (distance to inflection point) during the period of the count (Fig. 4). An empirical study of bias resulting from movement of birds used two consecutive, 5-min counts on circular plots 30 m in radius in coniferous forests of Yosemite National Park, California, and compared density estimates from the 5-min counts with those from the combined 10-min count (Granholm, 1983). Ten-minute counts gave density estimates that were 22%, 47%, and 56% higher for the three species studied. Granholm interpreted the differences as resulting from movement by birds across plot boundaries and considered them to be typical of species found in the forests he sampled.

Errors in distance estimation were discussed earlier (see Section 5.2.2). Effects of errors in distance estimation by observers trained to estimate distances showed that densities were overestimated for 32 of 35 species, typically by about 20% and in one case by more than 30% (Scott et al., 1981).

Bias from double counting an individual should be on the same order of magnitude, but in the opposite direction, as that resulting from failure to detect a bird. Incorrect identification combines the biases from missing a bird (the correct species) and double counting one (the incorrect species). No study reports quantitative results of bias from double counting or incorrect identification.

Accuracy of density estimates from point counting, as with all other methods, still needs thorough study with a variety of species in many

TABLE VI

Density Estimates by Point Counts of Birds on Two Sites in Oak-pine Woodlands of Central California[a]

Species	N[b]	Site #1 Density from			N	Site #2 Density from		
		Actual counts	Adjusted counts	Percent difference		Actual counts	Adjusted counts	Percent difference
Ash-throated Flycatcher	63	10.0	318.3	+3083	59	11.3	350.1	+2998
Scrub Jay	58	28.3	334.2	+1081	73	15.5	334.2	+2056
Plain Titmouse	101	69.0	334.2	+384	117	21.2	318.3	+1401
Bushtit	64	90.2	318.3	+253	43	49.5	366.1	+640
Bewick's Wren	58	37.1	318.3	+758	61	28.3	318.3	+1025

[a] Densities were estimated by using actual counts and actual counts adjusted by adding 0.25 bird per point count in the first grouping interval (0- to 10-m radius) to simulate effects of failing to detect birds. Eighty 8-min point counts were done on each site from April 14 through May 5, 1980.
[b] N = Total count of individuals used to estimate density.

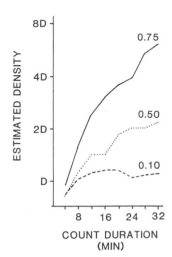

FIGURE 4. Simulated effects of bird movement on density estimates with increasing duration of count period. Fractional numbers designate proportions of an effective detection radius moved during a 4-min count period (adapted from Scott and Ramsey, 1981b).

habitats. Because violation of certain assumptions results in larger errors in density estimates from point counts than from line transects, Järvinen (1978) concluded that point counts are probably less accurate than line transects for estimating densities of birds. I agree.

Density estimates from variable-radius point counts and from spot mapping in two desert riparian sites in Arizona were comparable (Szaro and Jakle, 1982). Estimates from spot mapping were within the 95% confidence limits of estimates from point counts for all species compared except the White-winged Dove (*Zenaida asiatica*). Estimates in one comparison were the same; in 15 comparisons the mapping estimates were less, averaging 80% of the point-count estimates; and in eight the point-count estimates were less, averaging 86% of the mapping estimates. Nineteen comparisons were not possible because fewer than ten records were obtained during point counts (Burnham et al., 1980, recommended at least 40 records for line transects, which is a better standard for point counts as well).

The most thorough study of the accuracy of variable-distance point counts to estimate densities of birds was done by DeSante (1981), using eight species of passerines in a 36-hectare plot of coastal scrub in Marin County, California. DeSante used total mapping to determine densities, estimating that 64% of all territorial birds were color banded and that 43% of all nests were found. Eight-minute point counts were done four times by each of four observers at 13 points distributed evenly over the plot (total n = 208). DeSante did not report the minimum count accepted to estimate densities from point counts. Results were mixed. Point-count estimates ranged from 29% to 98% of those from total mapping ($\bar{x} = 75.0\%$). In my opinion, these results do not generate much confidence in the accuracy of point counts for estimating densities of birds. Furthermore, separate estimates by the four observers gave even less encouraging results, suggesting that pooled errors of several observers tend to balance out. This may mean that effective use of point counting will either require the use of several observers or extensive prior training such as that recommended by Kepler and Scott (1981). This reduces the efficiency of the method, which is one of its primary recommendations. Furthermore, point counts of five of the species consisted mainly of singing males, causing DeSante (1981, p. 184) to caution that the method, although "reasonably accurate during the height of the breeding season, may be quite inaccurate at other times of the year."

5.4. Other Methods

5.4.1. Capture–Recapture

Several methods use ratios of marked to unmarked animals to estimate population sizes. When these are combined with careful estimates of the area sampled, densities may be estimated from the results. Many different models have been developed to estimate population sizes according to a variety of assumptions about population closure and animal response to trapping. These methods have received little use in studies of bird populations, probably because they are extraordinarily time consuming. For this reason, I have elected not to discuss them in any detail here. The extensive literature on these methods may be traced by the reader through a number of excellent recent papers (Cormack, 1968, 1979; Otis et al., 1978; Jolly, 1979; Nichols et al., 1981; Pollock, 1981; Seber, 1982; White et al., 1982). White et al. (1978) document program CAPTURE, available free (Utah Cooperative Wildlife Research Unit, Utah State University, Logan UT 84322). This is an essential tool for anyone using capture–recapture methods that assume a closed population.

5.4.2. Indirect Measures of Density

Nests, tracks, feces, and other indirect signs of the presence of birds may provide indices to their abundance, but this is true for so few species that these measures are generally impractical (see review by Bull, 1981). Nest locations, however, can provide vital supplemental information for interpreting results from spot mapping. Mapping in combination with nest searching was even advocated as a standard of absolute density by Berthold (1976). This is not practical with most species, however, because it is too difficult and time consuming to locate all nests. Furthermore, such activity may disturb the birds and their habitat (Enemar, 1959; Oekle, 1977).

5.4.3. Aerial Counts

Large flocks of birds can sometimes be viewed from the air, in which case photography appears to offer the only reasonably accurate means for obtaining counts. Norton-Griffiths (1978) and Caughley (1980) review methodology and Ferguson and Gilmer (1980) discuss recent technology. Aerial counts usually cannot be used to estimate densities of birds, because birds are easily missed and because one normally cannot relate flocks sampled to total area occupied.

5.4.4. Migration Counts

Although counts of migrating birds cannot be used to estimate densities, carefully executed counts show promise as a means to examine trends in populations. In some instances, high correlations have been found between results of migration counts and spot mapping (O'-Connor, 1979; Svensson, 1979) and line transects (Järvinen and Väis-änen, 1979). An excellent review of this topic, together with a description of a new method for analyzing results of migration counts, is given by Hussell (1981).

6. CONCLUSIONS

Development of efficient and reasonably accurate methods to estimate densities of most species of birds cannot be considered accomplished. In previous sections, I scarcely introduced the extensive literature documenting bias in bird counts and the density estimates derived from them. Either we must assume that all biases miraculously compensate for one another or we must accept the fact that density estimates by standard methods are biased (inaccurate). This might not be so bad if estimates for all species were biased in the same direction and proportion (we could at least assume accurate ratio and ordinal scales of measurement). This is not true, however, as shown clearly by case studies cited here and by many others in the literature not discussed in this chapter. We must add to this the widely acknowledged fact that most methods fail to estimate densities of floaters. Finally, because transects and point counts are not limited to territorial birds, they should detect transients, subadults, and floaters, consistently giving higher density estimates than spot mapping. Just the reverse is generally true. In my opinion, an objective assessment fails to show that densities of most species of birds in most habitats can be measured accurately on an absolute scale or even on a ratio scale by any known method, with the possible exception of exhaustive total mapping.

If density estimates depart too far from true densities, the significance is far-reaching. It means that all studies requiring measurement at an absolute scale and most requiring measurement at a ratio scale are suspect. For example, I believe that studies of energy consumption by birds as part of a larger study of trophic dynamics should not be attempted with any method other than total mapping to estimate densities of birds (see the exemplary study by Holmes and Sturges, 1975), yet some observers express annual energy consumption to four significant digits (Alatalo, 1978, and references therein). Most methods fall short of giving estimates needed to compute indices of species diversity—a major effort of avian ecologists in the recent past. Many analyses of species diversity used density estimates from spot mapping (often from results of the American Breeding Bird Census, as reported in *Audubon Field Notes* and *American Birds*). Spot-map estimates may be reasonably good for many species, but I have never understood how a diversity index could be given for a bird community in which some species were listed simply as present and others (e.g., owls) were ignored altogether.

Because biases of bird counts differentially affect various sex and age classes, and because biases differ among seasons and among habitats, it is risky to assume that we can properly compare populations intraspecifically (1) at different times of the year, (2) in different habitats, or (3) in different seral stages on the same site. The same is true of interspecific comparisons, for which comparisons among species in the same habitat are also risky.

Line transects and point counts are appealing as methods because they are generally considered to be more efficient than mapping methods, and because they can give confidence limits for density estimates. This view can be challenged, however, because transects and point counts typically give counts of some (most?) species too low to allow a density estimate, using the minimum count criterion of 40 recommended by Burnham et al. (1980). The comparison of density estimates from point counts and spot mapping by Szaro and Jakle (1982) used 56 8-min point counts per site (7.5 hr) and allowed only 55% of the possible comparisons because some species had counts of less than ten. Even fewer comparisons would have been possible with a minimum count of 40 as the criterion. If any species in either area was detected only once in 56 counts, 300 hr would be needed to bring its count to 40. Similar calculations using results of transects and point counts in oak-pine woodlands of central California (Verner and Ritter, 1982, in press) show that 480 hr of transect and 533 hr of point-count sampling would be needed to obtain a count of 40 for species with the lowest counts. Six to nine well-executed spot mapping studies, using ten visits of 6 hr each to 40-hectare sites, could be done with similar time investments. Confidence limits could be calculated, and I believe they would be smaller and the density estimates more accurate than with either transects or point counts.

All of this is not to say that we cannot learn a great deal about avian ecology from counts of birds. We can. Many questions can be answered with a good index of abundance. Selection of the best method depends on study objectives (Järvinen, 1976; Dawson and Verner, in preparation). Table I shows that three basic study objectives can be met with lists of species, which can be obtained from any of the three primary methods (mapping, transects, and point counts) discussed in this chapter. Transects and point counts are more efficient for this than mapping. However, because the relative efficiency of transects and point counts depends in part on the habitat and in part on the duration of point counts and the spacing between counting stations (Verner and Ritter, in press), a pilot study is recommended before a choice is made between these methods. Unadjusted counts from a well-designed study can be used to track annual trends in populations. I strongly recommend point counting as the preferred method for studying annual trends, primarily because (1) the time spent counting can be absolutely controlled, and (2) more sites can be sampled, permitting more representative sampling. Furthermore, point counts can be designed to be about as efficient as transects.

Five of eleven study objectives listed in Table I require measurement of abundance on a ratio scale. Thorough, well-executed spot mapping, especially on areas of 40 hectares or more, is probably the method of choice at this time for such studies. As indicated in previous sections, I do not believe the state-of-the-art in line transects or point counts can yet deliver density estimates capable of meeting those study objectives. Hopefully, however, further research on species' detectabilities and on ways to estimate areas sampled will soon give these methods the level of accuracy we need. Total mapping is the only method suitable for studies in terrestrial communities requiring an absolute scale of density estimation.

7. RECOMMENDATIONS

Conclusions reached here seem to leave us with three alternatives: (1) We may continue to estimate densities as we have in the past, thus perpetuating what I judge to be poor science; (2) we may abandon as hopeless our efforts to estimate densities of birds efficiently and seek other ways to investigate questions we now try to answer with density estimates; and (3) we may seriously address the real challenge of finding efficient ways to obtain reasonably accurate density estimates of birds. I sincerely hope that this chapter will stimulate many to take alternative 3 seriously. A first attempt to identify higher priority research needs resulted from a short workshop following an international symposium in 1980 on *Estimating Numbers of Terrestrial Birds* (Verner 1981b). In this section, I expand on certain points brought out by the workshop and offer additional suggestions as well.

7.1. Compare Methods against an Absolute Standard

Numerous studies have compared one imperfect method with another; some were briefly described in this chapter. These studies, although interesting and useful for certain purposes, fall short of the goal of learning how accurately a given method estimates densities of birds. Spot mapping has been used as the standard of comparison in most of these studies. Although *well-executed* spot mapping probably permits density estimation of territorial birds during the breeding season more accurately than any method other than total mapping, it still has not been shown to be an accurate method.

I believe that total mapping can deliver reasonably accurate density estimates if done in a thorough way: (1) most birds banded and their movements studied long enough to establish territories or home ranges and to gather information on sex ratios; (2) trapping and netting operations maintained throughout the study to obtain information on floaters and transients; and (3) intensive search for nests if during the breeding season. I recommend total mapping as the best alternative we have as a standard for comparing the accuracy of other methods used to estimate densities of birds. We might be pleasantly surprised by how accurate density estimates are from certain well-executed methods.

7.2. Study Bias

The number of permutations possible, given the numbers of bird species, habitat conditions, seasons, and factors known to bias estimates of density, totally defy our ability to study them all. Burnham (1981, p. 325) correctly pointed this out, but stated that "Only general guidelines are needed, or feasible, and it should be possible for experienced ornithologists to provide such guidelines, in most cases, without further studies." This may be true for certain sources of bias, such as those associated with weather conditions or study design. But I have found few well-documented studies of bias in density estimates of birds, certainly not enough to formulate generalizations or correction coefficients that could be used to correct biased estimates in the variety of studies done regularly in avian ecology. Indeed, many topics mentioned in this chapter were left with the observation that I could find no studies relating to them. Much room remains for study of bias in bird counts and the density estimates derived from them. This should lead the way to a better understanding of how to eliminate some biases by standardizing study design or by testing and training observers. Biases not controllable in these ways must be studied so that coefficients of correction can be applied to the variety of cases involved.

First priority should go to biases shown to have the greatest impact on density estimates from standard methods (Table II). The following list gives my opinion of the problems that should be given higher priority: (1) effects of vegetation structure and composition on detectability of ground, shrub, and tree-nesting and foraging species; (2) magnitude and effects of interspecific differences in detectability; (3) effects of intraspecific differences in detectability; (4) effects of birds missed, bird movement, and errors in distance estimation in line transects and point counts; and (5) effects of edge clusters (hence plot size) in spot mapping.

Many of these problems will require two or more observers in carefully controlled studies designed specifically to learn about one or more sources of bias. Examples of such studies are those of Laake (1978), Preston (1979), Emlen and DeJong (1981), Hutto and Mosconi (1981), Kepler and Scott (1981), and Scott and Ramsey (1981a). Much more attention should be given to quantifying the efficiency of counts by species, season, and habitat, perhaps along lines suggested by Seierstad et al. (1965, 1967) and Emlen (1971, in preparation).

7.3. Facilitate Cross-study Comparisons through Standardization

Past studies have differed in many ways, and it is unreasonable to assume that this will change much in the future. Differences, however, sometimes make it imprudent or even impossible to compare results of different studies. We need to explore ways to achieve more standardization of study designs for the various methods used to count birds. Mapping studies, for example, could be better standardized on plot size (compare Scherner, 1981), number of visits (O'Connor and Marchant, 1981; Dawson and Verner, in preparation), and recognition of the importance of simultaneous registrations and bird movements (Tomialojć, 1980). Better guidelines for selecting line lengths, grouping intervals, and minimum counts of individuals, by species, for estimating densities by variable-distance line transects could lead to more standardization of that method. Point counts would be more comparable if all had the same duration. Better standards for locating sampling areas and for setting a minimum distance between them (mapping plots, transects, or counting stations) are needed for all methods.

We also need better ways to standardize, after the fact, studies that used different methods or study designs. Rarefaction procedures, recently introduced into the ornithological arena (James and Rathbun, 1981; James and Wamer, 1982), provide a better way than species diversity to compare diversities of different assemblages of birds. Observers have standardized count data by effort (e.g., species or individuals detected per hour), by kilometer of transect sampled, or by area sampled. However, studies that differ in all these regards strain credulity when standardized for comparison.

7.4. Study Floaters

Floaters are probably closely tied to the land somewhere, either as members of an organized "underworld" as described by Smith (1978) in the Rufous-collared Sparrow, or as other territorial birds wandering off their territories. Their study is important for reasons other than learning to adjust for them when estimating densities of birds. They are important to an understanding of avian social behavior in general and mating systems in particular; knowledge of floaters is also important for understanding population dynamics. Moreover, they are of interest solely on their own merit. I see the study of floaters as a major neglected area of avian field biology.

7.5. Quantify Statistical Distribution of Counts

Development of models to fit count data depends on a study of the statistical distribution of counts, based on very large data sets. The work of Burnham et al. (1980, and references therein) is exemplary in this regard.

7.6. Upgrade the Quality of Field Work

In my opinion, the literature dealing with density estimates of birds includes an appalling number of studies using poor designs and poor techniques. Sample sizes are inadequate; sampling procedures are unbalanced; analyses that assume independent samples are used when samples are obviously not independent; even recommended standards and guidelines, although cited, are often ignored. Greater care in planning and executing studies would unquestionably improve the accuracy of density estimates. This should include prior consultation with biometricians familiar with problems of estimating densities of birds, testing the skill (field identification of birds) and sensory capabilities (sight, hearing) of field personnel, and appropriate training of field personnel.

REFERENCES

Alatalo, R. V., 1978, Bird community energetics in a boreal coniferous forest, Holarctic Ecol. 1:367–376.

Anderson, B. W., and Ohmart, R. D., 1977, Climatological and physical characteristics affecting avian population estimates in southwestern riparian communities using transect counts, in: Importance, Preservation and Management of Riparian Habitat: A Symposium (R. R. Johnson and D. A. Jones, tech. coords.), USDA Forest Service, Rocky Mountain Forest and Range Exp. Stn., Fort Collins, Colorado, Gen. Tech. Rep. RM-43:193–200.

Anderson, D. R., and Pospahala, R. S., 1970, Correction of bias in belt transects of immotile objects, J. Wildl. Manage. 34:141–146.

Anonymous, 1970, Recommendations for an international standard for a mapping method in bird census work, Swedish Nat. Sci. Res. Council, Stockholm, Bull. Ecol. Res. Comm. 9:49–52.

Atkinson-Willes, G. L., 1963, Wildfowl in Great Britain, Monograph of the Nature Conservancy, No. 3, Her Majesty's Stationery Office, London.

Bell, B. D., Catchpole, C. K., Corbett, K. J., and Hornby, R. J., 1973, The relationship between census results and breeding populations of some marshland passerines, Bird Study 20:127–140.

Berthold, P., 1976, Methoden der Bestandserfassung in der Ornithologie: Übersicht und kritische Betrachtung, J. Ornithol. 117:1–69.

Best, L. B., 1975, Interpretational errors in the "mapping method" as a census technique, Auk 92:452–460.

Best, L. B., 1981, Seasonal changes in detection of individual bird species, in: *Estimating Numbers of Terrestrial Birds* (C. J. Ralph and J. M. Scott, eds.), Stud. Avian Biol. **6**:252–261.

Best, L. B., and Petersen, K. L., 1982, Effects of stage of the breeding cycle on Sage Sparrow detectability, Auk **99**:788–791.

Blondel, J., 1969, Méthodes de dénombrement des populations d'oiseaux, in: *Problèmes d'Ecologie: l'Echantillonnage des Peuplements Animaux des Milieux Terrestres* (M. Lamotte and F. Bourlière, eds.), Masson, Paris, pp. 97–151.

Bock, C. E., and Root, T. L., 1981, The Christmas Bird Counts and avian ecology, in: *Estimating Numbers of Terrestrial Birds* (C. J. Ralph and J. M. Scott, eds.) Stud. Avian Biol. **6**:17–23.

Brown, J. L., 1969, Territorial behavior and population regulation in birds, Wilson Bull. **81**:293–329.

Bull, E. L., 1981, Indirect estimates of abundance of birds, in: *Estimating Numbers of Terrestrial Birds* (C. J. Ralph and J. M. Scott, eds.), Stud. Avian Biol. **6**:76–80.

Buckland, S. T., 1982, A note on the Fourier series model for analysing line transect data, Biometrics **38**:469–477.

Burnham, K. P., 1981, Summarizing remarks: Environmental influences, in: *Estimating Numbers of Terrestrial Birds* (C. J. Ralph and J. M. Scott, eds.), Stud. Avian Biol **6**:324–325.

Burnham, K. P., Anderson, D. R., and Laake, J. L., 1980, Estimation of density from line transect sampling of biological populations, Wildl. Monogr. **72**:1–202.

Burnham, K. P., Anderson, D. R., and Laake, J. L., 1981, Line transect estimation of bird population density using a Fourier series, in: *Estimating Numbers of Terrestrial Birds* (C. J. Ralph and J. M. Scott, eds.), Stud. Avian Biol. **6**:466–482.

Caughley, G., 1980, *Analysis of Vertebrate Populations* (reprinted with corrections), Wiley, New York.

Chessex, C., and Ribaut, J. P., 1966, Evolution d'une avifaune suberbaine et test d'une méthode de recensement, Nos Oiseaux **28**:193–211.

Conant, S., Collins, M. S., and Ralph, C. J., 1981, Effects of observers using different methods upon the total population estimates of two resident island birds, in: *Estimating Numbers of Terrestrial Birds* (C. J. Ralph and J. M. Scott, eds.), Stud. Avian Biol. **6**:377–381.

Conner, R. N., and Dickson, J. G., 1980, Strip transect sampling and analysis for avian habitat studies, Wildl. Soc. Bull. **8**:4–10.

Cormack, R. M., 1968, The statistics of capture-recapture methods, Ocean. Mar. Biol. Annu. Rev. **6**:455–506.

Cormack, R. M., 1979, Models for capture-recapture in: *Sampling Biological Populations* (R. M. Cormack, G. P. Patil, and D. S. Robson, eds.), Int. Co-op. Publ. House, Fairland, Maryland, Stat. Ecol. Ser. **5**:217–255.

Dawson, D. G., 1981a, Counting birds for a relative measure (index) of density, in: *Estimating Numbers of Terrestrial Birds* (C. J. Ralph and J. M. Scott, eds.), Stud. Avian Biol. **6**:12–16.

Dawson, D. G., 1981b, The usefulness of absolute ("census") and relative ("sampling" or "index") measures of abundance, in: *Estimating Numbers of Terrestrial Birds* (C. J. Ralph and J. M. Scott, eds.), Stud. Avian Biol. **6**:554–558.

Dawson, D. G., and Bull, P. C., 1975, Counting birds in New Zealand forests, Notornis **22**:101–109.

Dawson, D. K., 1981, Sampling in rugged terrain, in: *Estimating Numbers of Terrestrial Birds* (C. J. Ralph and J. M. Scott, eds.), Stud. Avian Biol. **6**:311–315.

DeSante, D. F., 1981, A field test of the variable circular-plot censusing technique in a California coastal scrub breeding bird community, in: Estimating Numbers of Terrestrial Birds (C. J. Ralph and J. M. Scott, eds.), Stud. Avian Biol. 6:177–185.

Dickson, J. G., 1978, Comparison of breeding bird census techniques, Am. Birds 32:10–13.

Diehl, B., 1974, Results of a breeding bird community census by the mapping method in a grassland ecosystem, Acta Ornithol. 14:362–376.

Diehl, B., 1981, Bird populations consist of individuals differing in many respects, in: Estimating Numbers of Terrestrial Birds (C. J. Ralph and J. M. Scott, eds.), Stud. Avian Biol. 6:225–229.

Duke, G. E., 1966, Reliability of censuses of singing male woodcocks, J. Wildl. Manage. 30:697–708.

Eberhardt, L. L., 1968, A preliminary appraisal of line transects, J. Wildl. Manage. 32:82–88.

Ekman, J. 1981, Problems of unequal observability, in: Estimating Numbers of Terrestrial Birds (C. J. Ralph and J. M. Scott, eds.), Stud. Avian Biol. 6:230–234.

Emlen, J. T., 1971, Population densities of birds derived from transect counts, Auk 88:323–342.

Emlen, J. T., 1977, Estimating breeding season bird densities from transect counts, Auk 94:455–468.

Emlen, J. T. and DeJong, M. J., 1981, The application of song detection threshold distance to census operations, in: Estimating Numbers of Terrestrial Birds (C. J. Ralph and J. M. Scott, eds.), Stud. Avian Biol. 6:346–352.

Enemar, A., 1959, On the determination of the size and composition of a passerine bird population during the breeding season, Vår Fågelvårld, Suppl. 2:1–114.

Enemar, A., 1962, A comparison between the bird census results of different ornithologists, Vår Fågelvårld 21:109–120.

Enemar, A., Klaesson, P., and Sjöstrand, B., 1979, Accuracy and efficiency of mapping territorial Willow Warblers Phylloscopus trochilus: A case study, Oikos 33:176–181.

Enemar, A., Sjöstrand, B., and Svensson, S., 1978, The effect of observer variability on bird census results obtained by a territory mapping technique, Ornis Scand. 9:31–39.

Ferguson, E. C., and Gilmer, D. S., 1980, Small-format cameras and fine-grain film used for waterfowl population studies, J. Wildl. Manage. 44:691–694.

Ferry, C., Frochot, B., and Leruth, Y., 1981, Territory and home range of the blackcap (Sylvia atricapilla) and some other passerines, assessed and compared by mapping and capture-recapture, in: Estimating Numbers of Terrestrial Birds (C. J. Ralph and J. M. Scott, eds.), Stud. Avian Biol. 6:119–120.

Franzreb, K. E., 1976, Comparison of variable strip transect and spot-map methods for censusing avian populations in a mixed-coniferous forest, Condor 78:260–262.

Franzreb, K. E., 1981, A comparative analysis of territorial mapping and variable-strip transect censusing methods, in: Estimating Numbers of Terrestrial Birds (C. J. Ralph and J. M. Scott, eds.), Stud. Avian Biol. 6:164–169.

Frochot, B., Reudet, D., and Leruth, Y., 1977, A comparison of preliminary results of three census methods applied to the same population of forest birds, Polish Ecol. Stud. 3:71–75.

Gates, J. M., 1966, Crowing counts as indices to cock pheasant populations in Wisconsin, J. Wildl. Manage. 30:735–744.

Gill, B. J., 1980, Abundance, feeding and morphology of passerine birds at Kowhai Bush, Kaikoura, New Zealand, N. Z. J. Zool. 7:235–246.

Granholm, S. L. 1983, Bias in density estimates due to movement of birds, Condor 85:243–248.

Haukioja, E., 1968, Reliability of the line survey method in bird census, with reference to Reed Bunting and Sedge Warbler, Ornis Fenn. **45**:105–113.

Hildén, O., 1981, Sources of error involved in the Finnish line transect method, in: Estimating Numbers of Terrestrial Birds (C. J. Ralph and J. M. Scott, eds.), Stud. Avian Biol. **6**:152–159.

Hogstad, O., 1967, Factors influencing the efficiency of the mapping method in determining breeding bird populations in conifer forests, Nytt. Mag. Zool. (Oslo) **14**:125–141.

Holmes, R. T., and Sturges, F. W., 1975, Bird community dynamics and energetics in a northern hardwood ecosystem, J. Anim. Ecol. **44**:175–200.

Hussell, D. J. T., 1981, The use of migration counts for monitoring bird population levels, in: Estimating Numbers of Terrestrial Birds (C. J. Ralph and J. M. Scott, eds.), Stud. Avian. Biol. **6**:92–102.

Hutto, R. L., and Mosconi, S. L., 1981, Lateral detectability profiles for line transect bird censuses: some problems and an alternative, in: Estimating Numbers of Terrestrial Birds (C. J. Ralph and J. M. Scott, eds.), Stud. Avian Biol. **6**:382–387.

James F. C., and Rathbun, S., 1981, Rarefaction, relative abundance, and diversity of avian communities, Auk **98**:785–800.

James, F. C, and Wamer, N. O., 1982, Relationships between temperate forest bird communities and vegetation structure, Ecology **63**:159–171.

Järvinen, O., 1976, Estimating relative densities of breeding birds by the line transect method. II. Comparison between two methods, Ornis Scand. **7**:43–48.

Järvinen, O., 1978, Estimating relative densities of land birds by point counts, Ann. Zool. Fennici **15**:290–293.

Järvinen, O., and Väisänen, R. A., 1975, Estimating relative densities of breeding birds by the line transect method, Oikos **26**:316–322.

Järvinen, O., and Väisänen, R. A., 1976, Estimating relative densities of breeding birds by the line transect method. IV. Geographical constancy of the proportion of main belt observations, Ornis Fenn. **53**:87–91.

Järvinen, O., and Väisänen, R. A., 1977, Constants and Formulae for Analysing Line Transect Data, Mimeo., Helsinki.

Järvinen, O., and Väisänen, R. A., 1979, Numbers of migrants and densities of breeding birds: A Comparison of long-term trends in north Europe, Anser **18**:103–108.

Järvinen, O., Väisänen, R. A., and Enemar, A., 1978a, Efficiency of the line transect method in mountain birch forest, Ornis Fenn. **55**:16–23.

Järvinen, O., Väisänen, R. A., and Walankiewicz, W., 1978b, Efficiency of the line transect method in central European forests, Ardea **66**:103–111.

Jensen, H., 1974, The reliability of the mapping method in marshes with special reference to the internationally accepted rules, Acta Ornithol. **14**:378–385.

Johnson, D. H., 1981, Summarizing remarks: Estimating relative abundance (Part I), in: Estimating Numbers of Terrestrial Birds (C. J. Ralph and J. M. Scott, eds.), Stud. Avian Biol. **6**:58–59.

Johnson, R. R., Brown, R. T., Haight, L. T., and Simpson, J. M., 1981, Playback recordings as a special avian censusing technique, in: Estimating Numbers of Terrestrial Birds (C. J. Ralph and J. M. Scott, eds.), Stud. Avian Biol. **6**:68–75.

Jolly, G. M., 1979, A unified approach to mark-recapture stochastic models, exemplified by a constant survival rate model, in: Sampling Biological Populations (R. M. Cormack, G. P. Patil, and D. S. Robson, eds.), Int. Co-op. Publ. House, Fairland, Maryland, Stat. Ecol. Ser. **5**:277–282.

Jolly, G. M., 1981, Summarizing remarks: Comparison of methods, in: *Estimating Numbers of Terrestrial Birds* (C. J. Ralph and J. M. Scott, eds.), *Stud. Avian Biol.* **6:**215–216.

Karr, J. A., 1981, Surveying birds in the tropics, in: *Estimating Numbers of Terrestrial Birds* (C. J. Ralph and J. M. Scott, eds.), *Stud. Avian Biol.* **6:**548–553.

Kelker, G. H., 1945, *Measurement and Interpretation of Forces, That Determine Populations of Managed Deer*, Ph.D. dissertation, University of Michigan, Ann Arbor.

Kendeigh, S. C., 1944, Measurement of bird populations, *Ecol. Monogr.* **14:**67–106.

Kepler, C. B., and Scott, J. M., 1981, Reducing count variability by training observers, in: *Estimating Numbers of Terrestrial Birds* (C. J. Ralph and J. M. Scott, eds.), *Stud. Avian Biol.* **6:**366–371.

Laake, J. L., 1978, *Line Transect Estimators Robust to Animal Movement*, M. S. thesis, Utah State University, Logan.

Laake, J. L., Burnham, K. P., and Anderson, D. R., 1979, *User's Manual for Program TRANSECT*, Utah State University Press, Logan.

Larson, T. A., 1980, *Ecological Correlates of Avian Community Structure in Mixed-conifer Habitat: An Experimental Approach*, Ph.D. dissertation, Illinois State University, Normal.

Mackowicz, R., 1977, The influence of the biology of the River Warbler (*Locustella fluviatilis* Wolf.) on the effectiveness of the mapping method, *Polish Ecol. Stud.* **3:**89–93.

Mannes, P., and Alpers, R., 1975, Errors of census methods concerning birds breeding in boxes, *J. Ornithol.* **116:**308–314.

Manuwal, D. A., 1983, Avian abundance and guild structure in two Montana coniferous forests, *Murrelet* **64:**1–11.

Marchant, J. H., 1981, Residual edge effects with the mapping bird census method, in: *Estimating Numbers of Terrestrial Birds* (C. J. Ralph and J. M. Scott, eds.), *Stud. Avian Biol.* **6:**488–491.

Merikallio, E., 1946, Über regionale Verbreitung und Anzahl der Landvögel in Süd- und Mittelfinnland, besonders in deren östlichen Teilen, im Lichte von quantitativen Untersuchungen. I. Allgemeiner Teil, *Ann. Zool. Soc. "Vanamo"* **12:**1–140.

Merikallio, E., 1958, Finnish birds; Their distribution and numbers, *Fauna Fennica* **5:**1–181.

Murray, B. G., Jr., 1979, *Population Dynamics: Alternative Models*, Academic Press, New York.

Nichols, J. D., Noon, B. R., Stokes, S. L., and Hines, J. E., 1981, Remarks on the use of mark-recapture methodology in estimating avian population size, in: *Estimating Numbers of Terrestrial Birds* (C. J. Ralph and J. M. Scott, eds.), *Stud. Avian Biol.* **6:**122–136.

Nilsson, S. G., 1977, Estimates of population density and changes for titmice, nuthatch, and Tree Creeper in southern Sweden—An evaluation of the territory mapping method, *Ornis Scand.* **8:**9–16.

Norton-Griffiths, M., 1978, Counting animals, in: *Handbook No. 1, Serengeti Ecol. Monitoring Programme* (J. J. R. Grimsdell, ed.), African Wildlife Leadership Foundation, Nairobi, Kenya.

O'Connor R. J., 1979, BTO migration studies, *BTO News* **101:**1–2.

O'Connor, R. J., and Hicks, R. K., 1980, The influence of weather conditions on the detection of birds during Common Birds Census fieldwork, *Bird Study* **27:**137–151.

O'Connor, R. J., and Marchant, J. H., 1981, *A Field Evaluation of Some Common Birds Census Techniques*, Report from British Trust for Ornithology to Nature Conservancy Council, Huntingdon, England.

Oelke, H., 1977, Methoden de Bestandserfassung von Vögeln, Nestersuche—Rivierkartierung, Orn. Mitt. **29**:151–166.

O'Meara, T. E., 1981, A field test of two density estimators for transect data, in: Estimating Numbers of Terrestrial Birds (C. J. Ralph and J. M. Scott, eds.), Stud. Avian Biol. **6**:193–196.

Osborne, P. E., 1982, The Effects of Dutch Elm Disease on Bird Populations, Ph.D. dissertation, Oxford University, Oxford, England.

Otis, D. L., Burnham, K. P., White G. C., and Anderson, D. R., 1978, Statistical inference from capture data on closed animal populations, Wildl. Monogr. **62**:1–135.

Pollock, K. H., 1981, Capture-recapture models: A review of current methods, assumptions, and experimental design, in: Estimating Numbers of Terrestrial Birds (C. J. Ralph and J. M. Scott, eds.), Stud. Avian Biol. **6**:426–435.

Preston, F. W., 1979, The invisible birds, Ecology **60**:451–454.

Ramsey, F. L., and Scott J. M., 1978, Use of Circular Plot Surveys in Estimating the Density of a Population With Poisson Scattering, Tech. Report 60, Department of Statistics, Oregon State University, Corvallis.

Ramsey, F. L., and Scott, J. M., 1981, Tests of hearing ability, in: Estimating Numbers of Terrestrial Birds (C. J. Ralph and J. M. Scott, eds.), Stud. Avian Biol. **6**:341–345.

Rappole, J. H., Warner, E. W., and Ramoso, M., 1977, Territoriality and population structure in a small passerine community, Am. Midl. Natur. **97**:110–119.

Redmond, R. L., Bicak, T. K., and Jenni, D. A., 1981, An evaluation of breeding season census techniques for Long-billed Curlews (Numenius americanus), in: Estimating Numbers of Terrestrial Birds (C. J. Ralph and J. M. Scott, eds.), Stud. Avian Biol. **6**:197–201.

Reynolds, R. T., Scott, J. M., and Nussbaum, R. A., 1980, A variable circular-plot method for estimating bird numbers, Condor **82**:309–313.

Robbins, C. S., 1970, Recommendations for an international standard for a mapping method in bird census work, Aud. Field Notes **24**:723–726.

Robbins, C. S., 1978, Census techniques for forest birds, in: Proceedings of Workshop on Management of Southern Forests for Nongame Birds (R. M. DeGraaf, tech. coord.), USDA Forest Service, Southeastern Forest and Range Exp. Stn., Asheville, North Carolina, Gen. Tech. Rep. SE-**14**:142–163.

Robbins, C. S., 1981a, Effect of time of day on bird activity, in: Estimating Numbers of Terrestrial Birds (C. J. Ralph and J. M. Scott, eds.), Stud. Avian Biol. **6**:275–286.

Robbins, C. S., 1981b, Bird activity levels related to weather, in: Estimating Numbers of Terrestrial Birds (C. J. Ralph and J. M. Scott, eds.), Stud. Avian Biol. **6**:301–310.

Robinette, W. L., Jones, D. A., and Loveless, C. M., 1974, Field tests of strip census methods, J. Wildl. Manage. **38**:81–96.

Rodgers, R. D., 1981, Factors affecting Ruffed Grouse drumming counts in southwestern Wisconsin, J. Wildl. Manage. **45**: 409–418.

Root, T. L., 1983, A comparative study of distribution and abundance data of some wintering North American birds, Abstracts of Presented Papers, Annual Meeting of the Cooper Ornithological Society **53**:25.

Scherner, E. R., 1981, Die Flächengrösse als Fehlerquelle bei Brutvogel-Bestandsaufnahmen, Ökol. Vögel **3**:145–175.

Schweder, T., 1977, Point process models for line transect experiments, in: Recent Developments in Statistics (J. R. Barba, F. Brodeau, G. Romier, and B. Van Cutsem, eds.), North-Holland Publishing Co., New York, pp. 221–242.

Scott, J. M., and Ramsey, F. L., 1981a, Effects of abundant species on the ability of observers to make accurate counts of birds, Auk **98**:610–613.

Scott, J. M., and Ramsey, F. L., 1981b, Length of count period as a possible source of bias in estimating bird numbers; in: *Estimating Numbers of Terrestrial Birds* (C. J. Ralph and J. M. Scott, eds.), *Stud. Avian Biol.* **6**:409–413.

Scott, J. M., Ramsey, F. L., and Kepler, C. B., 1981, Distance estimation as a variable in estimating bird numbers, in: *Estimating Numbers of Terrestrial Birds* (C. J. Ralph and J. M. Scott, eds.), *Stud. Avian Biol.* **6**:334–340.

Seber, G. A. F., 1982, *Estimation of Animal Abundance and Related Parameters*, 2nd ed., Griffin, London.

Seierstad, S., Seierstad, A., and Mysterud, I., 1965, Statistical treatment of the "inconspicuousness problem" in animal population surveys, *Nature* **206**:22–23.

Seierstad, S., Seierstad, A., and Mysterud, I., 1967, Estimation of survey efficiency for animal populations with unidentifiable individuals, *Nature* **213**:524–525.

Shannon, C. E., and Weaver, W., 1963, Species diversity in lacustrine phytoplankton, I. The components of the index of diversity from Shannon's formula, *Am. Natur.* **103**:51–60.

Shields, W. M., 1979, Avian census techniques: an analytical review, in: *The Role of Insectivorous Birds in Forest Ecosystems* (J. G. Dickson, R. N. Conner, R. R. Fleet, J. C. Kroll, and J. A. Jackson, eds.), Academic Press, New York, pp. 23–51.

Sinclair, A. R. E., 1973, Population increases of the buffalo and wildebeast in the Serengeti, *E. Afr. Wild. J.* **11**:93–107.

Skirvin, A. A., 1981, Effect of time of day and time of season on the number of observations and density estimates of breeding birds, in: *Estimating Numbers of Terrestrial Birds* (C. J. Ralph and J. M. Scott, eds.), *Stud. Avian Biol.* **6**:272–274.

Smith, S. M., 1978, The "underworld" in a territorial sparrow: Adaptive strategy for floaters, *Am. Natur.* **122**:571–582.

Snow, D. W., 1965, The relationship between census results and the breeding population of birds on farmland, *Bird Study* **12**:287–304.

Svensson, S., 1974, Interpersonal variation in species map evaluation in bird census work with the mapping method, *Acta Ornithol.* **14**:322–338.

Svensson, S., 1979, Census efficiency and number of visits to a study plot when estimating bird densities by the territory mapping method, *J. Appl. Ecol.* **16**:61–68.

Svensson, S., 1980, Comparison of recent bird census methods, in: *Bird Census Work and Nature Conservation* (H. Oelke, ed.), *Proc. VI Intern. Conf. on Bird Census Work*, Univ. Göttingen, West Germany, pp. 13–22.

Szaro, R. C., and Jakle, M.D., 1982, Comparison of variable circular-plot and spot-map methods in desert riparian and scrub habitats, *Wilson Bull.* **94**:546–550.

Tiainen, J., Martin, J. L., Pakkala, T., Piiroinen, J., Solonen, T., Vickholm, M., and Virolainen, E., 1980, Efficiency of the line transect and point count methods in a south Finnish forest area, in: *Bird Census Work and Nature Conservation* (H. Oekle, ed.), *Proc. VI Intern. Conf. on Bird Census Work*, Univ. Göttingen, West Germany, pp. 107–113.

Tilghman, N. G., and Rusch, D. H., 1981, Comparison of line-transect methods for estimating breeding bird densities in deciduous woodlots, in: *Estimating Numbers of Terrestrial Birds* (C. J. Ralph and J. M. Scott, eds.), *Stud. Avian Biol.* **6**:202–208.

Tomialojć, L., 1980, The combined version of the mapping method, in: *Bird Census Work and Nature Conservation* (H. Oelke, ed.), *Proc. VI Intern. Conf. on Bird Census Work*, Univ. Göttingen, West Germany, pp. 92–106.

Tomialojć, L. 1981, On the census accuracy in the line transect, in: *Bird Census and Mediterranean Landscape* (F. J. Purroy, ed.), *Proc. VII Intern. Conf. on Bird Census Work*, Univ. of Leon, Spain, pp. 13–17.

Tramer, E. J., 1969, Bird species diversity: Components of Shannon's formula, *Ecology* **50**:927–929.

Verner, J., 1981a, Measuring responses of avian communities to habitat manipulation, in: *Estimating Numbers of Terrestrial Birds* (C. J. Ralph and J. M. Scott, eds.), *Stud. Avian Biol.* **6**:543–547.

Verner, J., 1981b, Appendix VI: Report of the working group to identify future research needs, in: *Estimating Numbers of Terrestrial Birds* (C. J. Ralph and J. M. Scott, eds.), *Stud. Avian Biol.* **6**:584.

Verner, J., and Ritter, L. V., 1982, Estimating the densities of breeding birds, in: *Abstracts of Presented Posters and Papers, 100th Stated Meeting of The American Ornithologists' Union*, Chicago, Illinois, p. 95.

Verner, J., and Ritter, L. V., A comparison of transects and point counts in oak-pine woodlands of California, *Condor*, in press.

von Haartman, L., 1971, Population dynamics, in: *Avian Biology*, Volume 1 (D. S. Farner and J. R. King, eds.), Academic Press, New York pp., 391–459.

Walankiewicz, W., 1977, A comparison of the mapping method and I.P.A. results in Bialowieza National Park, *Polish Ecol. Stud.* **3**:119–125.

White, G. C., Burnham, K. P., Otis, D. L., and Anderson, D. R., 1978, *User's Manual for Program CAPTURE*, Utah State University Press, Logan.

White, G. C., Anderson, D. R., Burnham, K. P., and Otis, D. L., 1982, *Capture-recapture and Removal Methods for Sampling Closed Populations*, Los Alamos National Laboratory, Los Alamos, New Mexico.

Wiens, J. A., 1969, An approach to the study of ecological relationships among grassland birds, *Ornithol. Monogr.* **8**:1–93.

Yapp, W. B., 1956, The theory of line transects, *Bird Study* **3**:93–104.

Yui, M., 1977, The census method of woodland bird populations during the breeding season, in: *Studies on Methods of Estimating Population Density, Biomass and Productivity in Terrestrial Animals* (M. Morista, ed.), *Japanese IBP Synthesis*, Volume 17, University Tokyo Press, Japan, pp. 97–112.

Zach, R., and Falls, J. B., 1979, Foraging and territoriality of male Ovenbirds (Aves: Parulidae) in a heterogeneous habitat, *J. Anim. Ecol.* **48**:33–52.

CHAPTER 9

CIRCADIAN ORGANIZATION OF THE AVIAN ANNUAL CYCLE

ALBERT H. MEIER and ALBERT C. RUSSO

1. INTRODUCTION

Organization of the avian annual cycle involves an interplay of exogenous and endogenous timing factors. In many species of the temperate zones, increasing daylengths in spring induce migratory and reproductive preparation. However, following breeding in summer when daylengths are still long, the reproductive system regresses and a complete feather molt occurs. Molting in turn is followed by metabolic and behavioral changes associated with fall migration to the wintering quarters. The entire sequence of conditions from reproductively photosensitive birds in spring to phototrefractory birds in late summer occurs when daylengths are long (greater than 12 hr). Re-establishment of photosensitivity in many species (i.e., of *Zonotrichia*) depends on an intervening period of short daylengths that occur naturally in late fall and winter. Thus, an endogenous system accounts for seasonal changes in responsiveness to daylength and this system in turn is gradually adjusted by daylengths.

Seasonal changes in reproductive and migratory conditions impose numerous demands on the neuroendocrine system. Photoperiodic stimulation of the reproductive system is mediated by follicle-stimulating

ALBERT H. MEIER and ALBERT C. RUSSO • Department of Zoology and Physiology, Louisiana State University, Baton Rouge, Louisiana 70803 hormone (FSH) and luteinizing hormone (LH) that stimulate gameto-genesis and gonadal steroidogenesis (review, Farner and Follett, 1979). Final stages in preparation for reproduction, particularly in the female, depends on social stimuli (courting) and may be modified by environ-mental factors such as food availability and inclement weather (review, Wingfield, 1980). Gonadal hormones are released in larger amounts during courting and have dominant roles in nesting and the initiation of egglaying, which involves a complex interaction of neural and hor-monal activities. The presence of eggs and young in the nest promote additional changes in the neuroendocrine system such as increased amounts of prolactin release that stimulates incubation and brooding behavior (review, Drent, 1975). Refractoriness is characterized by a severe reduction in FSH and LH secretion that leads to gonadal regres-sion (review, Sharp, 1980a).

Migration occurs at specific seasons and time of day and involves behavioral and metabolic changes that are regulated by neuroendocrine mechanisms. Migratory locomotor activity is supported largely by mo-bilization of fat stores that must be accumulated prior to and continually replaced during the migratory period. Mobilization of fat is not com-pletely understood but probably involves synergistic activities of growth hormone, glucagon, corticosteroid hormone (corticosterone), and thy-roid hormones. Fat synthesis is also incompletely understood, but in-volves prolactin stimulation at specific times of day (review, Meier and Fivizzani, 1981).

Although specific hormones may have more important roles at some times of year than at others, it is now evident that seasonal con-ditions depend on complexes of hormonal and neural activities acting synergistically with one another. The determination of these complexes and their sequential seasonal changes demands a high order of organ-ization. Recent research indicates that circadian rhythms may be the basic units for such an organization.

Circadian rhythms are evident at all levels of organization. They persist in constant environmental conditions with periods approxi-mating 24 hours. The daily light–dark (LD) cycle is the principle syn-chronizer or entrainer of the endogenous rhythms. Each cell is thought to have a circadian clock; however, cellular rhythms, as with all other rhythms, are imprecise and tend to free run with periods at variance from 24 hr unless synchronized. It is the important function of the neuroendocrine system to synchronize the cellular rhythms for the construction of tissue and organism rhythms so that the entire animal is organized internally and appropriately synchronized with its peri-odic environment.

A principle goal of this paper is to review research on circadian systems as they pertain to the avian annual cycle, to offer conclusions concerning the physiological relevance of the research, and to speculate on possible mechanisms for the purpose of stimulating further work. Those areas where circadian systems are clearly involved include photoperiodism, photosensitivity and photorefractoriness, timing of locomotor activity including migratory restlessness, migratory orientation, premigratory fattening, and egglaying. This partial list of the more important events in which circadian mechanisms have central roles should alert avian biologists to their potential.

2. CIRCADIAN MIGRATORY ACTIVITIES

2.1. Locomotor Activity Rhythms

Because migratory restlessness is essentially a specialized form of locomotor activity, an understanding of migration depends in part on an understanding of the regulation of activity. The exclusion of activity to specific times of day or night is a priori evidence that regulation exists. Because locomotor activity requires metabolic support, it also seems clear that metabolism is integrated with the behavior and that both are regulated through higher brain centers.

In many vertebrate species including birds, locomotor activity is divisible into two daily peaks (Beer, 1961; Aschoff, 1967). Usually one occurs early and the other late during the day. They appear to be expressions of two separate circadian oscillations that free run in continuous light in gradual changing relations with one another (McMillan et al., 1970; Gwinner, 1974). In a nocturnal migrant, such as the White-throated Sparrow Zonotrichia albicollis (Davis and Meier, 1973), the late afternoon peak apparently shifts to the dark at the onset of migratory restlessness.

Reentrainment of the circadian activity rhythm following a shift of the LD cycle may involve either an advance or delay of the rhythm depending on the phase relations of the exogenous cycle and endogenous rhythm. Resetting often occurs in the direction of shortest time differential. Based in part on the resetting phenomenon, it was proposed that the circadian rhythm of locomotor activity is an expression of two coupled circadian neural oscillations (Daan and Pittendrigh, 1976; Pittendrigh and Daan, 1976; Boulos and Rusak, 1982). Light is thought to advance the phase of one oscillation and to delay the other. This attractive hypothesis can account for splitting of activity that may occur under free-running conditions (Gwinner, 1974) and allow for a repositioning of the afternoon peak to night at the onset of nocturnal migratory restlessness. It might also account for the variations in lengths of circadian activity periods in birds kept on constant light (LL) of different intensities (Aschoff's rule.)

At least one photoreceptor is necessary for LD entrainment of the activity rhythm and two are certainly conceivable in transmitting different effects on a dual oscillation mechanism. Although the eyes are the most obvious sites for entraining photoreceptors, LD entrainment occurs in enucleated sparrows with similar properties of entrainment as found in eyed birds (McMillan *et al.*, 1975a,b,c,d). In addition, the period of the free-running rhythm also varies with light intensity. The evidence clearly indicates the presence of an extra retinal receptor, but the absence of an additional occular photoreceptor has not yet been established. Because the pineal organ has a photosensory appearance and function, early attention was focused on this organ as a possible entraining mechanism. It is now clear that the pineal is not required for LD entrainment of the activity rhythm. Pinealectomized as well as enucleated birds entrain to LD cycles of low light intensities (Gaston and Menaker, 1968). Thus, there is convincing evidence for an encephalic photoreceptor; however, a specific site has not yet been proposed. The methodology for such a demonstration must distinguish between receptor sites, pacemaker areas, and neural pathways. Localization of a photoreceptor site may also be compromised if there are two or more such sites.

2.2. Pacemaker Systems for Locomotor Rhythms

Although circadian periodicity is apparently a property of every cell, the synchronization of cellular activities for the expression of organism rhythms, such as the locomotor activity rhythm, indicates the existence of pacemaker systems. One potential pacemaker system is the pineal gland. Experiments in House Sparrows (*Passer domesticus*) demonstrate that whereas the locomotor rhythm persists in intact birds maintained in constant darkness (DD) or dim light, the rhythm disappears following pinealectomy (Gaston and Menaker, 1968). On the other hand, a locomotor rhythm is expressed in pinealectomized sparrows kept on a LD regimen and is subject to resetting by altering the time of light onset. A principal difference in the activity rhythm of pinealectomized and intact sparrows kept on LD cycles is that the onset of activity occurs several hours earlier (before light onset) in pinealectomized sparrows. Similar results have been reported in several other species (Gaston, 1971; McMillan, 1972).

The results are not so conclusive in other instances. Free-running rhythms of locomotor activity persist in pinealectomized European Starlings (*Sturnus vulgaris*) (Rutledge and Angle, 1977; Gwinner, 1978) and Japanese quail (*Coturnix coturnix*) (Simpson and Follett, 1980; MacBride and Ralph, 1972; MacBride, 1973) exposed to DD in which both audio and visual periodic cues were prevented. In starlings, the activity was more diffuse and the period was shorter in the pinealectomized birds than in the intact (Gwinner, 1978).

Pineal culture studies have demonstrated that the chick pineal does have the essential properties of a self-sustained oscillation as it must if it is to serve as a circadian pacemaker. There is a circadian rhythm of N-acetyltransferase (NAT) activity that persists under constant light conditions (Binkley *et al.*, 1978; Deguchi, 1979; Kasal *et al.*, 1979). Because NAT is the rate-limiting enzyme in formation of melatonin, it may be concluded that there is also a rhythm of melatonin synthesis. Light exerts a direct inhibitory influence on NAT activity so that the NAT rhythm is dampened in bright LL. In addition, an LD cycle can impose in pineal culture a cycle of NAT activity inversely related to light intensity (Wainwright and Wainwright, 1979; Takahashi, 1982).

A dampening effect of light on chick pineal NAT activity tested *in vitro* is most effective at a wavelength of 500 nm, and the sensitivity curve resembles the absorption spectrum of rhodopsin in the retina (Deguchi, 1981). On the basis of studies using agents that influence membrane potential, Deguchi (1979, 1981) suggested that light reduces NAT activity by hyperpolarizing pinealocyte membranes in a manner similar to that in retinal photoreceptor cells.

Thus, the chicken (*Gallus domesticus*) pineal gland in culture contains a self-sustained oscillation of melatonin synthesis under constant light conditions and a photoreceptor mechanism for direct influence of light on melatonin synthesis. Coupled with the findings that the source of circulating melatonin in the blood is from the pineal in chickens (Ralph *et al.*, 1974; Pelham, 1975) and that melatonin injections can synchronize circadian locomotor activity in pinealectomized starlings (Gwinner and Benziger, 1978), these evidences would seem to build a strong case that circadian rhythmicity and even LD entrainment of locomotor activity depends on the pineal. However, this surmise must be viewed in the context of clear evidence that pinealectomy does not abolish locomotor rhythmicity nor interfere with LD entrainment in chickens (MacBride and Ralph, 1972). One explanation for this quandary is that there is another pacemaker for locomotor rhythmicity in addition to the pineal.

Unlike the mammalian pineal gland, extirpation of the superior cervical ganglion does not prevent rhythmicity of LD entrainment of a melatonin synthesis rhythm in birds (Ralph et al., 1975; Binkley, 1976). Thus, the avian pineal either has its own photic and rhythmic machinery or else it is under humoral regulation. Furthermore, pineal maintenance of rhythmicity in House Sparrows held in DD does not depend on neural efferents (Zimmerman and Menaker, 1975). Thus, neural connectives are not required for possible pineal influences on the circadian locomotor system.

Because humoral elements are presumably involved in pineal activity and function, experiments were done with pinealectomized House Sparrows to determine whether pineal transplants could restore circadian locomotor rhythmicity. Not only did pineal transplants to the anterior chamber of the eye restore rhythmicity but more importantly the phase of the rhythm depended on the phase of the donor (Zimmerman and Menaker, 1975, 1979). Restoration of a rhythm by itself is interpretable as a permissive influence for the expression of a rhythm without regard to an underlying oscillator. Resetting the phase of a locomotor rhythm in the host to that found in the donor at the time of pineal transfer indicates a more significant pacemaker role of the pineal gland in House Sparrows.

Because of their central pacemaker role in mammalian pacemaker systems, the suprachiasmatic nuclei (SCN) have also been investigated in birds. In the Java Sparrow (Padda oryzivora) where pinealectomy does not remove locomotor rhythmicity, lesions of the SCN do abolish rhythmicity in birds held in continuous dim light (Kawamura et al., 1982). However, SCN–lesioned Java Sparrows do entrain to LD cycles. In House Sparrows, also, SCN lesions abolish rhythmicity, but do not prevent LD entrainment (Takahaski and Menaker, 1979). Thus, pinealectomy and SCN ablations have similar effects in House Sparrows, and neither the SCN nor the pineal can maintain rhythmicity without the other.

Because LD entrainment of the locomotor rhythm occurs in both pinealectomized and SCN–lesioned birds, it would seem to follow that the entrainment pathway can bypass these pacemaker centers. However, another possibility not yet examined, apparently, is that entrainment passes through both the pineal and the SCN so that both need to be removed in order to abolish LD entrainment. Thus, there could be dual photoreceptor systems as well and the loss of one might not prevent LD entrainment. A retinohypothalamic pathway and retinal input to the SCN have been demonstrated in several species (Hartwig, 1974; Bons, 1976). Such a dual system might account for the inconsistencies reported and would support the dual oscillatory hypothesis for entrainment of locomotor activity (Daan and Pittendrigh, 1976; Pittendrigh and Daan, 1976; Menaker and Zimmerman, 1976).

2.3. Circadian Hormone Rhythms

Inasmuch as locomotor activity must have metabolic support, it seems almost axiomatic that hormones involved in mobilization of carbohydrates and lipids should also undergo circadian rhythms of secretion. The corticosteroid hormone (corticosterone) has been studied most intensively in this regard. Circadian rhythms in plasma concentration have been reported in a large number of vertebrate species, including many birds (Assenmacher and Jallageas, 1980). Assenmacher argued that the daily increase in plasma corticosterone concentration prepares for and maintains locomotor activity. Given the undeniable stimulatory influences that corticosteroids exert on metabolism and neural activity, the idea is certainly an attractive one. Nevertheless, the relation of plasma corticosterone concentration and locomotor activity is not always direct. In Japanese quail, high plasma corticosterone levels occur during the dark. Although it may be argued that the high levels prepare for locomotor activity that occurs during the day, it should be recognized that the highest levels are found when the birds are asleep(Boissin and Assenmacher, 1970; Russo, 1983). In pigeons (*Columba livia*) also, the daily increase occurs shortly after the onset of dark and is low during the day when the birds are active (Joseph and Meier, 1973). In the migratory White-throated Sparrow as well, high concentrations of corticosterone are not directly related with *Zugunruhe* (Meier and Fivizzani, 1975). These findings indicate that the corticosteroid and locomotor rhythms are often associated, but are not always directly coupled with one another.

In the White-throated Sparrow, the phase, form, and amplitude of the plasma corticosterone rhythm vary with daylength and season (Meier and Fivizzani, 1975). In summer photorefractory birds during the postnuptial molt, there is no apparent rhythm at all. An interesting feature is that the corticosterone rhythm changes continuously even when daylength is maintained constant (LD 16:8). An association of the corticosterone rhythm and seasonal conditions is discussed in Section 4.

Another interesting feature of the plasma corticosterone rhythm is that it may not depend on a driving rhythm of ACTH secretion. Although not tested directly in birds (but see Assenmacher and Boissin,

1973), hypophysectomy and adrenal transplant studies in fish (Srivastava and Meier, 1972) and mammals (Meier, 1976; Ottenweller et al., 1978) demonstrated that both the free-running rhythm of plasma corticosteroid concentration and LD entrainment of the rhythm persist in hypophysectomized animals and depend on neural regulation of a rhythm of adrenal sensitivity to ACTH. The corticosteroid rhythm depends on serotonergic activity. Blockade of serotonin synthesis with para-chlorophenylalnine removes the corticosterone rhythm (Assenmacher and Boissin, 1973) and stimulation of serotonin synthesis with 5-hydroxytryptophan injections increase plasma corticosterone concentrations (Miller, 1979). These studies of corticosteroid rhythms suggest that circadian rhythms of other peripheral hormones may also be independent of driving rhythms of pituitary hormones.

Circadian rhythms of thyroid hormone concentration have been observed in plasma of chickens (Newcomer, 1974; Klandorf et al., 1978), ducklings (Pethes et al., 1978), Spotted Munia (Lonchura punctulata) (Chandola, 1978), and Japanese quail (Russo, 1983). As with the plasma corticosterone rhythm in Japanese quail (Boissin and Assenmacher, 1970), thyroid hormone rhythms in chickens can entrain to notably short days (ie., LD 8:8) (Klandorf et al., 1978). In addition, the plasma rhythms are composites of secretion and removal rates so that high plasma levels of thyroxim (T_4) may not indicate greater hormone activity. Plasma concentrations of triiodothyronine (T_3) may be more directly related to thyroid hormone activity since T_4 is converted to T_3 in actively metabolizing tissues. Thus in chickens, plasma T_4 levels are higher during rest and T_3 levels are higher during the active period (Klandorf et al., 1978). A relation between plasma thyroid hormone concentrations and locomotor activity is further indicated by the observations that both thyroid hormone rhythms and locomotor rhythms persist in chickens held on DD, even following pinealectomy, but are not apparent in chickens held on LL (Klandorf et al., 1978a; May, 1978). It seems likely that the daily variations of thyroid hormone concentrations are as much the consequences of, as determinants of, locomotor activity.

Apart from its own rhythmicity, thyroid hormone is necessary for the expression of other rhythms. In pigeons, the circadian rhythms of fattening and cropsac responses to prolactin are lost in birds held in LL unless treated with either T_4 or T_3 (John et al., 1972). Apparently LL decreases thyroid hormone secretion in pigeons, as it does in chickens (Klandorf et al., 1978a). It would be interesting to learn whether locomotor rhythmicity often lost or dampened in birds held on LL could be restored by treatment with thyroid hormone. Because of their inhibitory effects on monoamine oxidase activities in neurons, thyroid hormones have important potentiating influences on both serotonergic and catecholaminergic (dopaminergic and noradrenergic) activities. Photoperiodic control of thyroid hormone secretion is discussed in Section 3.

Several hormones, including growth hormone, undergo circadian variations directly related to their support of locomotor activity. Plasma concentrations of growth hormone are highest near the onset of locomotor activity in pigeons (McKeown *et al.*, 1973) when plasma free fatty acid levels are also high (March *et al.*, 1978). The fatty acids are probably mobilized by growth hormone (John *et al.*, 1973; Harvey *et al.*, 1977) and could serve as an energy source for locomotor activity early in the day.

Except for the studies relating to pineal and corticosteroid hormones, little is known about the mechanisms that regulate circadian rhythmicity of hormones and photoperiodic entrainment of their rhythms. This apparent lack of interest probably stems from the difficulties of hormone assays compared with monitoring locomotor activity. For those willing, research in this area could be fruitful. Seasonal changes in the phase relations of corticosteroid and prolactin hormone rhythms in the White-throated Sparrow occur not only with respect to the LD cycle, but also with respect to one another (Meier, 1976). These observations suggest some separateness in the regulation of these two hormone rhythms. Similarly, developmental alterations in the phases of corticosteroid and thyroid hormone rhythms are not parallel changes in Japanese quail (Russo, 1983). The physiological significance of different phase relations among hormone rhythms is discussed in Section 4.3. It is certainly premature to assume that pacemaker areas and photoreceptors for locomotor activity rhythms are also relevant for all hormone rhythms.

2.4. Circadian Rhythms in Orientation

Inasmuch as the object of locomotor activity is to move from one area to another, it is axiomatic that orientation be closely coupled with activity. Just as there are often two peaks of locomotor activity during a day, there are also two orientations associated with morning and evening activities that take a bird away from and back to its roosting area. Meier and Ferrell (1978) proposed that these activities and associated orientations are dissociable seasonally so that one of these activity peaks becomes migratory restlessness oriented toward the breeding grounds and that the other peak becomes migratory restlessness oriented toward the wintering quarters.

Orientation of locomotor activity with respect to the sun clearly involves a circadian system that allows maintenance of a specific direction despite movement of the sun (Kramer, 1951, 1953; von St. Paul, 1956; Able and Dillon, 1977). When trained to a specific compass direction and tested in the presence of a stationary light source, the starling's orientation shifts gradually during the course of a day (Kramer, 1953) so that it compensates for the apparent movement of the sun at the expected rate of about 15° per hour (Kramer, 1953). Whether time compensation is involved in stellar orientation by nocturnal migrants deserves further study, although Emlen and Emlen (1967) demonstrated that Indigo Buntings (*Passerina cyanea*) may use star patterns, which does not require time compensation (see Able, 1980).

The seasonal change in migratory orientation could result, conceivably, from a seasonal change in the environmental cue or from an endogenous change in the interpretation or response to a cue. On the basis of experiments in American Crows (*Corvus brachyrhynchos*), Rowan (1931) proposed that endogenous changes in the bird are responsible. Similar methodology and conclusions were reported by Emlen (1969) on the basis of studies with the Indigo Bunting, a nocturnal migrant. Birds placed into spring conditions in the fall by manipulation of daylength oriented northward (the spring direction) under a fall planetarium sky, whereas birds brought into fall condition in the spring oriented southward (the fall direction). Similar results have also been reported for the White-throated Sparrow (Miller and Weiss, 1978). Thus at least in some species, the seasonal change in migratory orientation results from endogenous seasonal changes in the bird. A circadian basis for this seasonal change in orientation is discussed in Section 4.

3. PHOTOPERIODISM AND ASSOCIATED ACTIVITIES

A role for circadian rhythms in photoperiodism was first proposed by Bünning (1960) on the basis of studies with plants and insects. Bünning suggested that light can act in two capacities in producing flowering and diapause. The daily photophase entrains a rhythm of photosensitivity so that a sensitive phase occurs 12–24 hr after light onset. If light is present during this latter portion of the day, it acts as an inducer of the photoperiodic effect.

This illuminating hypothesis was not applied to birds until the work by Hamner (1963, 1964), who employed protocol used successfully in plant and insect research. In one study, Hamner held House Finches (*Carpodacus mexicanus*) on a 72-hr "day" (LD 6:66) and provided short interruptions of light at one of several different intervals during the dark. Interruptions at 36 and 60 hr after light onset were stimulatory for gonad weights, whereas interruptions at 24 and 48 hr after light onset were ineffective. Apparently, the six-hour photophase entrained a rhythm of photosensitivity so that the concidence of light during a sensitive phase that reoccurs every 24 hr at 30 and 60 hr after light onset induced gonadal growth. In another study, House Finches were held on LD cycles composed of six-hour photophases followed by variable intervals of dark. Photocycles of 12, 36, and 60 hr were stimulatory for gonadal growth whereas 24- and 48-hr photocyles were not. Again, one may conclude that the 6-hr photophase entrains as rhythm of photosensitivity so that the succeeding 6-hr photophase coincides with a photosensitive phase of the rhythm and induces gonadal growth.

Since these pioneer studies in birds by Hamner, many other resonance studies have revealed similar mechanisms in photoperiodic species (Farner, 1964, 1965; Wolfson, 1965, Menaker, 1965; Follett and Sharp, 1969; Lofts and Lam, 1973; Turek, 1974; Meier, 1976). It is often stated that light acts as both an entrainer and as an inducer. The proof for entrainment resides in the observations that the photosensitive phase is set by a light–dark cycle and by the reoccurence of this phase every 24 hr during the dark.

3.1. External and Internal Coincidence Models

According to an external coincidence model (Fig. 1), photoinduction is envisioned as a direct effect of light on the production and release of gonadotrophic hormones acting on sensitive photoreceptors and mediated by sensitive neural pathways during the photosensitive phase. Perhaps the best evidence in support of such a direct inductive effect is the finding that gonadotropins are released in Japanese quail and two passerine species about 19 hr after light onset on the first day that the photophase is extended from a short nonstimulatory daylength to a long daylength (Follett et al., 1974; Follett and Robinson, 1980).

Because the external coincidence model is relatively easy to understand, it has also been generally accepted. However, this model may not be adequate to explain all the available evidence. It fails, for example, to account for gonadal stimulation in birds held in continuous dark. By definition, light must be present and coincide with a sensitive phase to induce gonadal growth.

Wolfson (1966) demonstrated continued gonadal recrudescence in White-throated Sparrows held in DD following initial exposure to long-day stimulation. Similar results have been reported in House Sparrows

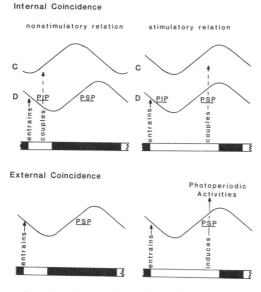

FIGURE 1. External and internal models in photoperiodism. In external coincidence, the daily photoperiod (onset of light indicated here) entrains a circadian neural oscillation that includes a rhythm of photosensitivity. If the photophase extends into the photostimulatory phase (PSP) of the photosensitivity rhythm, the light induces photoperiodic effects (i.e., gonadotropin release) directly along sensitive pathways. In internal coincidence, regulation of photoperiodic activities is accomplished by the interaction of two circadian neuroendocrine oscillations. One oscillation (D: driven oscillation) is entrained directly by some aspect of the daily photoperiod (light onset illustrated). The other oscillation (c) is coupled to the first in a nonstimulatory relation under short daylengths, perhaps by coincidence of light with a photoinhibitory phase (PIP). Coincidence of light during the PSP changes the phase of the coupled oscillation with the driven oscillation so that a temporal interaction of the two circadian systems produce photoperiodic effects.

(Vaugien and Vaugien, 1961) and White-crowned Sparrows (*Zonotrichia leucophrys*) Gwinner, 1980. Although these results have been construed as a prolonged carry-over effect of photostimulation (Farner and Gwinner, 1980), it may be noted that such a carry-over effect has not been given a physiological definition and other studies of birds held in DD are not so readily dismissed. Benoit *et al.* (1956,1959) reported several cycles of gonadal recrudescence and regression in domestic Mallards (*Anas platyrhynchos*) held continuously in DD. Recrudescence of testes also occurred in photorefractory starlings following transfer to DD (Rutledge and Schwab, 1974). In addition, the reproductive systems of both male and female Japanese quail developed to full maturity in birds held in DD following transfer of sexually immature quail raised on short nonstimulatory daylengths from hatch (Russo, 1983). The hens were laying eggs regularly after 16 weeks in DD.

Another conceptual construct for photoperiodism developed from several sources is known as the internal coincidence model (Tyshchenko, 1966; Pittendrigh, 1972; Gwinner, 1973; Pittendrigh and Daan, 1976; Dolnik, 1976; Meier, 1976). The basic difference from the external coincidence model is that there are two circadian oscillations in internal coincidence instead of one, and the temporal relation between the two oscillations determines whether or not gonadotropins are produced. The role of daylength is to adjust the temporal relations of the two oscillations.

Dolnik (1976) suggested that one of the oscillations might be entrained by light onset and the other by light offset. Thus, an increase in daylength would change the temporal relations in the phases of the two oscillations and produce a stimulatory pattern for gonadotropin release. A virtue of this paradigm for internal concidence, as that for external concidence, rests on its simplicity. Unfortunately, this design also has serious deficiencies in explaining all the information available (see Farner and Gwinner, 1980). For example, how does this system operate in birds treated with light interruptions during the night? It is also difficult to arrange such photostimulatory relations in House Sparrows in which very short daylengths (LD 2:22) as well as very long daylengths are stimulatory.

Despite the apparent deficiencies in this particular design, it would be reckless to abandon the basic idea involved in internal concidence. Other ways to account for photoperiodic adjustment of the phase relations of two oscillations are readily envisioned. One possibility is that the two oscillations are loosely coupled with one another. One oscillation is directly entrained by a daily photophase and this oscillation, in turn, couples the second oscillation. The phase relation of the two oscillations is determined by daylength. One visualization is that light advances the coupled oscillation so that it approaches a more stimulatory relation with the driving oscillation when the light occurs during the latter protion of the driving oscillation (Fig. 1) (see Meier, 1976).

The external coincidence model also offers no readily apparent explanation for observations that short daylengths are much more inhibitory for gonad weights than continuous darkness in gonadally developed birds transferred from long daylengths (Vaugien and Vaugien, 1961; Wolfson, 1966). Thus, short daylengths are not equatable with absence of light during a photosensitive phase. According to an internal concidence model (Fig. 1), one might postulate that just as there is a photosensitive phase about 14 hours after light onset when light adjusts the coupled oscillation into a stimulatory mode with the driven oscillation, so also there is photoinhibitory phase during the photosensitivity rhythm near light onset when light may adjust the coupled oscillation into an inhibitory mode, provided light is absent during the photosensitive phase.

Studies in White-throated Sparrows indicate that an internal co-

incidence model may be more appropriate for explaining photoperiodism. Whereas the phase of the photosensitivity rhythm apparently remains unchanged in sparrows held on long or short daylengths (Jenner
and Engels, 1952; Wolfson, 1965; Meier, 1976), the phase of the plasma
corticosterone rhythm does vary with respect to light onset (Meier and
Fivizzani, 1975). Apparently, the corticosterone and photosensitivity
rhythms are expressions of different oscillations. On LD 16:8, the peak
of plasma corticosterone concentration occurs 8 hr after light onset in
photosensitive birds. On LD 8:16, the corticosterone peak occurs at
light onset. On the other hand, the photosensitive phase occurs about
14 hr after light onset on both short and long daylengths. In sparrows
held on nonstimulatory daylengths (LD 6:18), daily corticosterone injections at 6–7 hr after light onset were stimulatory for gonad weights,
body fats stores, and nocturnal migratory restlessness, whereas injections at several other times of day were ineffective (Meier and Martin,
1971; Meier and Dusseau, 1973). According to an internal coincidence
model, it might be concluded that corticosterone injections entrain the
coupled oscillation of which it is an expression (see Section 4) so that
injections at the time of the endogenous corticosterone peak (8 hr after
light onset) on a stimulatory daylength produce a stimulatory relation
between the coupled and driven oscillations. Corticosterone injections
at other times produce nonstimulatory relations.

According to an external coincidence model, one might propose
that corticosterone injections entrain the photosensitivity rhythm so
that the photosensitive phase coincides with light (LD 6:18) when the
injections are made 18–24 hr before the photophase. However, this
interpretation appears untenable because the corticosterone and photosensitivity rhythms are not directly synchronized. Furthermore, if
one equated corticosterone injection time with the daily peak of plasma
corticosterone in sparrows held on LD 16:8 (8 hr after light onset), the
photosensitive phase would need to occur near light onset rather than
about 14 hr later.

Similar studies have been done with Japanese quail with equivalent
results (Russo, 1983). The phase of the corticosterone rhythm changes
with daylength, whereas the phase of the photosensitivity rhythm probably does not (Follett et al., 1974; Follett and Robinson, 1980). In addition, corticosterone injections also have stimulatory or inhibitory effects on reproductive indices as a function of injection time (Russo,
1983).

It should be noted that all models for photoperiodism are simply
guides for research and understanding. Neither internal nor external
coincidence models have been proved. Both models account for most

of the findings to date although most investigators claim that their results support an external coincidence model. However, the occurrence of gonadal recrudescence in the dark may prove a fatal blow for the external coincidence model where no satisfactory explanation is apparent. Gonadal development in the dark or on any daylength, long or short can be rationalized by internal coincidence.

An important consideration in understanding photoperiodism is that this system is often superimposed on an annual cycle of photosensitivity and photorefractoriness. Since seasonal changes in the interpretation of daylength appear to involve changes in the phase relations of circadian systems (internal coincidence), a fine tuning of such a system by daylength seems a reasonable option. Temporal synergism of circadian systems in construction of seasonality is discussed in Section 4.3.

3.2. Photoreceptor and Pacemaker Systems

Some additional questions that have arisen from the above demonstrations of circadian involvement in photoperiodism are as follows: How many photoreceptor systems are there and where are they located? How many pacemakers or centers of oscillatory activity are there and where are they located? Are the same mechanisms that are involved in reproductive photostimulation also involved in other photoperiodic events, such as vernal premigratory fattening?

Two photoreceptor systems are predictable on the basis of both the external and internal coincidence models. A photoreceptor involved in entrainment of a circadian oscillation including a rhythm of photosensitivity would be called for in both models. A second photoreceptor would be involved in photoinduction of gonadotropin release according to the external coincidence model and in phase adjustment of the two oscillations according to an internal coincidence model.

One of the more elegant early experiments that demonstrate two photosensory systems was done in House Sparrows (Menaker and Eskin, 1967). Sparrows were maintained on LD 14:10 with low light intensity (less than full moonlight) sufficient to entrain a rhythm of locomotor activity that occurred during tbe photophase, but not sufficient to stimulate the reproductive system. Reasoning that the photosensitivity rhythm may be coupled with the locomotor rhythm, bright light intervals (1.5 hr) were added near the onset of the dim photophase in one group and near the offset in another group. Only the latter group provided evidence of reproductive stimulation. Thus, the dim photophase entrains a rhythm of photosensitivity by way of one photoreceptor and bright light is necessary at a specific time to stimulate reproductive indices by way of another system.

Inasmuch as photostimulation of the reproductive system occurs in enucleated birds, an obligatory reproductive photosensor is not in the eye (Oishi et al., 1966; Menaker and Keatts, 1968; Gwinner et al., 1971; McMillan et al., 1975a,b,c,d; Yokoyama et al., 1978). Similarly, photostimulation occurs in pinealectomized birds (Sayler and Wolfson, 1968; Kobayashi, 1969; Menaker et al., 1970), although modifying effects of pinealectomy have been reported (Cardinali et al., 1971).

Following the initial demonstrations by Benoit (1964) on ducks, several other investigations have demonstrated photosensitive areas in the hypothalamus (Homma and Sakaibara, 1971; Menaker and Underwood, 1976; Yokoyama et al., 1978). Hamma and Sakakibara used probes tipped with luminscent paint to localize photosensitive areas in Japanese quail held in DD. The probes were ineffective when placed in most regions, including the vicinity of the pineal, but did promote gonadal stimulation when located in the hypothalamus.

The presence of two photosensory areas in photoperiodism were also demonstrated in Japanese quail using combinations of short nonstimulatory daylengths and luminescent probes in the hypothalamus. The environmental LD regimen entrained the circadian rhythm of locomotor activity in gonadally stimulated quail containing hypothalamic probes as well as in nonstimulated quail without the probes (Homma et al., 1979). Thus the photoreceptor for entrainment of locomotor activity is not in the same area of the hypothalamus as the photoreceptor for photostimulation on the gonads. In another experiment utilizing nonstimulatory short daylengths, hypothalamic probes provided for only short intervals daily were most effective when provided during the dark at the probable time of the photoinducible phase (Homma et al., 1979). The locomotor rhythm as with the photosensitivity rhythm, was entrained by the environmental LD regimen.

Inasmuch as photoperiodic entrainment of the locomotor activity rhythm can be induced in enucleated sparrows (McMillan et al., 1975a), it might seem plausible that the eye is not responsible for photoperiodic entrainment of the locomotor rhythm nor, by association, the photosensitivity rhythm. However, alternate explanations are possible and deserve consideration. If there are two circadian oscillations coupled together as proposed in the internal coincidence model, then removal of the photoreceptor (eyes) for entrainment of the driven oscillation, which includes the rhythm of photosensitivity, might allow for the coupling of this oscillation, which includes the rhythm of photosensitivity, to the one entrained by way of the hypothalamic photoreceptor (normally the coupled oscillation). Such a possibility is consistent with observations and arguments for a role of the eyes in photoperiodism (Benoit, 1964; Konishi and Homma, 1983). It should further be noted that there are evidences of pineal photoreceptor and pacemaker activities that might entrain a rhythm of photosensitivity in the absence of eyes.

An assemblage of the available information indicates that the photoreceptor mechanism for entrainment of photosensitivity may also have a role in entrainment of the locomotor rhythm. However, it is not yet apparent whether pacemaker systems involved in locomotor rhythmicity (SCN and pineal) are also a part of the photoperiodic mechanism. Although removal of the pineal does not usually influence photoperiodism, it should be noted that both the pineal and the SCN have been implicated in locomotor rhythmicity and thus by association the photosensitivity rhythm. A demonstration of these pacemaker activities, as in LD entrainment of the locomotor rhythm, may require removal of both organs.

Demonstrations of photoreceptor and pacemaker activities are made difficult by the presence of at least two photoreceptors and at least one (external coincidence) or more than one (internal coincidence model) pacemaker in reproductive photoperiodism. A single center (i.e., pineal and hypothalamic photoreceptor) may have both pacemaker and photoreceptor capabilities. In addition, lesions might destroy pathways or centers for production of gonadotropin-releasing hormones and be misinterpreted as influences on circadian systems.

3.3. Photoperiodic Stimulation of Nonreproductive Activities

Although photoperiodic stimulation of vernal migratory events (i.e., fattening and migratory restlessness) has been demonstrated in numerous species, possible roles for circadian rhythms in these events have received little attention. In the White-throated Sparrow, 2-hr light interruptions between 14 and 18 hours after onset of a 6-hr photophase was most stimulatory for fattening as well as gonadal growth (Jenner and Engels, 1952). Also in White-throated Sparrows, 6-hr light interruptions during the dark of a 72-hr photocycle (LD 6:66) were stimulatory for both gonadal growth and fattening when provided at 36–40 hr after onset of the entraining photophase, but not when given at 24–30 hr after light onset (see Meier, 1976).

As reproductive stimulation is in large measure the consequence of hormonal activity (gonadotropins), so fattening also can be considered an end result of hormone action. Prolactin has been implicated in premigratory fattening in the White-throated Sparrow, and thus also in the photoperiodic stimulation of vernal migratory metabolism and behavior (discussed in Section 4.3). Long daylengths have stimulatory influences on plasma prolactin concentrations in turkeys (*Meleagris gallopavo*) (Scanes et al., 1979; Burke et al., 1981) and ducks (Gourdji, 1970).

Thyroidal responsiveness to daylength varies with species, seasons, and apparently with the length of time the birds are maintained on long daylengths. In addition, there are considerable positive and negative interactions between the thyroid and gonads (Assenmacher, 1973; Chandola and Thapliyal, 1973; Sharp and Klandorf, 1981). A general finding is that long daylengths are inhibitory. In Japanese quail, long daylengths are inhibitory for uptake of ^{131}I by thyroid (Baylé and Assenmacher, 1967; Follett and Riley, 1967) and for plasma thyroxin concentration (Russo, 1983). Long daylengths are also inhibitory for thyroid ^{131}I uptake in castrated ducks (Jallageas and Assenmacher, 1972). An interesting finding in chickens and Willow Ptarmigan (*Lagopus lagopus*) is that LL, which is usually as stimulatory as long daylengths for gonadal function, may not be stimulatory for thyroid function (Klandorf et al., 1982). These results and others suggest some separation in photoperiod control of the gonads and the thyroid.

It is apparently generally assumed that photoperiodic stimulation of premigratory fattening and hormone (e.g., prolactin and thyroxin) secretion involve the same mechanisms as those involved in photoperiodic stimulation of reproduction. However, the basis for such an assumption is weak. In fact, many observations indicate some separation of metabolic and reproductive responsiveness and refractoriness to long daylengths (see Section 4). One possibility is that each oscillation is composed of a family of individual rhythms that may become uncoupled and assume different relations and thereby produce different net effects. It seems clear that our understanding of photoperiodism is still rudimentary. It would be premature to impose many limits on the possibilities at this juncture.

3.4. Egg Laying

In species studied intensively (chickens, ducks, quail, and turkey), egg laying follows a pattern in which a sequence of eggs are layed, one each day. This daily cycle of oviposition has features in common with circadian rhythms. It persists in LL with a periodicity approximating 24 hr and is entrained by a 24-hr LD cycle (Fraps, 1959; van Tienhoven, 1961; Arrington, et al., 1962; Opel, 1966).

There are several specific, timed events during the egg laying cycle. The first is the LH surge that occurs in response to a neurogenic stimulus (Zarrow and Bastian, 1953; Fraps and Conner, 1954; van Tienhoven *et al.*, 1954) and is timed by the daily photoperiod. In the chicken and Khaki Campbell duck, the LH surge is set by the onset of dark and occurs during the latter half of darkness (LD 14:10). In Japanese quail, the LH surge is set by the onset of light and occurs early during the photophase.

The LH surge, augmented by FSH (Kamiyoshi and Tanake, 1983), stimulates ovulation about 3–7 hr later. Following ovulation, there is a 24–26 hr maturational period for the developing egg in the oviduct before oviposition. Preparation of the oviduct depends on the synergistic activities (Hughes *et al.*, 1981) of estradiol and progesterone that are produced by the primary follicle during an interval of about 6 hr prior to ovulation. Estradiol may also be released for a period following ovulation. Mobilization of calcium and nutrients (vitellogenesis) for oviducal maturation is regulated primarily by estradiol.

The pause of a day or more that occurs in chickens between sequences of egg laying has stimulated considerable interest and several hypotheses (Fraps, 1954; Bastian and Zarrow, 1955; Nalbandov, 1959). The most recent hypothesis is one advanced by Kamiyoshi and Tanaka (1983) that ascribes a primary role for a minor circadian release of LH that results in a slight increase in progesterone secretion, which in turn induces an acute release of LH. The amount of progesterone produced in response to the minor circadian peak of LH is thought to decrease from day to day so that a longer time is required to reach a threshold for acute LH release until there is insufficient progesterone produced and the LH surge does not occur.

This hypothesis does take into account the findings that progesterone has a primary role in causing LH release and ovulation (Ralph and Fraps, 1959, 1960; see also Kamiyoshi and Tanake, 1983). It also accounts for the finding that progesterone can stimulate LH release at times of day other than a hypothetical sensitive period (Wilson and Sharp, 1975) as proposed by earlier models. However, it does not explain why there would be a gradual diminution of progesterone response to the priming LH release and it does not consider the possibility of a circadian release of progesterone from the adrenal cortex as a release for LH.

3.5. Pigeon Cropsac Responsiveness

Prolactin has a primary role in support of incubation and brooding (Bailey, 1952; Crispens, 1957; Lehrman, 1955). In pigeons, prolactin both supports brooding behavior and stimulates the cropsac mucosal layer to become a part of the cropsac milk (Riddle *et al.*, 1932). Because the cropsac response to prolactin was an important assay system for prolactin for many years, it is amazing that no one reported circadian responsiveness until relatively recently (Burns and Meier, 1971; Meier *et al.*, 1971a,b). Both systematic and intradermal injections over the cropsac were very stimulatory at nine hours after light onset (LD 12:12), but ineffective at light onset. In pigeons held in LL, prolactin was also most effective 18 hr after systemic or intradermal injections of corticosterone. This relation corresponds well with the interval between the peak of plasma corticosterone concentration at three hours after dark onset (LD 12:12) (Joseph and Meier, 1973) and the peak of cropsac sensitivity to prolactin at nine hours after light onset (LD 12:12). The rhythm of cropsac responsiveness free runs in LL without corticosterone injections, but gradually dampens out unless the birds are given thyroid hormone (Meier *et al.*, 1971b; John *et al.*, 1972). Because prolactin can also induce fattening in pigeons at a time when it has little influence on the cropsac, it follows that prolactin may have different activities depending on the phase of its rhythm (Meier *et al.*, 1971a). A possible role for circadian prolactin rhythms in regulation of incubation and brooding behavior invites investigation.

4. PHOTOSENSITIVITY AND PHOTOREFRACTORINESS

4.1. Links in the Annual Cycle

The sequential nature of reproductive and migratory events during the annual cycle has attracted considerable interest, and many attempts have been made to identify particular hormones that might lead toward the next stage or might prevent such progression. Rowan (1931) suggested that gonadal hormones have such prominent directive roles, not only with regard to reproduction, but migration as well. Subsequent research supports this hypothesis, but the overall mechanism is by no means evident since gonadal hormones have variable effects during the year.

Gonadal hormones have a permissive role in photoperiodic stimulation of vernal migratory indices in *Zonotrichia* species. If castration is carried out early in winter before there has been any photoperiodic stimulation and possibly before photosensitivity has been fully re-established, long daylengths in spring do not induce fattening or nocturnal restlessness (Weise, 1967; Gwinner *et al.*, 1971). Testosterone replacement in castrates restores migratory responsiveness to long day-lengths (Stetson and Erickson, 1972). However, castration does not pre-vent migratory responses to long daylengths if it is carried out later in winter or early spring (Miller, 1960; Morton and Mewaldt, 1962).

Gonadal hormones have been shown to delay the onset of photo-refractoriness (postnuptial molt and gonadal regression) in several spe-cies (Lofts, 1970), although plasma concentrations of FSH and LH de-crease at approximately the same time in castrate birds as in intact birds at the onset of gonadal regression (Wilson and Follett, 1974; Mattocks et al., 1976; Nicholls and Storey, 1976). This positive feedback influence of gonadal hormones prior to photorefractoriness is replaced in some species later during the photorefractory period by a strong negative feedback influence on gonadotropin release. Gonadotropin release in-creases sharply following castration in photorefractory Red Grouse (La-gopus lagopus) and Willow Grouse held on either long or short day-lengths (Sharp and Moss, 1977; Sharp, 1980a). Furthermore, implants releasing large amounts of testosterone prevent re-establishment of pho-tosensitivity in White-throated Sparrows held on short daylengths (Turek et al., 1980).

A preponderance of the evidence suggests that gonadal hormones are not directly equatable in a regulatory sense with either photosen-sitivity or photorefractoriness. A principal effect is to delay or prevent the transformation from sensitivity to refractoriness and from refrac-toriness to sensitivity. In addition, they appear to allow for the matu-ration of the photoperiodic mechanism involved in vernal migration. In view of the findings that photosensitivity, photorefractoriness, and migratory expressions may be regulated by altering the phase relations of circadian systems (see Section 4.3), it seems likely that the hormones exert their activities by maintaining or changing the phases of these systems. Such an influence by testosterone on the free-running rhythm of locomotor activity has been reported (Gwinner, 1974).

Thyroid involvement in the annual cycle has also attracted much attention.Unfortunately, the findings have been contradictory and dif-ficult to interpret. For example, histological evidences often judged to be indicators of secretion may only indicate increased hormone storage. In addition, blood concentrations of thyroxin are composites of secretion and utilization rates. Thus, high concentrations may be evidence of either greater secretion, less tissue utilization, or almost any possible combination of secretion and utilizations. Despite the fact that hormone concentrations, including thyroid hormone levels, often vary during a day as much as they do seasonally, investigators continue to ignore time of day when performing seasonal studies. Another important limitation in thyroid research is lack of appreciation of the enormous range of thyroid hormone activities and their synergistic roles with both neural and hormonal mechanisms. It is not surprising that thyroid hormone activity has been linked with adaptation to cold (Wilson and Farner, 1960; John and George, 1978), induction of molt (Voitkevitch, 1966; Payne, 1972), control of reproduction (Thapliyal and Pandha, 1965; Jallageas and Assenmaker, 1980), and stimulation of migratory restlessness (Merkel, 1938; de Graw et al., 1979). The thyroid, as with the heart, may well be linked with every seasonal condition.

Perhaps the most useful and interesting research on the thyroid relates to the reciprocal relation it often has with the reproductive system and the role it may exert in initiating photorefractoriness. Induced hypothyroidism enhances reproductive development and thyroxin injections may precipitate gonadal regression prematurely (Thapliyal and Pandha, 1965; Chandola et al., 1974). In addition, thyroidectomy prevents photorefractoriness in European Starlings if carried out in spring before photostimulation (Wieselthier and van Tienhoven, 1972). Assenmacher and colleagues (Assenmacher et al., 1975; Jallageas and Assenmacher, 1979, 1980) have presented data in Peking ducks and arguments supporting thyroid involvement in the initiation of photorefractoriness. Thyroid hormone is thought to decrease gonadal hormone activity and thereby allow for gonadal regression and molt. However, the universality of such a mechanism for all species has been questioned (Smith, 1982). Given the importance of phase relations of hormone rhythms in seasonal responsiveness to daylength (see Section 4.3), it is notable that thyroid hormones, as with gonadal hormones, influence phases of circadian rhythms. (Meier, 1975; Kovacs and Peczely, 1983).

4.2. Circannual Cycles

Annual cycles of molt and reproductive and migratory indices have been demonstrated in many species held on constant daylengths (Merkel, 1963; Lofts, 1964; Zimmerman, 1966; King, 1968; Gwinner, 1968, 1972; Berthold, 1974). These cycles often have periods somewhat less than a year and may persist for many years. An interesting feature is that dissynchrony among the several seasonal indices may gradually develop suggesting a partial independence of the mechanisms involved in molt, reproduction, and migration (Berthold, 1975; Gwinner, 1975).

These studies dramatize the central role of an endogenous system that accounts for an orderly sequence of seasonal conditions. The regularity of this cycle in some species has sometimes prompted the suggestion that circannual periodicity may be analogous to the circadian rhythm. This suggestion has been criticized on the basis that information of a cyclic nature may be obtainable from the daily LD regimen and that a circannual cycle must persist under LL or DD to be considered analogous to a circadian rhythm (Farner and Follett, 1966; Hamner, 1971). In those few instances where circannual cycles were tested in birds held in LL or DD, the findings have been negative (Hamner, 1971; Rutledge and Schwab, 1974), equivocal (Benoit et al., 1956, 1959), or positive (Gwinner, 1973).

Numerous investigators have also proposed that the period of a circannual cycle might be determined by a count of the number of circadian days (frequency demultiplication hypothesis; FDH). However, this hypothesis is sustained primarily by extrapolation from research in insects (see Saunders, 1976). When tested directly in Garden Warblers, *Sylvia borin*, and European Starlings, the results did not support the hypothesis (Gwinner, 1981). Gwinner placed birds on LD cycles of 11:11, 12:12, and 13:13, reasoning that those on LD 13:13 would require a longer period to complete a circannual cycle than those birds held on the shorter "days" and that the circannual cycles should have the same number of circadian days under each LD regimen. Neither of these expectations for the FDH were fulfilled.

It now appears probable that although there is an endogenous seasonal timing mechanism, this mechanism is not equivalent to the circadian rhythm. Thus, a circannual cycle may be expressed in European Starlings on short or medium (LD 12:12) daylengths, but not on long daylengths (Schwab, 1971; Hamner, 1971). In birds such as starlings, gonadal recrudescence can occur on short daylengths and short daylengths can also break photorefractoriness. Thus, cycling is possible on short daylengths, but the birds are trapped in photorefractoriness when they are maintained on long daylengths. Inasmuch as photosensitivity is re-established in some species such as *Quelea quelea* (Lofts and Murton, 1968) when held on long daylengths as well as short daylengths, it seems likely that such species would be able to show cyclicity on long daylengths.

4.3. Seasonality Based on Temporal Synergisms of Circadian Systems

Because circadian rhythms are involved in photostimulation in photosensitive birds as well as in maintenance of photorefractoriness (Hamner, 1968; Murton et al., 1970; Turek, 1972), it seems axiomatic that circadian rhythms would have important roles in the endogenous seasonal mechanism. Based on a series of studies by our laboratory with White-throated Sparrows, we have proposed that the endogenous seasonal determining photosensitivity and photorefractoriness involves a temporal interaction of circadian neuroendocrine oscillations (Meier, 1976; Meier and Fivizzani, 1981; Meier et al., 1981). A principal experimental basis for this hypothesis is that hormones have different activities as functions of time of day when hormone injections are given. Prolactin injections carried out daily during the latter half of the photophase (LD 16:8) induced large increases in body fat stores in lean photorefractory sparrows; however, prolactin injections made early during the photophase were completely ineffective (Meier and Davis, 1967). Midday injections also stimulated nocturnal migratory restlessness as well as fattening, whereas injections given early or late during the photophase did not (Meier, 1969).

Assays of pituitary prolactin content support a circadian involvement of prolactin in regulating seasonal conditions. Circadian variations were observed in photostimulated sparrows during the vernal migratory period as well as in photorefractory birds undergoing postnuptial molt. However, the phase of the rhythm differed at the two seasons. The release of pituitary prolactin (sharp decrease in pituitary content) occurred during the afternoon in photosensitive birds and late during the dark in photorefractory birds (Meier et al., 1969).

Although on normal LD regimens the circadian rhythm of responsiveness to prolactin is entrained by the daily photoperiod, in birds held in LL injections of corticosterone can entrain the response rhythms. Daily injections of prolactin at six different times (at 4-hr intervals) relative to corticosterone injections had variable effects. Prolactin injections at 12 hr after corticosterone injections (12-hr relations) induced in lean photorefractory sparrows marked increases in fat stores that reached the high levels found during the migratory periods after only five days of injections. The 4-hr relation of corticosterone and prolactin injections was also stimulatory for fat stores, but all other relations were ineffective. In fact, the 8-hr relation induced further loss of fat in lean photorefractory birds (Meier and Martin, 1971). Similar results were also found in photosensitive sparrows in winter maintained in LL and treated with timed daily injections of corticosterone and prolactin (Meier et al., 1971a).

Inasmuch as there were evidences of gonadal stimulation as well as fattening in the birds treated with corticosterone and prolactin in the 12-hr relation, it seemed possible that the 12-hr treatment may have induced spring conditions. Accordingly, the 8-hr relation that was inhibitory for fattening and reproductive indices may have induced summer conditions, and the 4-hr relation may have induced fattening associable with fall migratory conditions. As a test for this hypothesis, sparrows were injected with the hormones in the 12-, 8-, and 4-hr relations at several seasons and tested after one week for nocturnal restlessness and orientation (Martin and Meier, 1973). There was no nocturnal activity at all in the eight-hour group. However, both the 12-hr and 8-hr treatments induced nocturnal restlessness and fattening. Moreover, the orientation of birds tested in cages under the open night sky was directed northward (toward the breeding grounds) in the 12-hr group and southward (toward the wintering quarters) in the 4-hr group.

Assays of plasma corticosterone, as with those of prolactin, indicated that the phase of the circadian hormone rhythm changes seasonally. Even when photosensitive sparrows were placed on long daylengths (LD 16:8) in winter and held on that daylength until fall, the phase, amplitude, and pattern varied continuously with the sequence of changes in migratory and reproductive conditions (Meier and Fivizzani, 1975). The rhythm dampened out completely during the postnuptial molt several weeks after initiation of photorefractoriness. Because thyroid hormones are important for maintenance of corticosterone rhythms in pigeons and some other vertebrate species (John et al., 1972; Meier, 1975), the loss of the corticosterone rhythm might be a consequence of hypothyroidism. Thyroid activity is low during the postnuptial molt in the White-crowned Sparrow (Wilson and Farner, 1960).

Some of the temporal activities of prolactin are probably at the tissue level. Prolactin apparently directly stimulates hepatic lipogenesis and enzymes involved in lipogenesis (Wheeland et al., 1976). It can also suppress oviducal responses to exogenous gonadotropins (Meier and MacGregor, 1972). Nevertheless, it seems most unlikely that regulation of a multitude of metabolic and behavioral seasonal conditions would be funneled through the timed activities of just two hormones. It appeared more probable to us that many of the effects induced by the hormone injections were indirect influences by way of the nervous system.

Inasmuch as corticosteroids stimulate serotonin synthesis (Sze et al., 1976; Telegdy and Vermes, 1975) and prolactin increases dopamine synthesis (Hokfelt and Fuxe, 1972; Weisel et al., 1978), it seemed possible that timed injections of drugs that have similar activities on neurotransmitter synthesis might also induce complexes of seasonal conditions. Accordingly, 5-hydroxytryptophan (5-HTP), a rate-limiting precursor substrate for serotonin, was substituted for corticosterone and dihydroxyphenylalanine (DOPA), a rate-limiting presursor for dopamine, was substituted for prolactin.

Timed daily injections of 5-HTP and DOPA produced changes comparable to those induced by timed hormone treatment. Dihydroxyphenylalanine was stimulatory for reproductive and migratory (spring conditions) when injected 12 hours after 5-HTP injections and inhibitory for these indices when injected at 5–8 hr after 5-HTP (Miller and Meier, 1983a). Perhaps the most dramatic effect of timed drug treatment was the resetting of the annual cycle by re-establishing photosensitivity during late summer in photorefractory sparrows (Miller and Meier, 1983b). The birds were placed in LL and injected with saline or timed injections of 5-HTP and DOPA for ten days. Afterward they were returned to long daylengths (LD 16:8) and examined periodically for five months. The saline-injected controls remained in the fall photorefractory conditions as expected. However, the 12-hr relation of 5-HTP and DOPA treatment induced spring conditions including gonadal growth, fattening, and nocturnal activity oriented northward under the open fall sky. After several weeks, these spring conditions were replaced by summer conditions including loss of nocturnal activity, involution of the gonads, and a complete postnuptial molt. These summer conditions in turn were replaced by fall migratory conditions including fattening and nocturnal restlessness oriented southward under the open spring sky. Thus, the drug treatment induced photosensitivity that naturally requires an interval of about eight weeks of short daylengths in White-throated Sparrows.

Re-establishment of photosensitivity during the late summer in photorefractory White-throated Sparrows can also be induced by the 12-hr relation of corticosterone and prolactin injections (Fig. 2) (Ferrell, 1979; Meier et al., 1981). Spring migratory and reproductive conditions appear first and these are followed in an appropriate time sequence by summer and fall conditions, as described for drug-induced resetting of the annual cycle. Later during the photorefractory period in October and November, neither timed drug or hormone treatment in the 12-hr relation resets the annual cycle (unpublished data). One explanation for this resistance later during photorefractoriness involves the recovery of both the thyroid and adrenal cortex from an annual low in functional capacities during early photorefractoriness (see Meier and Fivizzani, 1975). Both thyroid and adrenal cortical hormones have augmenting effects on serotonergic and catecholaminergic activities and might make it more difficult to adjust such neural activities by exogenous stimuli.

We believe that the complex of metabolic and behavioral condi-

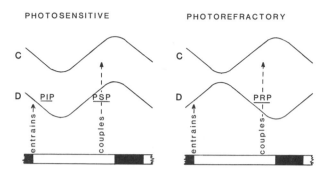

FIGURE 2. Internal coincidence model for photosensitivity and photorefractoriness. In photosensitive birds, coincidence of light during a photostimulatory phase (PSP) of the driven oscillation (D) associates the coupled oscillation (C) in a stimulatory relation. In photorefractory birds, coincidence of light during a photorefractory phase (PRP) of D maintains C in a refractory relation.

tions associated with a specific season are the consequences of inter-action between two circadian neuroendocrine oscillations that have many other circadian neural and hormonal expressions. The temporal interaction of these oscillations and their numerous expressions produce the season conditions. Changes in the phase relations of these oscil-lations and their expressions account for the sequence of seasonal con-ditions and for the alternation of photosensitivity and photorefracto-riness. Products of temporal interaction such as thyroid and gonadal stimulation may provide a driving force for changes in the temporal relations of the circadian systems.

Injections of corticosterone and prolactin are thought to reset the neuroendocrine oscillations to which the endogenous hormone rhythms are coupled. Thus, the hormone injections alter the phase relations of the two oscillations including their circadian neural and hormonal expressions. It seems probable that timed injections of other hormones may also influence the phases of the neural oscillations. Melatonin might be one such candidate based on its stimulatory effect on brain serotonin content in rats (Anton-Tay et al., 1968). Any hormone rhythm might well be expected to feed back on its controlling neural oscillation in such a manner as to sustain the overall rhythmicity.

5. CONCLUSIONS AND PROJECTIONS

The annual cycle involves an orderly sequence of complex inte-grated behavioral and physiological conditions. Regulation of this cycle depends on a high level of organization that is provided by temporal interaction of circadian systems. Temporal interaction and synergisms may occur at several levels of organization. At the cell level, specific target cells may have a rhythm of sensitivity (response rhythm) to one hormone (stimulus) that is entrained or driven by a circadian rhythm of another hormone. If the daily peak of the stimulus rhythm coincides with the peak of the response rhythm, the effect is greatest and lead to cumulative conditions such as premigratory fattening. On the other hand, if the stimulus peak occurs at a time when there is little or no responsiveness, the stimulus would be relatively ineffective (Fig. 3). Such a system allows one stimulus to be involved in several activities as a function of the phase relations of the stimulus rhythm with several response rhythms. For example, prolactin stimulates fattening in pigeons at one time of day and cropsac proliferation at another time (John et al., 1972). Thus, the potential activities of each hormone are enhanced by this dimension of biological time and might help to account for involvement of individual hormones in several separate activities.

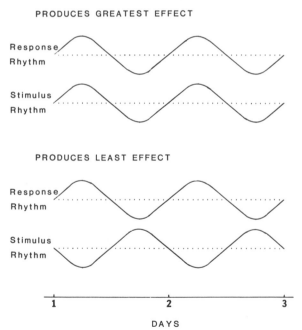

FIGURE 3. Temporal synergism of circadian stimulus and response rhythms. Coincidence of greatest stimulus with greatest responsiveness of target tissue produces the greatest cumulative net effect. Multitudes of these temporal interactions may produce seasonal complexes of physiological and behavioral conditions.

In a similar manner, circadian interaction in the nervous system may influence not only the daily sequence of various activities, but also the strength of these activities. Coincidence of inhibitory afferent neural activity at a hypothalamic nucleus with circadian stimulatory afferent activity may block development of efferent activity. If the circadian inhibitory activity occurs in another phase relation, efferent activity would be less impeded.

Our studies of White-throated Sparrows indicate that an interaction of circadian systems can determine responsiveness to daylength. Photosensitivity and photorefractoriness appear to be consequences of differences in the phase relations of two circadian systems. The phase relations of these neural oscillations can be set by timed daily injections of hormones (corticosterone and prolactin) or drugs (5-HTP and DOPA) given in specific phase relations. Although reset by exogenous stimuli that influence serotonin and dopamine synthesis, each neural oscillation probably includes many other neural activities. It seems probable that additional hormones and neurotransmitter affectors will also be found to reset these oscillations.

Inasmuch as every cell probably has an endogenous circadian rhythm, it may be unrealistic to limit the neuroendocrine system to just two interacting circadian systems. There may be subsidiary circadian rhythms that couple and uncouple with the principal regulating oscillations. In addition, the phase relation that a subsidiary-coupled oscillation takes with a principal oscillation may vary as a result of hormonal activities, such as those of thyroid and gonadal hormones.

If photosensitivity and photorefractoriness are determined in part by an interaction of two circadian neuroendocrine oscillations, it seems reasonable that photoperiodism should involve a part of this system. Both external and internal coincidence models of photoperiodism would appear consistent with an internal coincidence model for seasonality. However, the external coincidence model would seem to be more restrictive in terms of regulation. A direct photic stimulation of photoperiodic events along sensitive pathways (external coincidence) would not have the flexibility for producing fine tuning and possible separation of photoperiodic events that an interaction of two oscillating systems (internal coincidence) would seem to have.

As a working model, we would propose that the same circadian neuroendocrine oscillations involved in photosensitivity and photorefractoriness are also involved in photoperiodism. When the phase relations of the oscillations are set in a photosensitive relation, they require further tuning by the presence of light during a critical phase of the driven oscillation (photosensitivity rhythm) to couple the oscillations in a stimulatory relation for photoperiodic events.

Serotonergic mechanisms appear to have important roles in seasonality, and manipulation of brain serotonin synthesis by means other than timed daily injections of precursor substrates and hormones can influence the endogenous seasonal mechanism. In the Golden Hamster, increased diet tryptophan increases brain serotonin content, changes the phase relations of plasma corticosterone and prolactin rhythms, and converts scotorefractory hamsters in summer (resistance to inhibitory effects of short daylengths on the reproductive system) to scotosensitive hamsters (the natural condition in winter) (Wilson and Meier, 1983). In the White-throated Sparrow, parachlorophenylalanine (inhibitor of brain serotonin synthesis) given alone can convert a small percentage of photorefractory birds to photosensitive birds (Miller and Meier, 1983a,b).

Involvement of serotonin in regulation of phase relations of circadian rhythms and in control of the reproductive system (overall effects of high brain serotonin content are inhibitory) suggests several areas of research that would have potential interest. One such area relates to influence of feeding on reproduction. Tryptophan availability is the principal rate-limiting factor in brain serotonin synthesis and this availability depends on the amount of competition with other amino acids in the blood for active transport mechanisms in the brain and on the amounts and ratios of tryptophan and other amino acids in the food (Fernstrom and Wurtman, 1972). Inasmuch as amino acid content in new vegetation is considerably greater than in old vegetation, the greater competition among amino acids in animals feeding on new vegetation would seem to favor lower levels of brain serotonin and greater stimulation of the reproductive system. Tryptophan levels are extremely low in new sprouts (Orr and Watt, 1968). Both temperature (ambient) and barometric pressure have been shown also to influence serotonergic activity (see Meier and Fivizani, 1981). Another area that deserves study is the possible mediation by serotonin of the influences of melatonin. Although a comparable study has apparently not been done in an avian species, melatonin induces increases in brain serotonin content in rats and ferrets (Anton-Tay, 1968; Yates and Herbert, 1979).

REFERENCES

Able, K. P., 1980, Mechanisms of orientation, navigation, and homing in: *Animal Migration, Orientation, and Navigation* (S. A. Gauthreaux, ed.), Academic Press, New York, pp. 283–373.

Able, K. P., and Dillon, P. M., 1977, Sun compass orientation in a nocturnal migrant, the white-throated sparrow, Condor **79**:393–395.

Anton-Tay, F., Chou, E., Anton, S., and Wurtman, R. J., 1968, Brain serotonin concentration: Elevation following intraperitoneal administration of melatonin, Science **162**:277–278.

Arrington, L. C., Abplanalp, H., and Wilson, W. O., 1962, Experimental modification of the laying pattern in Japanese Quail, Br. Poult. Sci. **3**:105–113.

Aschoff, J., 1967, Circadian rhythms in birds, Proc. Int. Ornithol. Congr. **14**:81–105.

Assenmacher, I., 1973, The peripheral endocrine glands, in: Avian Biology, Volume III (D. S. Farner and J. R. King eds.), Academic Press, New York, pp. 185–285.

Assenmacher, I., and Boissin, J., 1973, Mecanismes aminergiques et controle du cycle circadien de la corticostéronémie chez les oiseaux, Colloques INSERM (Brain-adrenal interactions) **22**:257–282 (1974).

Assenmacher, I., and Jallageas, M., 1980, Adaptive aspects of endocrine regulation in birds, in: Hormones, Adaptation and Evolution (S. Ishii, ed.), Japan Sci. Soc. Press and Springer-Verlag, Berlin, p. 91.

Assenmacher, I., Astier, H., Daniel J. Y., and Jallageas, M., 1975, Experimental studies on the annual cycles of thyroid and adrenocortical functions in relation to the reproductive cycle of drakes, J. Physiol. (Paris) **70**:507–520.

Bailey, R. E., 1952, The incubation-patch of passerine birds, Condor **54**:121–136.

Bastian, J. W., and Zarrow, M. X., 1955, A new hypthesis for the asynchronous ovulatory cycle of the domestic hen (Gallus domesticus), Poult. Sci. **34**:776–788.

Baylé, J. D., and Assenmacher, I., 1967, Controle hypothalamo-hypophysaire du fonctionnement thyroidien chez la Caille, C. R. Acad. Sci **264**:125–128.

Beer, J. R., 1961, Winter feeding patterns of the House Sparrow, Auk **78**:63–71.

Benoit, J., 1964, The structural component of the hypothalamo-hypophyseal pathway, with particular reference to photostimulation of the gonads in bird, Ann. N. Y. Acad. Sci. **117**:23–34.

Benoit, J., Assenmacher, I., and Brard E., 1956, Apparition et maintien de cycles sexuels non saisonniers chaz le Canard domestique place pendant plus de trois ans a l'obscurite totale, J. Physiol. (Paris) **48**:388–391.

Benoit, J., Assenmacher, I., and Brard, E., 1959, Action d'un éclairment permanent prolonge sur l'evolution testiaulaire du canard Pekin, Arch. Anat. Micro. Morph. Exp. **48**:5–11.

Berthold, P., 1974, Circannuale peroidik bei grasmucken (Sylvia). III. Periodik der mauser, der nachtururahe und des korpergewichtes bei mediterranen arten mit unterschiedlichen zugverhalten, J. Ornithol. **115**:251–272.

Berthold, P., 1975, Migration: Control and metabolic physiology, in: Avian Biology, Volume V (D. S. Farner and J. R. King, eds.), Academic Press, New York, pp. 77–128.

Berthold, P., 1978, Concept of endogenous control of migration in warblers, in: Animal Migration, Navigation and Homing (K. Schmidt-Koenig and W. T. Ketton, eds.), Springer-Verlag, Berlin, New York, pp. 275–282.

Binkley, S., 1976, Pineal gland biorhythmic N-acetyltransferase in chickens and rats, Fed. Proc. **35**:2347–2352.

Binkley, S. A., Reibman, J. B., and Reilly, K. B., 1978, The pineal gland: A biological clock in vitro, Science **202**:1198–1201.

Boissin, J. and Assenmacher, I., 1970, Circadian rhythms in adrenal cortical activity in the quail, J. Interdisc. Cycle Res. **1**:251–265.

Bons, N. 1976, Retinohypothalamic pathway in duck (Anas platyrhynchos), Cell Tissue Res. **168**:343–360.

Boulos, Z., and Rusak, B., 1982, Phase-response curves and the dual-oscillator model of circadian pacemakers, in: *Vertebrate Circadian Systems* (J. Aschoff, S. Daan, and G. A. Groos, eds.), Springer-Verlag, Berlin, pp. 215–223.

Bünning, E., 1960, Circadian rhythms and time-measurement in photoperiodism, *Cold Spring Harbor Symp. Quant. Biol.* **25**:249–256.

Burke, W. H., Dennison, P. T., Silsby, J. L., and El Halawani, M. E., 1981, Serum prolactin levels of turkey hens in relation to reproductive function, in: *Recent Advances in Avian Endocrinology* (G. Pethes, P. Péczely, and R. Rudas, eds.), *Adv. Physiol. Sci.* **33**:109–116.

Burns, J. T., and Meier, A. H., 1971, Daily variations in pigeon cropsac responses to prolactin, *Experientia* **27**:527–574.

Cardinali, D. P., Luello, A. E., Tramezzani, J. H., and Rosner, J. M., 1971, Effects of pinealectomy on the testicular function of the adult male duck, *Endocrinology* **89**:1082–1093.

Chandola, A., 1978, Daily rhythmicity of thyroid function in Spotted Munia, *Gen. Comp. Endocrinol.* **34**:93–98.

Chandola, A., and Thapliyal, J. P., 1973, Gonadal hormones and thyroid function in spotted munia, *Lonchura punctulata*, *Gen. Comp. Endocrinol.* **21**:305–319.

Chandola, A., Thapliyal, J. P., and Paynaskar, J., 1974, Effects of thyroidal hormones an avian response to photoperiod in a tropical finch (*Ploceus phillipinus*), *Gen. Comp. Endocrinol.* **24**:437–441.

Crispens, C. G., 1957, Use of prolactin to induce broodiness in two wild turkeys, *J. Wildlife Manage.* **21**:462–468.

Daan, S., and Pittendrigh, C. S., 1976, A functional analysis of circadian pacemakers in nocturnal rodents II. The variability of phase response curves, *J. Comp. Physiol.* **106**:252–266.

Davis, K. B., and Meier, A. H., 1973, Seasonal and daily variations of sodium, potassium and chloride levels in the plasma and brain of the migratory white-throated sparrow, *Zonotrichia albicollis*, *Physiol. Zool.* **47**:13–21.

de Graw, W. A., Kern, M. D., and King, J. R., 1979, Seasonal change in blood composition of captive and free-living white-crowned sparrows, *J. Comp. Physiol.* **129(2)**:151–162.

Deguchi, T., 1979, Circadian rhythm of serotonin N-acetyltransferase activity in organ culture of chicken pineal gland, *Science* **203**:1245–1247.

Deguchi, T., 1981, Rhodopsin-like photosensitivity of isolated chicken pineal gland, *Nature (London)* **290**:702–704.

Dolnik, V., 1976, Fotoperiodism u ptits, in: *Fotoperioism zhivotnykh* (D. A. Scarlato, ed.), Academiya Nauk, SSSR, Leningrad, pp. 47–81.

Drent, R., 1975, Incubation, in: *Avian Biology*, Volume 5 (D. S. Farner and J. R. King, eds.), Academic Press, New York, pp. 333–420.

Emlen, S. T., 1969, Bird migration: Influence of physiological state upon celestial orientation, *Science* **165**:716–718.

Emlen, S. T., and Emlen, J. T., 1967, A technique for recording migratory orientation of captive birds, *Auk* **83**:361–367.

Ensor, D. M., and Phillips, J. M., 1970, The effect of salt loading on the pituitary prolactin levels of the domestic duck (*Anas phatyrhynchos*) and juvenile herring or lesser black-backed gulls (*Larus argntatus* or *Larus fuscus*), *J. Endocrinol.* **48**:167–172.

Farner, D. S., 1964, The photoperiodic control of reproductive cycles in birds, *Am. Sci.* **52**:137–156.

Farner, D. S., 1965, Circadian systems in the photoperiodic response of vertebrates, in: *Circadian Clocks* (J. Ascoff, ed.), North-Holland, Amsterdam, pp. 357–369.

Farner, D. S., and B. K. Follett, 1979, Reproductive periodicity in birds, in: *Hormones and Evolution*, V.2., (E. J. W. Barrington, ed.), Academic Press, New York, pp. 829–872.

Farner, D. S., and Gwinner, E., 1980, Photoperiodicity, circannual, and reproductive cycles, in: *Avian Endocrinology* (A. Epple and M. H. Stetson, eds.), Academic Press, New York, pp. 331–366.

Fernstrom, J. D., and Wurtman, R. J., 1972, Brain serotonin content: Physiological regulation by plasma neutral amino acids, *Science* **178**:414.

Ferrell, B. R., 1979, Thyroid hormones and seasonality in white-throated sparrow and green anole, Ph.D. Dissertation, Louisiana State University, Baton Rouge.

Follett, B. K., and Riley, J., 1967, Effect of the length of the daily photoperiod on thyroid activity in the female Japanese quail (*Coturnix coturnix japonica*), *J. Endocrinol.* **39**:615–616.

Follett, B. K., and Robinson, J. E., 1980, Photoperiod and gonadotrophin secretion in birds, in: *Progress in Reproductive Biology*, Volume 5 (R. J. Reiter and B. K. Follett, eds.), Karger, Basel, pp. 407–420.

Follett, B. K., and Sharp, P. J., 1969, Circadian rhythmicity in photoperiodically induced gonadotrophin release and gonadal growth in the quail, *Nature (London)*, **223**:968–971.

Follett, B. K., Mattocks, P. W., and Farner, D. S., 1974, Circadian function in the photoperiodic induction of gonadotropin secretion in the white-crowned sparrow, *Zonotrichia leucophrys gambelli*, *Proc. Nat. Acad. Sci. U.S.A.*, **71**:1666–1669.

Fraps, R. M., 1954, Neural basis of diurnal periodicity in release of ovulation-inducing hormone in fowl, *Proc. Natl. Acad. Sci., U.S.A.*, **40**:348–356.

Fraps, R. M., 1959, Photoperiodism in the female domestic fowl, in: *Photoperiodism and Related Phenomena in Plants and Animals* (R. B. Withrow, ed.), Publ. No. 55, Am. Assoc. Adv. Sci., Washington, D.C., pp. 767–785.

Fraps, R. M., and Conner M. H., 1954, Neurohypophysial control of follicular maturation in the domestic fowl, *Nature* **174**:1148–1150.

Gaston, S., 1971, The influence of the pineal organ on the circadian activity rhythm in bird, in: *Biochronometry* (M. Menaker, ed.), National Academy of Sciences, Washington, D.C., pp. 541–548.

Gaston, S., and Menaker, M., 1968, Pineal function: The biological clock in the sparrow? *Science* **160**:1125–1127.

Gourdji, D., 1970, Prolactine et relations photosexuelles chez les viseaux, *Colloq. Int. Cent. Nat. Rech. Scient.* **172**:233–258.

Gwinner, E., 1968, Circannuale periodik als grundlage der jahreszietlechen funktionswandlels bei zugvogeln. Untersuchungen am fitis (*Phyllosopus trochilus*) und am uraldlauresanger (*P. sibilatrix*), *J. Ornithol.* **109**:70–95.

Gwinner, E., 1972, Adaptive function of circannual rhythms in warblers, *Proc. Int. Ornithol. Congr.* **15**:218–236.

Gwinner, E., 1973, Circannual rhythms in birds: Their interactions with circadian rhythms and environmental photoperiod, *J. Reprod. Fert.* (suppl.) **19**:5–18.

Gwinner, E., 1974, Testosterone induces splitting of circadian locomotor activity rhythms in birds, *Science* **185**:72–74.

Gwinner, E., 1975, Circadian and circannual rhythms in birds, in: *Avian Biology*, Vol. V., (D. S. Farner and J. R. King, eds.), Academic Press, New York, pp. 221–285.

Gwinner, E., 1978, Effects of pinealectomy on circadian locomotor activity rhythms in European starlings, *J. Comp. Physiol.* **126**:123–129.

Gwinner, E. 1980, Relationship between circadian activity patterns and gonadal function: Evidence for internal coincidence? in: *Acta XVII Congressus Internationalis Orni-*

thologici (R. Nohring ed.), Verlag der Deutochen Ornithologen-Gesellschaft, Berlin, pp. 409–416.

Gwinner, E., 1981, Circannuale rhythm von Tieren and Ihre Photoperiodische Synchronisation, *Naturwissenschaften* **68**:542–551.

Gwinner, E., and Benzinger, I., 1978, Synchronization of a circadian rhythm in pinealectomized European starlings by daily injection of melatonin, *J. Comp. Physiol.* **127**:209–213.

Gwinner, E., Turek, F., and Smith, S. D., 1971, Extraocular light perception in photoperiodic responses of the White-crowned Sparrow (*Zonotrichia Leucophrys*) and of the Golden-crowned Sparrow, (*Z. atricapilla*), *Z. Vergl. Physiol.* **75**:323–331.

Hamner, W. M., 1963, Diurnal rhythm and photoperiodism in testicular recrudescense of the House Finch, *Science* **142**:1294–1295.

Hamner, W. H., 1964, Circadian control of photoperiodism in the house finch demonstrated by night-interruption experiment, *Nature* **203**:1400–1401.

Hamner, W. M., 1968, The photorefractory period of the house finch, *Ecology* **49**:211–227.

Hamner, W. M., 1971, On seeking an alternative to the endogenous reproductive rhythm hypothesis in birds, in: *Biochronometry*, National Academy of Sciences, Washington, D.C., pp. 448–462.

Hartwig, H. G., 1974, Electron microscopic evidence for a retino-hypothalamic projection to the suprachiasmatic nucleus of *Passer domesticus. Cell. Tiss. Res.* **153**:89–99.

Harvey, S., Scanes, C. G., and Howe, T., 1977, Growth hormone effects on *in vitro* metabolism of avian adipose and liver tissue, *Gen. Comp. Endocrinol.* **33**:322–328.

Hokfelt, T., and Fuxe, K., 1972, Effects of prolactin and ergot alkaloids on tuveroinfundibular dopamine (DA) neurons, *Neuroendocrinology* **9**:100.

Homma, K., and V. Sakakibara, 1971, Encephalic photoreceptors and their significance in photoperiodic control of sexual activity in Japanese quail, in: *Biochronometry* (M. Menaker, ed.), National Academy of Science, Washington, D.C., pp. 333–341.

Homma, K., Sakaibara, V., and Ohta, M., 1977, Potential sites and action spectra for encephalic photoreception in the Japanese quail, in: *Abstracts of the First International Symposium on Avian Endocrinology* (B. K. Follett, ed.), Univ. College of North Wales, Bangor, pp. 25–26.

Homma, K. Ohta, M., and Sakakibara, Y., 1979, Photoinducible phase of the Japanese quail detected by direct stimulation of the brain, in: *Biological Rhythms and their Central Mechanism* (M. Suda, O. Hayaishi, and N. Nakagawa, eds.), Elsevier/North-Holland Biomedical Press, Amsterdam, pp. 85–94.

Hughes, M. R., Compton, J. R., Schrader, W. T., and O'Malley, B. M., 1981, Interaction of chick oviduct progesterone receptors with DNA, *Biochemistry* **20**:2481–2491.

Jallageas, M., and Assenmacher, I., 1972, Efft de la photoperiode et du taux d'androgene circulant sur la fonction thyroidienne du Canard, *Gen. Comp. Endocrinol.* **19**:331–340.

Jallageas, M., and Assenmacher, I., 1979, Further evidence for reciprocal interactions between the annual sexual and thyroid cycles in male peking ducks, *Gen. Comp. Endocrinol.* **37**:44–51.

Jallageas, M., and Assenmacher, I., 1980, Annual endocrine cycles in male teal (*Anas cresca*) and Peking Ducks *Anas platyrhynchos*), in: *Acta XVII Congressus Internationalis Ornithologisci*, Verlag der Deutschen Ornithologen-Gesellschaft, pp. 447–452.

Jenner, C. E., Engles, W. L., 1952, The significance of the dark period in the photoperiodic response of male juncos and white-throated sparrows, *Biol. Bull.* **103**:345–355.

John, T. M., and George, J. C., 1978, Circulating levels of thyroxin and triiodothyronine in the migratory Canada goose, *physiol. Zool.* **54**:361–370.

John, T. M., Meier, A. H., and Bryant, E. E., 1972, Thyroid hormones and the circadian fattening and cropsac responses to prolactin in the pigeon, Physiol. Zool. 45:34–42.

Joseph, M. M., and Meier, A. H., 1973, Daily rhythms of plasma corticosterone in the common pigeon, Columba livia, Gen. Comp. Endocrinol. 20:326.

Kamiyoshi, M., and Tanaka, K., 1983, Endocrine control of ovulatory sequence in domestic fowl, in: Avian Endocrinology: Environmental and Ecological Perspectives (S. Mikami, K. Homma, and M. Wada, eds.), Japan Sci. Soc. Press, Tokyo/Springer-Verlag, Berlin, pp. 167–177.

Kasal, C. Menaker, M., and Perez-Polo, R., 1979, Circadian clock in culture: N-acetyltransferase activity of chick pineal glands oscillates in vitro, Science 103:656–658.

Kovacs, K., and Peczely, P., 1983, Phase shifts in circadian rhythmicity of total free corticosterone and transcortin plasma levels in hypothyroid male Japanese quail, Gen. Comp. Endocrinol. 50(3):483–489.

Kawamura, H., Inacye, S. T., Ebihara, S., and Noguchi, S., 1982, Neurophysiological studies of the SCN in the rat and in the Java sparrow, in: Vertebrate Circadian Systems (J. Aschoff, S. Dean, and G. A. Groos, eds.), Springer-Verlag, Berlin, pp. 106–111.

King, J. R., 1968, Cycles of fat deposition and molt in White-crowned sparrows in constant environmental conditions, Comp. Biochem. Physiol. 23:827–837.

Klandorf, H., Sharp, P. J., and Duncan, I. J. H., 1978, Variations in levels of plasma thyroxine and tricodothyronine in juvenile female chickens during 24 hour and 16 hour lighting cycles, Gen.Comp. Endocrinol. 36:238–243.

Klandorf, H., Stokkan, K., and Sharp, P. J., 1982, Plasma thyroxine and triiodothyronine levels during development of photorefractoriness in willow ptarmigan Lagopus lagopus lagopus) exposed to different photoperiod, Gen. Comp. Endocrinol. 47:64–69.

Kobayashi, H., 1969, Pineal and gonadal activity in birds, in: Seminar on Hypothalamic and Endocrine Functions in Birds, International House of Japan, Tokyo, p. 72.

Konishi, H., and Homma, K., 1983, Role of eyes in external coincidence and modulation of steroidal feedback in male Japanese quail, in: Avian Endocrinology: Environmental and Ecological Perspectives (S. Mikami, K. Homma, and M. Wada, eds.), Japan Sci. Soc Press, Tokyo, pp. 179–190.

Kramer G., 1950, Weitere analyse der faktoren, welch die Zugaktivitat des gekafigten vogels orientieren, Naturwissenschften 37:377–378.

Kramer, G., 1951a, Experiments on bird orientation, Ibis 94:265–285.

Kramer, G., 1951, Eine neue Methode zur erforschung de Zugorientierung und die bisher erzielten Ergebnisse, in: Proc.Int. Ornithol. Congr.,(S. Hörstadius, eds.), 10th, pp. 269–280.

Kramer, G., 1953, Die Sonnenorientierung der Vogel, Verh. Dsch. Zool. Ges. Frieburg, suppl. 17:72–84.

Lehrman, D. S., 1955, The physiological basis of parental feeding behavior in the ring dove (Streplopelia risoria), Behavior 7:241–286.

Lofts, B., 1964, Evidence of an antonomous reproductive rhythm in an equatorial bird, (Quelea Quelea, Nature, (London), 201:523–524.

Lofts, B., 1970, Cytology of the gonads and feedback mechanisms with respect to photosexual relations in male birds, in: La Photoregulation de la Reproduction chex les Oiseaux et les Mammiferes (J. Benoit and I. Assenmacher, eds.), Colloq. Int. Cent. Nat. Rech. Sci. 172:301–323.

Lofts, B., and Lam, W. L., 1973, Circadian regulation of gonadotrophin secretion, J. Reprod. Fert. Suppl. 19:19–34.

Lofts, B., and Murton, R. K., 1968, Photoperiodic and physiological adaptations regulating avian breeding cycles and their ecological significance, *J. Zool.* **155**:327–394.

MacBride, S. E., 1973, Pineal biochemical rhythms of the chicken (*Gallus domesticus*): Light cycle and locomotor activity correlates, Ph.D. Dissertation, University of Pittsburg, Pittsburg, Pennsylvania.

MacBride, S. E., and Ralph, C. L., 1972, Is the pineal a biological clock in birds? *Am. Zool.* **12**:34.

March, G. L., McKeown, B. A., John, T. M., and George, J. C., 1978, Diurnal variation in circulating levels of free fatty acids and growth hormone during crop gland activity in the pigeon (*Columbia livia*), *Comp. Biochem. Physiol.* **59B**:143.

Martin, D. D., and Meier, A. J., 1973, Temporal synergism of corticosterone and prolactin in regulating orientation in the migratory white-throated sparrow (*Zonotrichia albicollis*), *Condor* **75**:369–374.

Mattocks, P. W., Farner, D. S., and Follett, B. K., 1976, The annual cycle of luteinizing hormone in the plasma of intact and castrated white-crowned sparrows, *Zonotrichia leucophrys gambelii*, *Gen. Comp. Endocrinol.* **30**:156–161.

May, J. D., 1978, Effect of fasting on T_3 and T_4 concentrations in chicken serum, *Gen. Comp. Endocrinol.* **34**:323–327.

McKeown, B. A., John, T. M., and George, J. C., 1973, Circadian-rhythm of plasma growth-hormone levels in pigeon, *J. Interdisc. Cycle Res.* **4**:221–227.

McMillan, J. P., 1972, Pinealectomy abolishes the circadian rhythm of migratory restlessness, *J. Comp. Physiol.* **79**:105–112.

McMillan, J. P., and Elliot, J. A., and Menaker, M. M., 1975a, On the role of eyes and brain photoreceptors in the sparrow: Aschoff's Rule, *J. Comp. Physiol.* **102**:257–262.

McMillan, J. P., Elliot, J. A., Menaker, M. M., 1975b, On the role of the eyes and brain photoreceptors in the sparrow: Rhythmicity in constant light, *J. Comp.Physiol.* **102**:263–268.

McMillan, J. P., Gauthreaux, S. A., and Helms, C. W., 1970, Spring migratory restlessness in caged birds: A circadian rhythm, *Bioscience* **20**:1259–1260.

McMillan, J. P., Keatts, H. C. and Menaker, M. M., 1975c, On the role of eyes and brain photoreceptors in the sparrow: Entrainment of light, *J. Comp. Physiol.* **102**:251–256.

McMillan, J. P., Underwood, H. A., Elliot, J. A., Stetson, M. H., and Menaker, M. 1975d, Extraretinal light perception in the sparrow. 4. Further evidence that the eyes do not participate in photoperiodic photoreception, *J. Comp. Physiol.* **97**:205–214.

Meier, A. H., 1969, Antigonadal effects of prolactin in the white-throated sparrow, *Zonotrichia albicollis*, *Gen. Comp. Endocrinol.* **13**:222–225.

Meier, A. H., Chronoendocrinology of vertebrates, in: *Hormonal Correlates of Behavior* (B. E. Eleftheriou and R. L. Sprott, eds.), Plenum Press, New York, pp. 469–549.

Meier, A. H. 1976, Chronoendocrinology of the white-throated sparrow, in: *Proc. 16th Intern. Ornith. Cong.* (H. J. Frith and J. H. Calaby, eds.), Griffin Press Ltd, Wetley, S. Australia, pp. 355–368.

Meier, A. H., and Davis, K. B., 1967, Diurnal variations of the fattening response to prolactin in the white-throated sparrow, *Zonotrichia albicollis*, *Gen. Comp. Endocrinol.* **8**:110–114.

Meier, A. H., and Dusseau, J. W., 1973, Daily entrainment of the photoinducible phases for photostimulation of the reproductive system of the sparrows, *Zonotrichia albicollis* and *Passer domesticus*, *Biol. Reprod.* **8**:400–410.

Meier, A. H., and Ferrell, B. R., 1978, Avian endocrinology, in: *Chemical Zoology* (X. M. Florkin, B. T. Schier, and A. H. Brush, eds.), *Academic Press*, New York, pp. 214–260.

Meier, A. H., and Fivizzani, A. J., 1975, Changes in the daily rhythm of plasma corticosterone concentration related to seasonal conditions in the white-throated sparrow, *Zonotrichia albicollis*, *Proc. Soc. Exp. Biol. Med.* **150:**356–362.

Meier, A. H., and Fivizzani, A. J., 1981, Physiology of migration in: *Animal Migration* (S. Gautreaux, ed.), Academic Press, New York, pp. 255–282.

Meier, A. H., and MacGregor, R., 1972, Temporal organization in avian endocrinology, *Am. Zool.* **12:**257–271.

Meier A.H., and Martin, D. D., 1971, Temporal synergism of corticosterone and prolactin controlling fat storage in the white-throated sparrow, *Gen Comp. Endocrinol.* **17:**311–318.

Meier, A. H., Burns, J. T., and Dusseau, J. W., 1969, Seasonal variations in the diurnal rhythm of pituitary prolactin content in the white-throated sparrow, *Zonotrichia albicollis, Gen. Comp. Endocrinol.* **12:**282–289.

Meier, A. H., Burns, J. T., Davis, K. B., and John, T. M., 1971a, Circadian variations in sensitivity of the pigeon cropsac response to prolactin, *J. Interdis. Cycle Res.* **2:**161–172.

Meier, A. H., John, T. M., and Joseph, M. M., 1971b, Corticosterone and the circadian cropsac response to prolactin, *Comp. Biochem. Physiol.* **40:**459–465.

Meier, A. H., Ferrell, B. R., and Miller, L. J., 1981, Circadian components of the circannual mechanism in the white-throated sparrow, in: *Proc. XVII Intern. Ornith. Congr. Berlin* (R. Nöhning, ed.), pp. 458–462.

Menaker, M., 1965, Circadian rhythms and photoperiodism in *Passer domesticus* in: *Circadian Clocks* (J. Aschoff, ed.), North Holland Publ., Amsterdam, pp. 385–395.

Menaker, M, and Eskin, A., 1967, Circadian clock in photoperiodic time measurement: A test of the Bunning hypothesis, *Science* **157:**1182–1185.

Menaker, M., and Keatts, H., 1968, Extraretinal light perception in the sparrow. II. Photoperiodic stimulation of testicular growth, *Proc. Natl. Acad. Sci. U.S.A.* **60:**146–151.

Menaker, M, and Underwood, H., 1976, Extraretinal photoreception in birds, *Photochem. Photobiol.* **23:**299–306.

Menaker, M, and Zimmerman, N., 1976, Role of the pineal in the circadian system of birds, *Am. Zool.* **16:**45–55.

Menaker, M., Roberts, R. J., Elliot, J., and Underwood, H., 1970, Extraretinal light perception in the sparrow. III. The eyes do not participate in the photoperiodic perception, *Proc. Natl. Acad. Sci. U.S.A.* **67:**320–325.

Merkel, F. W., 1938, Zur Physiologie der Zugunruhe bei Vogeln, *Ber. Ver. Sches. Orn.* **23:**1–72.

Merkel, F. W., 1963, Long-term effects of constant photoperiods on European robins and whitethroats, *Proc. Intern. Ornith. Congr.* **13:**950–959.

Miller, J. B., 1960, Migratory behavior in the white-throated sparrow, *Zonotrichia albicollis* at Madison, Wisconsin, Ph.D. Dissertation, University of Wisconsin, Madison, Wisconsin.

Miller, L. J., 1979, Circadian neurotransmitter activity regulating seasonal conditions in three avian species, Ph.D. Dissertation, Louisiana State University, Baton Rouge.

Miller, L. J., and Meier, A. H., 1983a, Circadian neurotransmitter activity resets the endogenous annual cycle in a migratory sparrow, *J. Interdis. Cycle Res.* **14:**85–94.

Miller, L. J., and Meier, A. H., 1983b, Temporal synergism of neurotransmitter-affecting drugs influences seasonal condition in sparrow, *J. Interdis. Cycle Res.* **14:**75–84.

Miller, L. J. and Weiss, C. M., 1978, Effects of altered photoperiod on migratory orientation in white-throated sparrow (*Zonotrichia albicollis*), *Condor* **80:**94–96.

Morton, M. L., and Mewaldt, L. R., 1962, Some effects of castration on a migratory sparrow, *Zonotrichia atricapilla, Physiol. Zool.* **35:**237–247.

Murton, R. K., Lofts, B., and Westwood, N. J., 1970, Manipulation of photorefractoriness in the house sparrow (Passer domesticus) by circadian light regimes, Gen. Comp. Endocrinol. 14:107–113.

Nalbandov, A. V., 1959, Neuroendocrine reflex mechanism: Bird ovulation, in: Comparative Endocrinology (A. Gorbman, ed.), Wiley, New York, pp. 161–173.

Newcomer, W. S., 1974, Diurnal rhythms of thyroid function in birds, Gen. Comp. Endocrinol. 24:65–73.

Nicholls, T. D., and Storey, C. R., 1976, The effects of castration on plasma levels in photosensitive and photorefractory canneries (Serinus canaris), Gen. Comp. Endocrinol. 29:170–174.

Oishi, T., Konishi, T., and Kato., M., 1966, Investigation on photorecepting mechanism to control gonadal development in Japanese quail, Environ. Cont. Biol. 3:87–90.

Opel, H. 1916, The timing of oviposition and ovulation in the quail (Coturnix coturnix japonica), Br. Poult. Sci. 7:29–38.

Orr, M. L., and Watt, B. K., 1968, Amino acid control of food, Home Economics Research Report No. 4, U. S. Department of Agriculture.

Ottenweller,J. E., Meier, A. H., Ferrell, B. R., Horseman, N. D., and Proctor, A., 1978, Extrapituitary regulation of the circadian rhythm of plasma corticosteroid concentration in rats, Endocrinology 103:1875–1879.

Payne, R. B., 1972, Mechanisms and control of molt, in: Avian Biology, Volume 2 (D. S. Farner and J. R. King, eds.), Academic Press, New York, pp. 103–155.

Pelham, R. W., 1975, A serum melatonin rhythm in chickens and its abolition by pinealectomy, Endocrinology 96:543–546.

Pethes, G., Losonczy, S., and Rudas, P., 1978, Measurement of serum triiodothyronine by radioimmunoassay, Magyar Allatorvosok Lapja 33:177–182.

Pittendrigh, C. S., 1972, Circadian surfaces and the diversity of possible roles of circadian organization in photoperiodic induction, Proc. Nat. Acad. Sci. USA 69:2734–2737.

Pittendrigh, C. S. and Daan, S., 1976, A functional analysis of circadian pacemakers in nocturnal rodents. A. pacemaker structure: A clock for all seasons, J. Comp. Physiol. 106:333–355.

Ralph, C. L., and Fraps, R. M., 1959, Effects of hypothalamic lesions on progesterone induced ovulation in the hen, Endocrinology 65:819–824.

Ralph, C. L., and Fraps, R. M., 1960, Induction of ovulation in the hen by injection of progesterone into the brain, Endocrinology 66:269–272.

Ralph, C. L, Pelham, R. W., MacBride, S. E., and Reilly, D. P., 1974, Persistent rhythms of pineal and serum melatonin in cockerels in continuous darkness. J. Endocrinol. 63:319–324.

Ralph, C. C., Binkley S., MacBride, S. E., and Klein, D. C., 1975, Regulation of pineal rhythms in chickens: Effects of blinding, constant light, constant dark, and superior cervical ganglionectomy, Endocrinology 97:1371–1378.

Riddle, O., Bates, R. W., and Dykshorn, S. W., 1932, A new hormone of the anterior pituitary, Proc. Soc. Exp. Biol. Med. 29:1211–1212.

Rowan, W., 1931, The Riddle Of Migration, Williams & Wilkins, Baltimore, Maryland.

Russo, A. C., 1983, Circadian systems in development and photoperiodism in Japanese quail, Ph.D. Dissertation, Louisiana State University, Baton Rouge.

Rutledge, J. T., and Angle, M. J., 1977, Persistence of circadian activity rhythms in pinealectomized European starlings, J. Exp. Zool. 200:333–338.

Rutledge, J. T., and Schwab, R. G., 1974, Testicular metamorphosis and prolongation of spermatogenesis in starlings (Sturnus vulgaris) in the absence of daily photostimulation, J. Exp. Zool. 187:71–76.

Saunders, D. S., 1976, Insect Clocks, Pergamon Press, Oxford.

Sayler, A., and Wolfson, A., 1968, Influence of the pineal gland on gonadal maturation in the Japanese quail, Endocrinology 83:1237–1246.

Scanes, C. G., Sharp, P. U., Harvey, S., Goddin, P. J. M., Chadwick, A., and Newcomer, W. S., 1979, Variations in plasma prolactin, thyroid-hormones, gonadal-steroids and growth-hormone in turkeys during the induction of egg-laying and molt by different photoperiods. Br. Poult. Sc. 20(2):143–148.

Schwab, R. G., 1971, Circannual testicular periodicity in the European starling in the absence of photoperiodic change, in: Biochronometry (M. Menaker, ed.), National Academy of Science, Washington, D.C., pp. 428–447.

Sharp, P. J., 1980a, The role of the testes in the initiation and maintenance of photore-fractoriness, in: Acta XVII Congressus Internationalis Ornithologici (R. Nohring, ed.), Verlag der Deutschen Ornighologen-Gesellschaft, Berlin, pp. 468–472.

Sharp, P. J. 1980b, The endocrine control of ovulation in birds, in XVII Congressus Internationalis Ornithologicus (R. Nohring, ed.), Verlag der Deutschen Ornithologen-Gesellschaft, Berlin, pp. 245–250.

Sharp, P. J., and Klandorf, H., 1981, The interaction between daylength and the gonads in the regulation of levels of plasma thyroxine and triiodothyronine in the Japanese quail, Gen. Comp. Endocrinol. 45:504–512.

Sharp, P. J., and Moss, R., 1977, The effects of castration on concentrations of luteinizing hormone in the plasma of photorefractory red grouse (Lagopus, lagopus scoticus), Gen. Comp. Endocrino. 32:289–293.

Simpson, S. M., and Follett, B. K., 1980, Investigations on the possible roles of the pineal and anterior hypothalamus in regulating circadian activity rhythms in Japanese quail, in: Acta XVII Congr. Intern. Ornith. (R. Nohring, ed.), Verlag der Deutschen Or-nithologen-Gesellschaft, Berlin, pp. 435–438.

Smith, J. P. 1982, Changes in blood levels of thyroid hormones in two species of passerine birds, Condor 84(2):160–167.

Srivastava, A. K., and Meier, A. H., 1972, Daily variation in concentration of cortisol in intact and hypophysectomized Gulf Killifish, Science 177:185–187.

Stetson, M. H., and Erickson, J. E., 1972, Hormonal control of photoperiodically induced fat deposition in white-crowned sparrows, Gen. Comp. Endocrinol. 19:355–362.

Sze, P. Y., Neckers, L., and Towle, A. C., 1976, Glucocorticoids as a regulatory factor for brain tryptophan-hydroxylase, Neurochemistry 26:169–173.

Takahashi, J. S., 1982, Circadian rhythms of the isolated chicken pineal in vitro, in:Vertebrate Circadian Systems (J. Aschoff, S. Dean, and G. A., Groos, eds.), Springer-Verlag, Berlin, pp. 158–162.

Takahashi, J. S., and Menaker, M., 1979, Brain mechanisms in avian circadian systems, in: Biological Rhythms and Their Central Mechanism (M. Suda, O. Hayaishi, and H. Nakagawa, eds.), Elsevier-North Holland, Amsterdam, pp. 95–109.

Telegdy, G., and Vermes, I., 1975, Effect of adrenocortical hormones on activity of the seronotinergic system in limbic structures in rats, Neuroendocrinology 18:16–26.

Thapliyal, J. P., and Pandha, S. K., 1965, Thyroid-gonad relationship in spotted munia, Uroloneher punctulata, J. Exp. Zool. 158:253–261.

Turek, F. W., 1972, Circadian involvement in termination of the refractory period in two sparrows, Science 178:1112–1113.

Turek, F. W., 1974, Circadian rhythmicity and initiation of gonadal growth in sparrows, J. Comp. Physiol. 92:59–64.

Turek, F. W., Wolfson, A., and Desjardins, C., 1980, Testosterone treatment blocks the termination of the gonadal photorefractory condition in the white-throated sparrows maintained on short days, Gen. Comp. Endocrinol. 41:365–371.

Tyshchenko, V. P., 1966, Dvukhostsillyatornaya model' fiziologicheskogo mekhanizma fotoperiodicheskoi reakstii nasekomykh, Zh. Obschchei Biol. 27:209–222.

van Tienhoven, A., 1961, Endocrinology of reproduction in birds, in: Sex and Internal Secretions, Volume 2 (W. C. Young, ed.), Williams and Wilkins, Baltimore, Maryland, pp. 1088–1169.

van Tienhoven, A., Nalbandov, A. V., and Nortion, H. W., 1954, Effect of dibenamine or progesterone-induced and 'spontaneous' ovulation in the hen, Endocrinology 54:605–611.

Vaugien, M., and Vaugien, L., 1961, Le Moineau domestique peut developer son activite sexuelle et la maintenic dans l'abscurite complete, C. R. Seances Acad. Sci. 253:2762–2764.

Voitkevitch, A. A., 1966, The Feathers and Plumage of Birds, October House Inc., New York.

von St. Paul, V., 1956, Compass directional training of Western meadowlarks, Auk 73:203–210.

Wainwright, S. D., and Wainwright, L. K., 1979, Chick pineal serotonin acetyltransferase: A diurnal cycle maintained in vitro and its regulation by light, Can. J. Biochem. 57:700–709.

Weise, C. M., 1967, Castration and spring migration in the white-throated sparrow, Condor 69:49–68.

Wheeland, R. A., Martin, R. J., and Meier, A. H., 1976, The effect of prolactin and CB 154 on in vitro lipogenesis in the Japanese Quail, Coturnix coturnix japonica and of photostimulation on enzyme patterns in the white-throated sparrow, Zonogrichia albicollis, Comp. Biochem. Physiol. 53:379–385.

Wiesel, F. A., Fuxe, K., Hokfelt, T., and Agnati, L. F., 1978, Studies on dopamine turnover in ovariextomized female rats. Effects of 17-B-estradiol benzoate, ethynodiolacetate and ovine prolactin, Brain Res. 148:399–411.

Weiseltheir, A. S., and van Tienhoven, 1972, The effect of thyroidectomy on testicular size and on the photorefractory period in the starling (Sturnus vulgaris), J. Exp. Zool. 179:331–338.

Wilson, A. C., and Farner, D. S., 1960, The annual cycle of the thyroid activity in white-crowned sparrows of eastern Washington, Condor 62:414–425.

Wilson, F. C., and Follett, B. K. 1974, Plasma and pituitary lutenizing hormone in intact and castrated tree sparrows (Spizella arborea), Gen. Comp. Endocrinol. 23:82–93.

Wilson, J. M. and Meier, A. H., 1983, Tryptophan feeding induces sensitivity to short daylengths in photorefractory hamsters, Neuroendocrinology 36:59–63.

Wilson S. C. and Sharp, P. J., 1975, Episodic release of luteinizing hormone in the domestic fowl, J. Endocrinol. 67:59–70.

Wingfield, J. C., 1980, Fine temporal temporal adjustment of reproductive functions in: Avian Endocrinology (A. Epple and J. H. Stetson, eds.), Academic Press, New York, pp. 368–389.

Wolfson, A., 1965, Circadian rhythm and the photoperiodic regulation of the annual reproductive cycle in birds, in: Circadian Clocks, North Holland Publ. Co., Amsterdam, pp. 370–378.

Wolfson, A., 1966, Environmental and neuroendocrine regulation of annual gonadal cycles and migratory behavior in birds, Recent Progr. Horm. Res. 22:177–239.

Yates, C. A., and Herbert, J., 1979, The effects of different photoperiods on circadian 5-HT rhythms in regional brain areas and their modulation by pinealectomy, melatonin and estradiol, Brain Res. 176:311–326.

Yokoyama, K., Osche, A., Darden, T. R., and Farner, D. S., 1978, The sites of encephalic photoreception in the photoperiodic induction of the growth of the testes in the white-crowned sparrow, *Zonotricia leucophrys gambelii, Cell Tissue Res.* **189:**441–467.

Zarrow, M. X., and Bastian, J. W., 1953, Blockage of ovulation in the hen with adrenolytic parasympatholytic drugs, *Proc. Soc. Exp. Biol. Med.* **84:**457–459.

Zimmerman, J. L., 1966, Effects of extended tropical photoperiod and temperature on the dickcissel, *Condor* **68:**377–387.

Zimmerman, N. H. and Menaker, M., 1975, Neural connections of sparrow pineal: Role in circadian control of activity, *Science* **190:**477–479.

Zimmerman, N. H., and Menaker, M., 1979, The pineal: A pacemaker within the circadian system of the house sparrow, *Proc. Natl. Acad. Sci. U.S.A.* **76:**999–1003.

AUTHOR INDEX

BIRD NAME INDEX

SUBJECT INDEX

Air consumption, during calls, 220–221
Amplitude modulation, 229, 237
  flow rate and, 230
Analysis, *see specific types*
Analysis of covariance, description of, 45
Analysis of variance, 32
  description of, 45
ANOVA, *see* Analysis of variance
Anticannibalism, reversed sexual
      dimorphism and, 68
Avian classification
  cladistic, 188, 189, 194
  eclectic, 191
  problems in, 187–211
Avian phonation
  mechanism of, 238
  syringeal structure and, 213–242

Behavior, morphology, 175
Bernoulli principle, 237
Bill
  adaptation of, 162
  dimorphism in, 71, 72–73
  size, diet and, 79
Biogeography
  bird counting and, 249, 250
  island, equilibrium model of, 9
  population, as a verbal model, 5
Birds
  altricial, growth of, 6

Birds (*cont'd*)
  call,
    amplitude modulation and, 230
    sonogram of, 226, 237
    trachea length and, 239
  counting of, *see* Counting, birds
  distribution,
    direct gradient analysis and, 35
    empiric model for display of, 34
  distribution of island, 16
  large, metabolic rate of, 5
  model for energetics, 6
  population, island, 16
  precocial growth of, 6
  size,
    climate and, 8
    low food availability and, 8
Box plot, 46
Breeding
  song development and, 116
Brooding
  prolactin release and behavior in, 304,
     321
  reversed sexual dimorphism and
     behavior in, 83
  role of sexes in, 74, 75

Canonical correlation analysis, 46, 158,
     175, 176, 178
Cauchy distribution, 41